Applied AI Techniques in the Process Industry

Preface

In the Industry 4.0 era, the process industry is embracing the broad adoption of artificial intelligence (AI) methods. Universities, institutions, and industrial enterprises are all eagerly seeking to leverage advanced AI techniques to enhance efficiency, optimize processes, and reduce costs.

Applied AI Techniques in the Process Industry aims to compile recent advances and progress in the field of AI that systematically address challenges from molecular design to process design and optimization. The book comprehensively summarizes state-of-the-art techniques such as computer-aided molecular design, high-throughput screening, variational inference, reduced-order modeling, and physics-informed learning. It also demonstrates the benefits of utilizing these methods, such as improved interpretability and predictive power. In essence, by addressing the following fundamental questions, the book helps consolidate the existing body of knowledge:

a) How to mathematically represent the structure–property relationship for molecular design?
b) How to couple molecular and process for multi-scale design and optimization?
c) How to integrate prior knowledge and data for modeling and inference?

This book not only enhances readers' comprehension of the core principles of AI but also aids them in discovering optimal solutions to both theoretical and practical challenges. Hence, it has the potential to attract a diverse readership, encompassing university students to seasoned engineers in the industry.

Chang He
Sun Yat-Sen University,
School of Chemical Engineering and Technology
Jingzheng Ren
The Hong Kong Polytechnic University,
Department of Industrial and Systems Engineering

1

AI for Property Modeling, Solvent Tailoring, and Process Design

Yuqiu Chen

University of Delaware, Department of Chemical and Biomolecular Engineering, 150 Academy Street, Newark, DE 19716, USA

1.1 AI-Assisted Property Modeling

The accurate prediction of physical properties is critical for the successful application of both chemicals and materials across various industries. Property modeling using artificial intelligence (AI) has emerged as a powerful and efficient approach in various scientific and engineering disciplines [1]. This methodology leverages advanced algorithms and machine learning (ML) techniques to predict and model the physical and chemical properties of materials, compounds, or systems [2]. One of the key advantages of employing AI in property modeling is its ability to handle complex relationships and patterns within large datasets, leading to more accurate predictions. ML models, such as neural networks and support vector machines (SVMs), can be trained on diverse datasets, enabling them to capture intricate correlations that traditional modeling methods might overlook [3]. In the field of materials science, AI-driven property modeling has proven valuable for predicting properties like conductivity, thermal conductivity, and mechanical strength [4]. Similarly, in chemistry, AI has been applied to forecast molecular properties, solubility, and reaction outcomes [5]. Given the broad scope of AI techniques, the author has chosen to focus on their application in property modeling, solvent tailoring, and process design for systems involving ionic liquids (ILs) and aqueous two-phase systems (ATPSs).

ILs are innovative fluids that have garnered significant attention from both academia and industries [6]. Over the past few decades, extensive research has been conducted on the applications of ILs in various fields such as electrochemistry [7], synthetic materials [2, 8], and pharmaceutical manufacturing [9, 10]. In these areas, ILs serve different roles, including as extractants/absorbents in separations, media and/or catalysts in lignocellulosic biomass pretreatment, and functional materials in batteries. With the growing interest in ILs within industrial settings, it is crucial to deeply understand their property behaviors for effective product and process design. For example, viscosity is a crucial transport property that plays a significant

Applied AI Techniques in the Process Industry: From Molecular Design to Process Design and Optimization, First Edition. Edited by Chang He and Jingzheng Ren.

role in fluid flow analysis, process optimization, and material characterization [11]. Understanding and controlling viscosity enable engineers, researchers, and industries to design efficient systems, optimize processes, develop high-quality products, and ensure the desired performance and functionality of materials and fluids in a wide range of applications. Another example is that heat capacity is crucial for thermodynamic calculations, energy balance analyses, temperature control systems, material selection, process optimization, and fundamental research. It plays a crucial role in various scientific, engineering, and industrial applications, enabling the efficient and effective design, operation, and optimization of processes involving heat transfer and energy conversion [12]. Consequently, the characterization of IL properties and the establishment of structure–property relationships for ILs and IL-based mixtures are equally important in investigating their applications.

ATPSs, also known as aqueous biphasic systems (ABSs), are formed when two or more water-soluble components, such as polymers, salts, ILs, alkaline, and alcohols, are mixed in water at appropriate concentrations and temperatures [13]. To date, various combinations of phase-forming agents (e.g. polymer–salt/alkaline, polymer–polymer, ionic liquid–salt, and alcohol–salt) have been proposed for the creation of ATPSs [14]. Many ATPSs and the combination of these two-phase systems with other techniques such as microfluidic apparatus [15], have exhibited great technical and economic advantages in biotechnological applications [16]. Due to the high water content in both phases, ATPSs can provide a biocompatible and nondenaturing environment for cells, proteins, and other biomolecules. Meanwhile, ATPSs generally present less damage to the extracted biomolecules as they allow rapid phase separation and compound partition, leading to much lower interfacial stress than that of organic-water solvent systems. In addition, ATPSs can offer high recovery percentages and high purity of biomolecules in a one-step process. Besides these, ATPSs show characteristics of high tailored space, and they are also easy to scale up [17]. A good understanding of the mechanism that can guide the phase formation of ATPS is obviously of great importance for enhancing the opportunity for ATPS adoption in the industry. Meanwhile, the ability to provide reliable predictions on the partition of biomolecules in ATPS is also essential, given the fact that it would largely reduce the time and cost to find high-performance ATPS for biomolecules. Therefore, a systematic modeling study on the phase equilibria behavior of ATPS and the partition of biomolecules in ATPS is highly desirable for the transition of ATPS separation technique from pure academic focus to industrial implementation.

Due to the high complexity of ATPS and IL-involved systems, empirical correlations and theory-driven models cannot simultaneously provide reliable physical and thermodynamic predictions. In this respect, AI techniques such as ML algorithms are potential alternatives to model thermodynamic and transport properties of complex systems [18] such as IL-based ATPS and ionic liquid–water mixtures. The advantages of using AI techniques for modeling complex systems include the following: (i) AI techniques, particularly ML, can handle complex patterns and nonlinear relationships in data, making them well-suited for modeling intricate chemical properties. (ii) AI models can provide accurate predictions for various chemical properties. This accuracy is crucial in industries such as pharmaceuticals, materials science, and environmental research, where precise chemical property information is essential. (iii) Chemical data often involves a high number of dimensions, with numerous variables influencing the properties of interest. AI algorithms, especially those used in deep learning, excel at extracting relevant features from high-dimensional datasets. (iv) AI techniques facilitate the integration of diverse data types, including experimental results, literature data, and computational simulations. This comprehensive approach enhances the accuracy and reliability of chemical property models. (v) The increasing availability of large datasets in chemistry, including databases of chemical compounds and their properties, aligns well with AI's ability to handle big data. AI can uncover meaningful patterns and correlations within massive datasets. (vi) AI techniques can be applied across a range of chemical domains from organic chemistry to materials science. Their adaptability makes them versatile tools for modeling diverse chemical properties.

Recently, ML algorithms, including multiple linear regression (MLR), k-nearest neighbor (KNN), decision tree (DT), random forest (RF), gradient boosting (GB), artificial neural network (ANN), multilayer perceptron (MLP), SVM, XGBoost, and lightweight gradient-boosting machine (LightGBM), have been employed as property prediction tools for complex systems. The advantages and disadvantages of these ML algorithms are described in the following text:

MLR is a supervised ML algorithm used for predicting the numerical value of a dependent variable based on the values of two or more independent variables. It is a foundational and widely used technique, especially when dealing with datasets where the outcome is influenced by multiple factors. MLR serves as the basis for more advanced regression techniques in ML.

The KNN algorithm is a versatile and intuitive ML algorithm used for both classification and regression tasks. KNN is a valuable tool in various applications, including classification problems in image recognition, recommendation systems, and anomaly detection. KNN is easy to understand and implement, making it a good choice for initial exploration of a dataset. It does not involve a training phase in the traditional sense, and the model is trained during the prediction phase. KNN can be applied to various types of datasets and is effective in capturing complex patterns. However, KNN may struggle with imbalanced datasets, where one class significantly outnumbers the others. In addition, calculating distances between data points for large datasets can be computationally intensive. Nonetheless, its simplicity and ease of implementation make it a valuable tool, especially in cases where interpretability and explainability are important.

Decision tree is a versatile and widely used ML algorithm that can be applied to both classification and regression tasks. It is a supervised learning algorithm that recursively partitions the data into subsets based on the features, ultimately leading to a decision or prediction. Decision trees are easy to interpret and visualize. The decision-making process is transparent, allowing users to understand how predictions are made. Decision trees can be prone to overfitting, especially if the tree is deep and complex. Regularization techniques like pruning or using a minimum number of samples per leaf can help mitigate this. Decision trees serve as the foundation for more advanced ensemble methods like RF and GB.

RF is an ensemble learning method in ML that operates by constructing a multitude of decision trees during training and outputting the class that is the mode of the classes (classification) or mean prediction (regression) of the individual trees. RF often provides higher accuracy compared to individual decision trees, especially for complex datasets. The ensemble approach and feature randomization help mitigate overfitting, making RF more robust. RF can be applied to both classification and regression tasks and can handle missing values in the dataset without the need for imputation. However, RFs are less interpretable compared to individual decision trees, as the combined effect of multiple trees can be complex. In addition, RF can be memory intensive, especially with a large number of trees or features. Nonetheless, RF is widely used in practice and is considered a powerful and versatile algorithm. It is effective for a variety of tasks, including classification, regression, and feature selection. Its robustness and high accuracy make it a popular choice in the ML community.

GB is a powerful ensemble learning technique used for both classification and regression tasks. It is a boosting algorithm that combines the predictions of multiple weak learners (typically decision trees) sequentially, with each tree correcting the errors of the previous ones. GB can capture complex, nonlinear relationships in the data and often achieves high accuracy, making it one of the state-of-the-art algorithms for many ML tasks. However, GB can be prone to overfitting, especially if the number of trees is too large, and it can be computationally intensive when training a large number of trees. Nonetheless, GB is widely used in practice and has proven effective in various ML tasks. It is particularly popular in Kaggle competitions and real-world applications where accurate predictions are essential.

ANN is a computational model inspired by the structure and function of the human brain. It consists of interconnected nodes, called artificial neurons or "neurons," organized into layers. Each neuron takes input, performs a computation, and produces an output that is passed on to other neurons. The advantages of ANN in modeling studies can be summarized as follows: (i) ANNs can learn organically, meaning their outputs are not limited solely by inputs, and they have the ability to generalize from their inputs. (ii) Nonlinear systems can find shortcuts to reach computationally expensive solutions. (iii) ANNs have high fault tolerance potential. (iv) ANNs can perform tasks beyond routing around nonoperational parts of the network. The advantages of ANNs are particularly evident when a large experimental dataset with a wide range of variables is available. MLP is a type of ANN and a fundamental ML algorithm. It falls under the category of supervised learning and is widely used for both classification and regression tasks. Figure 1.1 presents a structure of the three-layer ANN with an input vector size of 46×1.

SVM is a supervised ML algorithm used for both classification and regression tasks. The primary goal of SVM is to find a hyperplane that best separates the data into different classes in a high-dimensional space. For a binary classification task, SVM aims to find a hyperplane that separates data points of one class from another. For regression tasks, SVM seeks to find a hyperplane that best fits the data. In a

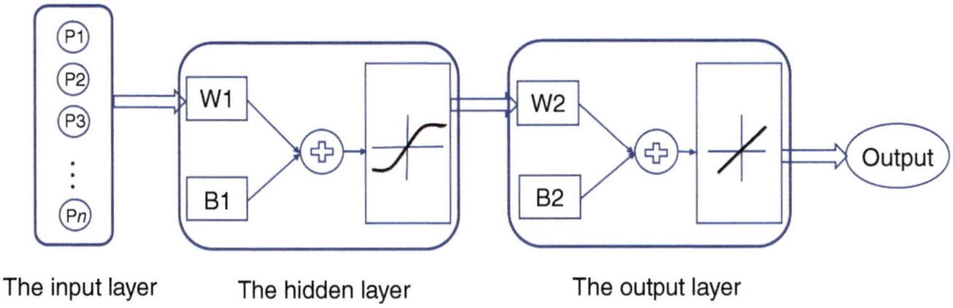

Figure 1.1 Structure of a three-layer ANN with an input vector size of $n \times 1$.
Source: [19]/with permission of Elsevier.

two-dimensional space, a hyperplane is a line. In higher dimensions, it becomes a hyperplane. Support vectors are the data points that are closest to the decision boundary (hyperplane). SVM can use a kernel function to map the input features into a higher-dimensional space. Common kernel functions include linear, polynomial, radial basis function (RBF), and sigmoid. The optimization objective is to maximize the margin between classes. SVM solves a quadratic optimization problem to find the optimal hyperplane and support vectors. SVM is effective in high-dimensional spaces, making it suitable for tasks with many features. It is widely used in various applications, including image recognition, text classification, and bioinformatics. Proper tuning of parameters, especially the choice of the kernel and regularization parameters, is crucial for achieving optimal performance with SVM.

XGBoost is a parallel regression tree model based on the boosting technique, where boosting refers to obtaining the final classifier by weighting the sum of existing weak classifiers. The XGBoost model is an improvement over the Gradient Boosted Decision Tree (GBDT) model. Compared to GBDT, XGBoost significantly enhances the speed of model training calculations and improves prediction and classification accuracy, making it an upgraded version of the GBDT algorithm. To prevent model overfitting and improve generalization ability, XGBoost incorporates regularization terms into the loss function of the GBDT model. The traditional GBDT loss function adopts a first-order Taylor expansion and uses the negative gradient value to replace the residual for fitting. In XGBoost, a second-order Taylor expansion is added to the loss function, capturing the second derivative to gather gradient direction information and improve model accuracy. Furthermore, XGBoost employs block-wise and sorted feature processing, enabling parallelization for finding the best-split points and thereby enhancing computation speed.

LightGBM is a GBDT algorithm framework renowned for its rapid training, minimal memory requirements, support for efficient parallel training, enhanced accuracy, and the ability to process extensive datasets swiftly. While XGBoost is a widely recognized GBDT tool, it grapples with significant memory usage. To tackle this challenge, Microsoft developed LightGBM as an optimization of the conventional GBDT algorithm. LightGBM distinguishes itself from XGBoost by implementing the histogram algorithm instead of the presorted algorithm. This choice results in decreased memory consumption and simplifies data separation.

However, it is worth noting that the histogram algorithm may not be as time efficient as the presorted algorithm when processing sparse data. LightGBM incorporates two novel techniques, Exclusive Feature Bundling (EFB) and Gradient-based One-Side Sampling, into the histogram-based GBDT algorithm. EFB enables the fusion and bundling of certain features, reducing the feature dimension without sacrificing accuracy. Additionally, LightGBM employs a leaf-wise strategy with a capped maximum depth, which serves to prevent overfitting without compromising computational efficiency.

To date, integrating ML algorithms into property modeling has been widely studied for different systems, including ATPS and IL-involved systems [20]. Yusuf et al. (2020) [21] reviewed the application of AI-based predictive methods in IL studies. Koutsoukos et al. (2021) [22] conducted a critical review of the use of ML algorithms as property prediction tools for the viscosity, density, melting point, and toxicity of pure ILs as well as the solubility of acid gases (H_2S, CO_2) in ILs. Therefore, the discussion in this work focuses on studies not covered in Yusuf's and Koutsoukos's review.

For pure ILs, Baskin et al. (2022) [20] combined three traditional ML algorithms and neural networks with seven different architectures with five types of molecular representations (in the form of either numerical molecular descriptors or simplified molecular input line entry system (SMILES) text strings) to construct quantitative structure-property relationship (QSPR) models for predicting six important physical properties of ILs: density, electrical conductivity, melting point, refractive index, surface tension, and viscosity. The results showed that (i) nonlinear ML algorithms perform much better than linear ones, (ii) neural networks perform better than traditional ML methods, and (iii) transformers that are actively used in natural language processing (NLP) perform better than other types of neural networks due to the advanced ability to analyze chemical structures of ILs encoded into SMILES text strings.

Dhakal and Shah (2021) [23] applied SVM and ANN to correlate ionic conductivity of imidazolium-based ILs. Both models were shown to successfully capture the entire range of ionic conductivity spanning 6 orders of magnitude over a temperature range of 275–475 K with relatively low statistical uncertainty, and the ANN-based model presented slightly better performance. The performance of the ANN algorithm in the modeling study of the conductivity of ILs was also validated by Datta et al. (2022) [24]. Dhakal and Shah (2022) [25] further applied three ML algorithms (MLR, RF, and XGBoost) to conduct models for the conductivity prediction of ILs, and XGBoost performed the best. Meanwhile, Karakasidis et al. (2022) [26] incorporated six different numerical ML algorithms, namely MLR, KNN, DT, RF, gradient boosting regressor (GBR), and MLP for electrical conductivity prediction. Results showed that all ML algorithms performed well on their predictions, while the best fit was obtained for the GBR algorithm. Yalcin et al. (2019) [27] employed MLR and Bayesian regularized ANNs to build semiempirical structure-property models for predicting the surface tension and liquid nanostructure of solvents containing a protic ionic liquid (PIL) with water and excess acid or base present. The results showed that all the models were successful in prediction.

For IL–H_2O mixture systems, Chen et al. (2022) [28] applied an ANN algorithm to build a group contribution (GC) model for predicting the viscosity of these binary mixtures at different compositions and temperatures, as demonstrated in Figure 1.2. The results show that the ANN–GC model with four or five neurons in the hidden layer is capable of providing reliable predictions on the viscosities of IL–H_2O mixtures. In addition, comparisons showed that the nonlinear ANN–GC model has much better prediction performance on the viscosity of IL–H_2O mixtures than that of the linear mixed model. The combination of the ANN algorithm and GC method was introduced by Fu et al. (2023) [30] to model the surface tension of IL–H_2O mixtures. The results show that the model with 4, 5, and 7 neurons in the hidden layer can provide reliable predictions for the surface tension of the IL–H_2O hybrid system. They also extended the ANN–GC model to the surface tension prediction of pure ILs, and the results showed that the model with 4 and 5 neurons in the hidden layer can provide reliable predictions for the surface tension of pure ILs. For IL–organic solvent mixture systems, Liu et al. (2023) [19] reported using three ML algorithms (i.e. ANN, XGBoost, and LightGBM) to build predictive models for the density and heat capacity of these binary mixtures. The results demonstrate that all three algorithms can provide accurate predictions, with the ANN model exhibiting the best performance. For using the same ML algorithms, Lei et al. (2023) [31] studied the surface tension and viscosity of IL–organic solvent mixtures. The modeling results indicate that all three algorithms can reliably predict the surface tension and viscosity of these binary mixture systems. Among them, XGBoost demonstrates the best performance for surface tension predictions, while the ANN model presents the best predictions for heat capacity. Most recently, Chen et al. (2024) [29] conducted a comprehensive modeling study on the viscosity, density, heat capacity, and surface tension of IL–IL binary mixtures. In their work, three ML algorithms, ANN, XGBoost, and LightGBM, were used. The results indicated that the ANN-based model exhibits the best prediction performance for viscosity, density, and heat capacity, while the XGBoost-based model provides the highest accuracy in surface tension predictions, as shown in Figure 1.2.

Unlike IL-involved systems, the modeling study on ATPS using ML algorithms is still limited. Mohsen et al. (2020) [32] reported using SVM and ANN to build models for the prediction of phase equilibria in poly(ethylene glycol) (PEG) + sodium phosphate ATPS. The average absolute error of SVM with the differential evolution (DE) algorithm for the testing dataset is 6.71%, which, for the considered process, is in an acceptable interval.

Chen et al. (2022) [33] combined the ANN algorithm with the GC method to conduct a comprehensive modeling study on the phase equilibria behavior of IL-based ATPSs, as illustrated in Figure 1.3. The ANN–GC model was trained and tested by 17 449 experimental binodal data points covering 171 IL-ABS at different temperatures (278.15–343.15 K). The results indicate that this ML-based model is capable of predicting the phase equilibria behavior of IL-ATPS in a general way. For the ATPSs involving biomolecules, Pazuki et al. (2010) [35] applied an ANN algorithm to model the partition coefficients of some well-known biomolecules (e.g. α-amylase, β-amylase, and albumin) in PEG-dextran ATPSs. The network topology is optimized,

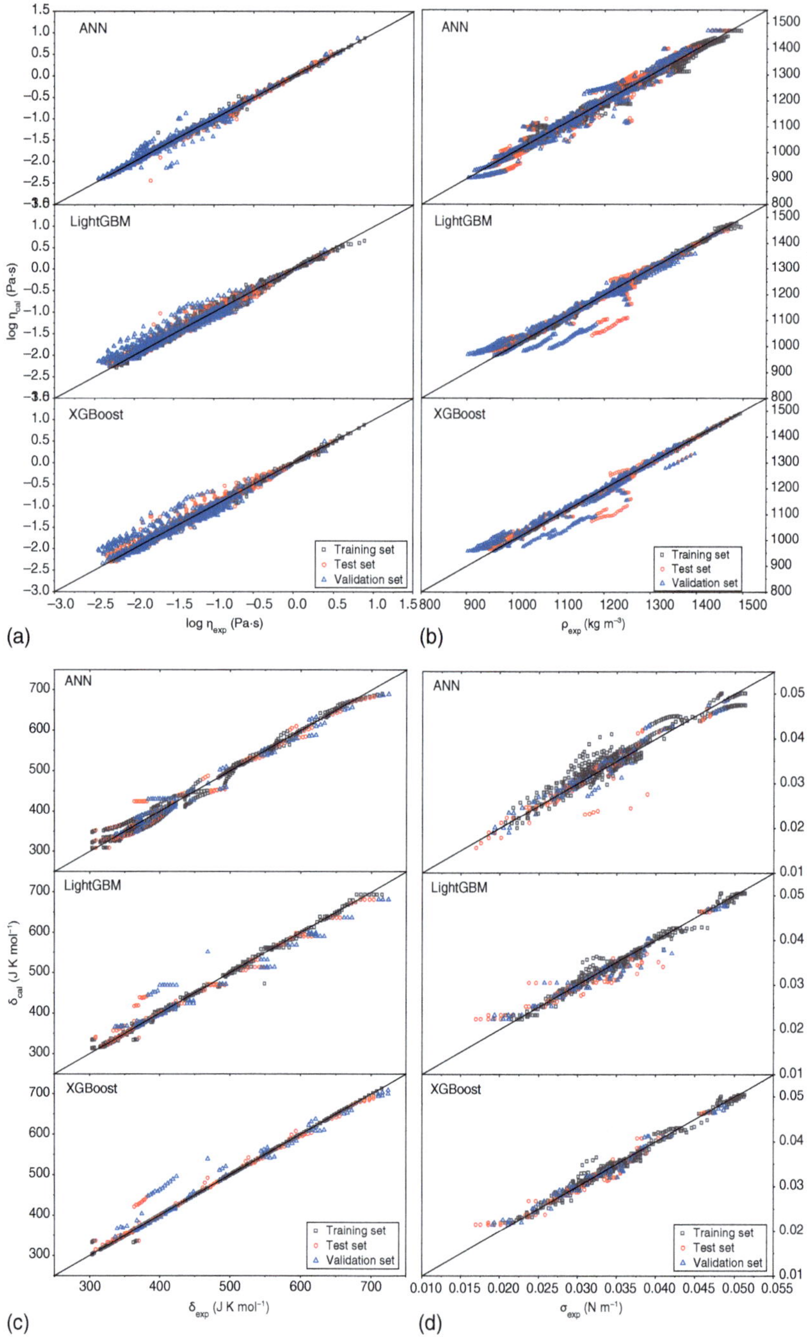

Figure 1.2 Plots of experimental and predicted values from the GC model using ANN, XGBoost, and LightGBM in the predictions of (a) viscosity, (b) density, (c) heat capacity, and (d) surface tension of IL–IL binary mixtures. Source: [29]/with permission of John Wiley & Sons.

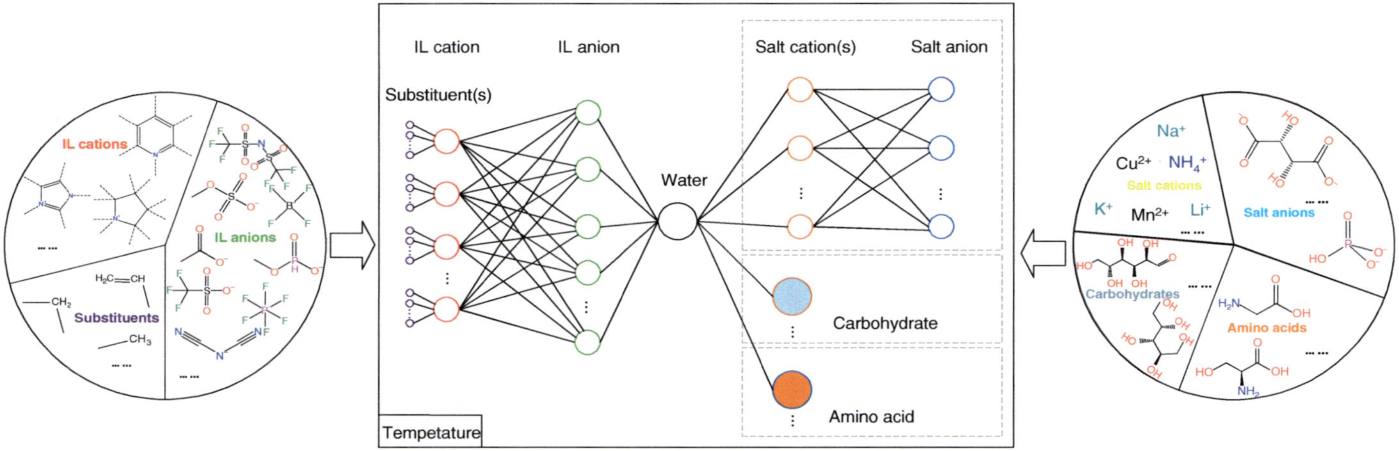

Figure 1.3 Conceptual structure of ionic liquid-based aqueous biphasic systems combining GC method. Source: [34]/with permission of American Chemical Society.

and the (6-1-1) architecture is found using the optimization of an objective function with the sequential quadratic programming (SQP) method for 450 experimental data points. The results obtained from the neural network of the partition coefficients of biomolecules in polymer–polymer ATPSs were compared with those from the modified Flory–Huggins model. Comparisons showed that the ANN is 50% more accurate than the Flory–Huggins model in obtaining partition coefficients of biomolecules in polymer–polymer ATPSs.

In 2023, Chen et al. [36] conducted an ANN modeling on the polymer-electrolyte aqueous two-phase systems involving biomolecules. In their work, 11 998 experimental binodal data points covering 276 polymer-electrolyte ATPS at different temperatures (273.15–399.15 K) and 626 experimental partition data points involving 22 biomolecules in 42 polymer-electrolyte ATPSs at different temperatures (283.15–333.15 K) were included. An ANN–GC model (ANN–GC model1) was built to predict the binodal curve behavior of polymer-electrolyte ATPS, while another ANN–GC model (ANN–GC model2) was developed to predict the partition of biomolecules in these biphasic systems. The modeling results show that ANN-GC model1 can give reliable predictions on the binodal curve behavior of polymer-electrolyte ATPS and ANN–GC model2, to some extent, is capable of predicting the partition coefficient of biomolecules in these biphasic systems in a widespread way. In addition, the obtained results also indicate that the tie-line length of polymer-electrolyte ATPS calculated from ANN–GC model1 can be directly used in ANN–GC model2 for predicting the partition coefficient of biomolecules in these ATPSs.

1.2 AI-assisted Solvent Tailoring

Solvent screening is crucial for optimizing chemical processes. Different solvents can significantly impact reaction rates, yields, and selectivity. The choice of solvent has a direct impact on the quality and purity of the final product. The environmental impact of a process is closely tied to the choice of solvents. Solvent properties influence safety considerations in a laboratory or industrial setting. Different solvents have varying costs, and solvent selection can significantly impact the overall cost of a process. Efficient solvent screening helps in identifying the most suitable solvent for sustainable and responsible manufacturing practices [37]. However, the vast number of available solvents with diverse chemical properties makes comprehensive screening challenging. Access to accurate and comprehensive solvent property data is crucial for effective screening [38]. In some cases, data for certain solvents may be limited, affecting the reliability of the screening process. Solvent performance can vary under different process conditions. Screening solvents under various temperature, pressure, and concentration conditions adds complexity to the screening process. Balancing diverse criteria (e.g. process performance, cost, and environmental impact) to identify the most suitable solvent can be complex. There are often trade-offs between different desirable solvent properties. Finding a solvent that meets all criteria without compromises can be challenging, requiring careful consideration of priorities [39].

Effective solvent screening is essential for achieving optimal process outcomes, but it requires a balance between the complexity of the screening process and the need for accurate and reliable results. However, traditional experimental solvent screening is generally time and resource intensive. Exploring a wide range of solvents through experimentation may not be feasible within practical timeframes [40]. On the other hand, the optimal design of compounds through manipulating properties at the molecular level is often the key to considerable scientific advances and improved process systems performance, as reported in Ref. [41]. Property models are particularly essential due to their role as the foundation for the development of computer-aided molecular design (CAMD) methods [42]. The application of AI techniques to optimize and discover novel solvents for various industrial processes has gained prominence [43]. This is due to AI algorithms, particularly ML models, that can analyze vast databases of chemical information, properties, and reactions. This aids in predicting and identifying solvent candidates with specific desirable properties. ML models can predict the properties of solvents, such as viscosity, boiling point, density, and environmental impact. This helps researchers select solvents that meet the requirements of a particular application. AI is often integrated with computational chemistry methods to predict molecular interactions and behaviors, aiding in the understanding of how solvents interact with other chemicals [44].

To date, some efforts have been made to integrate AI techniques into solvent design. Liu et al. (2021) [45] proposed a machine learning-based atom contribution (MLAC)-CAMD framework for solvent design. In their work, three-dimensional atomic descriptors are used to develop an MLAC method for fast and accurate predictions of molecular $\rho(\sigma)$ for the COSMO-SAC model. Then, the MLAC method is integrated with the CAMD problem, which is formulated as a mixed-integer nonlinear programming (MINLP) model and solved by a decomposition-based solution algorithm. The proposed framework was successfully used in the solvent design for the improved crystallization operation. Wang et al. (2022) [46] proposed a new CAMD approach for solvent design by combining ML algorithms with deterministic optimization. Variational autoencoder (VAE), a powerful generative ML method, is used to transfer a molecular structure into a continuous latent vector with an encoder and to convert the latent vector back to the molecule with a decoder. Solvent properties of interest are estimated by an FNN using the latent vector as input. The proposed design method successfully generated superior separation performance solvent candidates for the separation of 1-butene and butadiene. Sui et al. (2022) [47] combined eight ML algorithms with CAMD technique to design high cycle performance IL solvent for absorption heat transformer. The results showed that the optimum ILs screened by this ML-based CAMD method perform better than the currently investigated ILs.

Wang et al. (2021) [48] integrated the SVM algorithm into the IL solvent design for CO_2 separation from flue gas. By using this ML-based design method, 1-ethyl-3-methylimidazolium tricyanomethanide ([EMIM][TCM]) was identified as the most suitable solvent for CO_2 separation from flue gas, and the performance of this IL solvent was further validated via rigorous process simulation in Aspen Plus. The results showed that the process using [EMIM][TCM] has a 12.9% savings on total annualized cost compared to that of a reported IL, 1-ethyl-3-methylimidazolium bis(trifluoromethylsulfonyl)amide ([EMIM][Tf_2N]). Zhang et al. (2021) [49] presented an ML-based approach combining multiplayer Monte Carlo tree search and recurrent neural network for the tailor-made design of ILs. The application of this ML-based IL solvent design method was demonstrated through cases of CO_2 capture from (i) flue gas (CO_2/N_2) and (ii) from syngas (CO_2/H_2). The results showed that the IL solvent identified through this ML-based design method presents great efficiency for the studied CO_2 capture processes. Zhang et al. (2021) [50] proposed a computer-aided ionic liquid design (CAILD) approach for the optimal design of ILs for CO_2 capture, where three different ML models were applied to predict the CO_2 solubility. This ML-based IL solvent design problem was formulated as a MINLP problem with the objective of maximizing the CO_2 solubility of ILs under prespecified conditions. The CO_2 capture performance of the designed ILs was further confirmed using density functional theory calculations. The applicability of the proposed data-driven IL design method was then demonstrated via a case study of postcombustion carbon capture.

Most recently, Liu et al. [51] proposed a novel ML-based IL solvent design method that combines a syntax-directed variational autoencoder (SDVAE), deep factorization machine (DeepFM), and gradient-based particle swarm optimization (GBPSO). The SDVAE converts the molecular structure and chemical space of the ILs, and then DeepFM predicts the solubility of each coordinate in the chemical space representing an IL. Finally, GBPSO identifies the coordinates that represent ILs with ideal properties. In their work, the main optimization objective was a high solubility difference for CO_2 between its absorption and desorption conditions in commercial plant capture systems, which represents the CO_2 capture ability. The best IL generated has a predicted solubility difference that is 35.3% higher than that of the best one in the data set. A synthetic novel IL, [EMIM][TOS], from the generated results, was experimentally evaluated; it has a sufficiently high solubility difference to be a capture solvent with low energy consumption. This model has proved to be a high-efficiency molecular design model that can be used for sparse small data sets.

Kuroki et al. (2023) [52] proposed an electronic structure informatics approach to predict and develop ILs with high CO_2 solubility based on geometric and electronic factors of the constituent cations and anions. With a ML-assisted search for the best cation/anion combination, targeted organic syntheses, and precision measurements, the IL trihexyl(tetradecyl)phosphonium perfluorooctanesulfonate ([P66614][PFOS]) was experimentally proven to have a higher CO_2 solubility than trihexyl(tetradecyl)phosphonium bis(trifluoromethanesulfonyl)amide ([$P_{6,6,6,14}$][TFSA]). Some more works associated with the ML-based solvent design of ILs for carbon capture can be found in Sun's review study Sun et al. (2023).

Chen et al. [29] combined ANN algorithms with the CAMD technique for the design of an IL–IL binary mixed solvent for recovering H_2 from raw coke oven gas. The optimal solution was achieved in 30 seconds on an Intel(R) Xeon(R) E5-1620 3.70 GHz PC, demonstrating the high computational efficiency and integrability

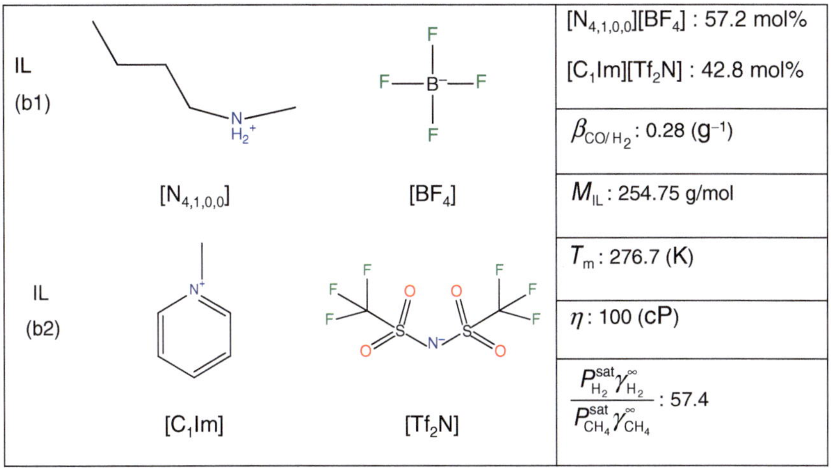

Figure 1.4 Molecular structure and some important properties of the binary mixed solvent combining 57.2 mol% $[N_{4,1,0,0}][BF_4]$ and 42.8 mol% $[C_1Im][Tf_2N]$. Source: [29]/with permission of John Wiley & Sons.

of the ANN algorithm in solvent design. Optimization results show that a binary mixed solvent combining 57.2 mol% butylethylammonium tetrafluoroborate ($[N_{4,1,0,0}][BF_4]$) and 42.8 mol% 1-methylimidazolium bis(trifluoromethanesulfonyl) amide ($[C_1Im][Tf_2N]$) meets all the design constraints and has the maximum H_2 absorption capacity ($= 0.28\,g^{-1}$), as presented in Figure 1.4. To the best of our knowledge, this is the only work using ML algorithms in the design of IL–IL mixed solvents that have been reported.

For ATPSs, Chen et al. (2021) [34] integrated the ANN algorithm and CAMD technique to tailor optimal IL-ATPSs for the recovery of hydrophilic ILs from dilute aqueous solutions. Two case studies were performed to test this ML-based ATPSs design method by formulating and solving their optimization-based MINLP problems. In both cases, the salting-out agents, i.e. $(NH_4)_2SO_3$ and KH_2PO_4, identified in this work have better ABS-forming ability than their counterparts K_2CO_3 and $(NH_4)_2SO_4$ reported in the literature. For the recovery of 10 wt% $[C_4Py][TfO]$ from aqueous solutions, the ATPS composed of $[C_4mIm][Cl]$-H_2O-$(NH_4)_2SO_3$ gives an IL recovery efficiency of 95.0 wt% and a salting-out agent input of 2.36 kg/kg IL recovery, and for the ATPS of $[C_4mIm][Cl]$-H_2O-K_2CO_3, they are 81.7 and 5.25, respectively. For the aqueous solutions containing 10 wt% $[C_4Py][TfO]$, the ATPS composed of $[C_4Py][TfO]$-H_2O-KH_2PO_4 gives an IL recovery efficiency of 95.6 wt% and a salting-out agent input of 1.81 kg/kg IL recovery, and for the ABS composed of $[C_4Py][TfO]$-H_2O-$(NH_4)_2SO_4$, they are 80.6 and 3.16, respectively. Most recently, Chen et al. (2023) [53] proposed an ML-based design method of high-performance ATPS for partitioning biomolecules, as presented in Figure 1.5.

In this design method, two ML models that combine the ANN algorithm and GC method are, respectively, employed to predict the phase equilibrium composition of polymer-electrolyte ATPS and the partition of biomolecules

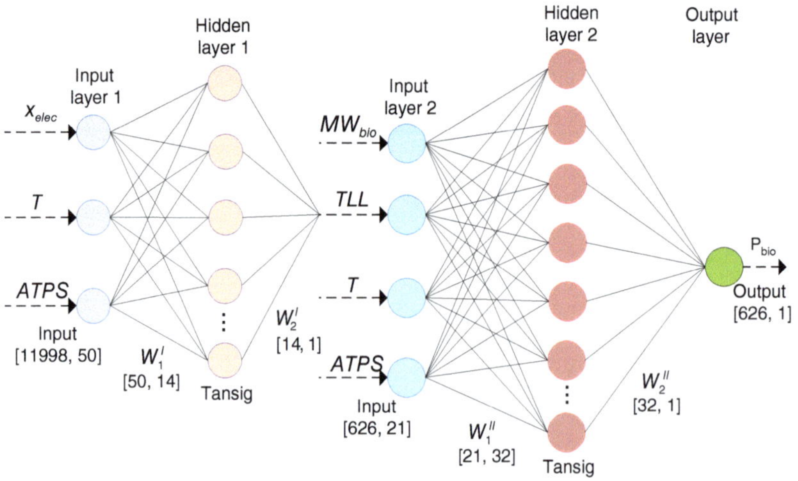

Figure 1.5 Diagram of the ML-based design of ATPS for the partitioning of biomolecules. Source: [54]/with permission of American Chemical Society.

in these aqueous systems. By integrating these two ANN–GC models into the computer-aided design technique, the optimal ATPS is identified by solving an optimization-based MINLP problem. Results of partitioning cefazolin and β-amylase were presented to demonstrate the viability of this ML-based ATPS design method. In the case of cefazolin, the partitioning performance of the tailored ATPS (PPG600 + $KNaSO_4$ + H_2O) is nearly seven times greater than that of the reported ATPS (PEG6000 + $Na_3C_6H_5O_7$ + H_2O). Meanwhile, the ATPS of PPG600 + $KNaSO_4$ + H_2O gives a cefazolin recovery of 95.0 wt% and an agent input of 0.154 kg/kg aqueous solution, and for the ATPS of PEG6000 + $Na_3C_6H_5O_7$ + H_2O, these values are 90.6 and 0.233, respectively. For the second case, the partition coefficient of β-amylase in ATPS (PPG400 + $KNaHPO_4$ + H_2O) (identified by ML-based design method) is about 13.5 times higher than that of the reported ATPS (PEG10000 + KH_2PO_4 + H_2O). In addition, the ATPS of PPG600 + $KNaSO_4$ + H_2O gives a β-amylase recovery of 97.3 wt% at a cost of 0.387 kg agent input/kg aqueous solution, and for the ATPS of PEG6000 + $Na_3C_6H_5O_7$ + H_2O, they are 66.3 and 0.252, respectively.

1.3 AI-Assisted Process Design

Process design is a critical phase in the lifecycle of any manufacturing or industrial operation. It involves the detailed planning and configuration of systems, equipment, and processes to convert raw materials into finished products. Efficient process design ensures the optimal use of resources, including raw materials, energy, and labor. Well-designed processes are crucial for ensuring the quality and consistency of the final product. Property models play a crucial role in process design across various industries, and they are essential for sizing and designing process equipment. Parameters like heat capacity, vapor pressure, and phase equilibrium data are critical for designing reactors, heat exchangers, distillation columns, and other equipment. Property models also contribute to understanding thermodynamic behavior and ensuring the safety and environmental compliance of industrial processes. Accurate property models are indispensable for successful and efficient process design. However, there are several challenges associated with acquiring precise property models. Industrial processes often involve complex mixtures and reactions. Developing accurate property models for complex systems can be challenging due to the interactions between different components. In addition, many real-world systems exhibit nonideal behavior, such as deviations from ideal gas or solution behavior. Modeling such nonidealities accurately can be challenging. Furthermore, developing and solving complex models computationally can be resource intensive and time consuming.

ML algorithms have significant potential in addressing these challenges associated with process design. This is due to the following factors: They can analyze vast amounts of process data to identify patterns, correlations, and insights that may not be apparent through traditional methods. ML algorithms can estimate properties such as viscosity, density, and thermodynamic parameters, providing accurate predictions for materials with limited experimental data. ML algorithms, particularly advanced neural networks, can capture complex relationships within systems, allowing for more accurate modeling of intricate processes. ML algorithms trained on existing data can provide estimations for properties of materials not yet extensively studied, facilitating the incorporation of new materials into process designs. Besides these, ML algorithms can integrate process models with economic considerations, optimizing designs based on both technical and economic criteria [54].

Recently, efforts have been made to integrate ML algorithms into the process design, especially for additive manufacturing. In addition to the studies discussed in Razvi's review work, Jiang et al. (2022) [55] recently proposed an ML-integrated design framework to establish process–structure–property (PSP) relationships for additive manufacturing. With the help of ML, the analysis and design processes based on PSP no longer need to establish complex surrogate models that are also unable to establish the relationships of PSP in a reversed direction. The relationships between process, structure, and property can be established simply through ML in whichever direction it is desired using the available additive manufacturing data. DNNs for point data and CNNs for distributions and image data were proposed as the specific ML techniques for the proposed framework.

Chen et al. (2023) [56] proposed an ML-based hybrid process design method for efficient recovery of hydrophilic IL from dilute aqueous solutions. The application of this ML-based hybrid process design method is illustrated through case studies of recovering two hydrophilic ILs, *n*-butylpyridinium trifluoromethanesulfonate ($[C_4Py][TfO]$) and 1-butyl-3-methylimidazolium chloride ($[C_4mIm][Cl]$), from their dilute aqueous solutions. For the recovery of 10 wt% $[C_4Py][TfO]$ from the aqueous solution, the proposed hybrid process could reduce the total annual cost (TAC) and energy consumption by 57% and 91%, respectively, compared with pure distillation processes. In the case of recovering 10 wt% [C_4mIm][Cl] from the aqueous solution, the reduction in TAC and energy savings of the hybrid process could reach 49% and 87%, respectively, compared with the pure distillation process.

Most recently, Chen et al. (2024) [29] integrated ML algorithms into process design involving IL–IL mixed solvent. In their work, the ANN algorithm was combined with the GC method to predict properties, including viscosity, density, surface tension, and heat capacity in the equipment design, as shown in Figure 1.6. The application of this ML-based process design method is demonstrated via a case study on H_2 recovery from raw coke oven gas, as shown in Figure 1.7. In that case, the best solution was found within 27 seconds on an Intel(R) Xeon(R) E5-1620 3.70 GHz PC, indicating the high computational efficiency and integrability of ML algorithms in process design.

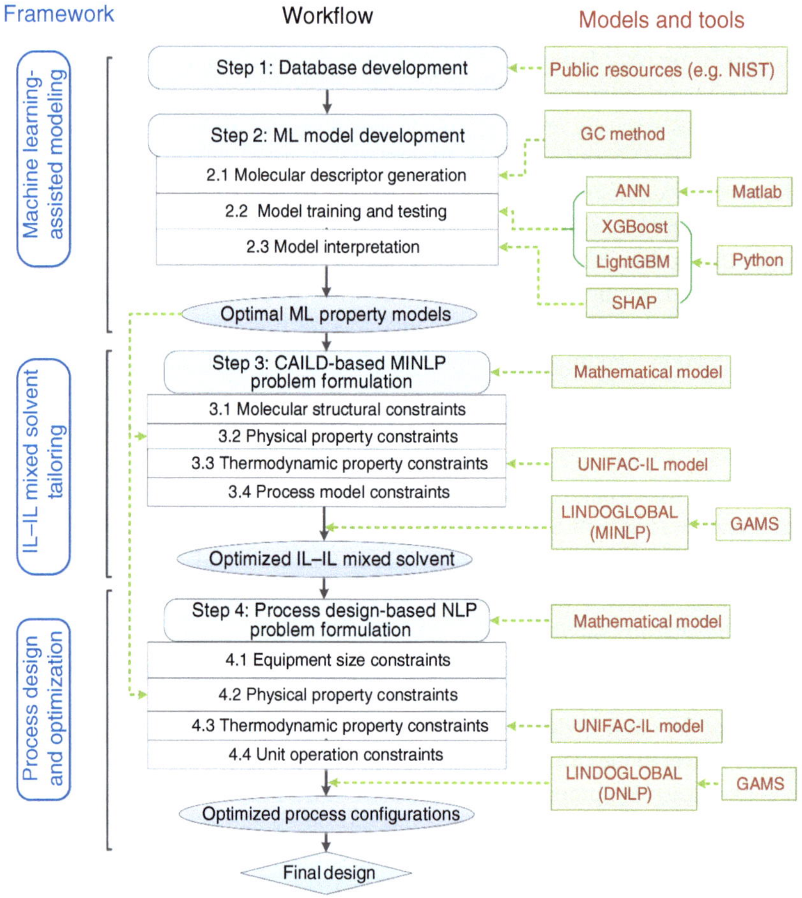

Figure 1.6 Framework of the ML-assisted modeling, solvent tailoring, and process design involving IL–IL mixed solvent. Source: [29]/with permission of John Wiley & Sons.

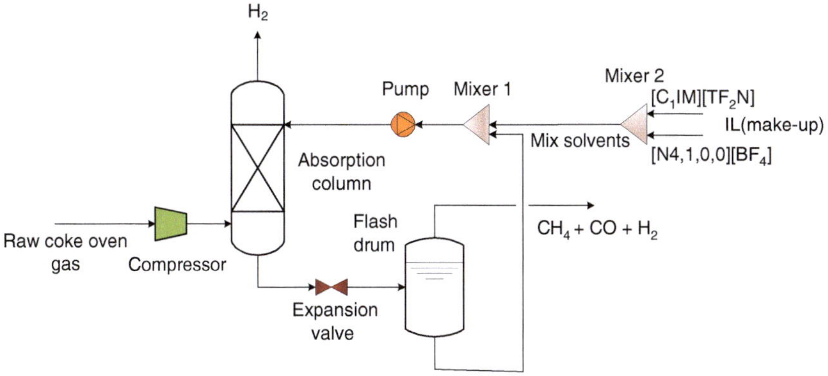

Figure 1.7 Process flowsheet for recovering H_2 from raw coke oven gas using a binary IL–IL mixed solvent. Source: [29]/with permission of John Wiley & Sons.

1.4 Conclusions

Despite the promising capabilities of AI techniques in property modeling, solvent tailoring, and process design, it also presents certain challenges as follows:

Property Modeling: (i) Data might be limited, especially for certain rare compounds or newly synthesized substances. This limitation can impact the training and performance of AI models. (ii) AI models used in chemical property modeling, particularly deep learning models, can be highly complex and difficult to interpret. Understanding how the model arrives at a specific prediction is crucial, especially in applications with potential safety implications. (iii) If the training data used for chemical property models is biased, the AI model may inherit and perpetuate these biases. This leads to inaccurate predictions and potentially unsafe recommendations. (iv) AI models developed for specific chemicals or conditions may not generalize well to different scenarios or substances. Ensuring the transferability of models across a range of chemicals is a challenge. Despite these challenges, the field continues to advance, with ongoing research focused on refining algorithms and expanding applications in diverse scientific domains. In conclusion, property modeling using AI holds significant promise for enhancing our understanding and prediction of various material and chemical properties. Continued research and advancements in AI methodologies are likely to contribute further to the accuracy and applicability of such models in scientific and engineering pursuits.

Solvent Tailoring: (i) AI models, especially ML algorithms, require large and high-quality datasets for training, but obtaining comprehensive and accurate data on solvent properties and behaviors remains a challenge in many cases. Limited or noisy data can lead to biased or inaccurate models, impacting the reliability of AI-driven solvent design. (ii) Solvent design involves complex interactions between molecules, and the chemical space is vast. Capturing these complexities accurately with AI models, especially for novel solvents, poses a significant challenge. Incomplete understanding of the chemical space may result in AI models that struggle to generalize well to diverse solvent systems. (iii) Many AI models, especially deep learning models, are often considered "black boxes" due to their complex architectures. Understanding how and why a model makes a particular prediction in solvent design is challenging. Lack of interpretability can hinder the trust and acceptance of AI-driven results, particularly in critical applications such as solvent design for pharmaceutical or industrial processes. (iv) AI models trained on specific datasets may struggle to generalize well to different conditions or diverse solvent systems not present in the training data. Generalization issues limit the applicability of AI models to a broader range of solvent design scenarios. (v) The integration of AI techniques with experimental data is crucial for validating and refining models. However, reconciling experimental data with computational predictions presents challenges. The lack of seamless integration may hinder the practical application of AI-driven solvent design in laboratory or industrial settings. (vi) Solvent systems can exhibit dynamic behaviors under different conditions. AI models may struggle to adapt to these dynamic changes effectively. Inaccuracy in capturing dynamic behaviors may limit the real-time applicability of AI models in dynamic solvent design scenarios. Addressing these challenges requires a multidisciplinary approach, combining expertise in chemistry, data science, and AI. Ongoing research aims to overcome these challenges and enhance the effectiveness of AI techniques in solvent design.

Process Design: (i) ML algorithms heavily rely on high-quality and abundant data. In some cases, obtaining sufficient and relevant training data can be challenging, and the available data may have errors or biases. Poor data quality can lead to inaccurate models and unreliable predictions. (ii) Many ML algorithms, especially complex ones like deep neural networks, are often considered "black boxes," making it challenging to interpret how they arrive at specific decisions. Lack of interpretability can hinder the understanding of model outputs, making it difficult for engineers to trust and act upon the results. (iii) ML models may overfit the training data capturing noise rather than underlying patterns. Ensuring that models generalize well to unseen data is a challenge. Overfit models may perform poorly on new data, leading to inaccurate predictions in real-world scenarios. (iv) ML models may overfit the training data capturing noise rather than underlying patterns, ensuring that models generalize well to unseen data is a challenge. (v) Validating ML models for complex processes can be challenging, and ensuring their reliability in a real-world environment is crucial. (vi) ML models may struggle to adapt to dynamic processes or changing conditions, as they may not capture evolving patterns effectively. Inaccurate predictions in dynamic environments may lead to suboptimal process control and performance. (vii) Quantifying and managing uncertainties in ML predictions can be challenging, particularly when dealing with complex and uncertain process conditions. (viii) Integrating ML approaches with traditional engineering methods may pose challenges in terms of compatibility and consistency. Inconsistent integration may lead to conflicting results and difficulties in adopting ML-based solutions. Addressing these challenges requires a comprehensive approach, including data preprocessing, model validation, transparency, continuous monitoring, and collaboration between domain experts and data scientists. As the field of AI techniques in process design continues to evolve, overcoming these challenges will be crucial for realizing its full potential.

References

1 Khaleel, M., Ahmed, A.A., and Alsharif, A. (2023). Artificial intelligence in engineering. *Brilliance: Research of Artificial Intelligence* 3 (1): 32–42.

2 Guo, K., Yang, Z., Yu, C.-H., and Buehler, M.J. (2021). Artificial intelligence and machine learning in design of mechanical materials. *Materials Horizons* 8 (4): 1153–1172.

3 Khanzode, K.C.A. and Sarode, R.D. (2020). Advantages and disadvantages of artificial intelligence and machine learning: a literature review. *International Journal of Library & Information Science (IJLIS)* 9 (1): 3.

4 DeCost, B.L., Hattrick-Simpers, J.R., Trautt, Z. et al. (2020). Scientific AI in materials science: a path to a sustainable and scalable paradigm. *Machine Learning: Science and Technology* 1 (3): 033001.

5 Baum, Z.J., Yu, X., Ayala, P.Y. et al. (2021). Artificial intelligence in chemistry: current trends and future directions. *Journal of Chemical Information and Modeling* 61 (7): 3197–3212.

6 Lei, Z., Chen, B., Koo, Y.-M., and MacFarlane, D.R. (2017). Introduction: ionic liquids. *Chemical Reviews* 117: 6633–6635.

7 Liu, K., Wang, Z., Shi, L. et al. (2021). Ionic liquids for high performance lithium metal batteries. *Journal of Energy Chemistry* 59: 320–333.

8 Kerche, E.F., Demétrio da Silva, V., da Cruz, J.A. et al. (2023). Aramid pulp modified with imidazolium ionic liquids as a potential replacement for carbon black in fluorinated rubber composites. *Polymer Composites* 44 (5): 2887–2897.

9 Silva, A.B., Jesus, A.R., Agostinho, D.A. et al. (2023). Using dicationic ionic liquids to upgrade the cytotoxicity and solubility of poorly water-soluble drugs. *Journal of Ionic Liquids* 3 (1): 100052.

10 Greaves, T.L. and Drummond, C.J. (2008). Protic ionic liquids: properties and applications. *Chemical Reviews* 108 (1): 206–237.

11 Rooney, D., Jacquemin, J., and Gardas, R. (2010). Thermophysical properties of ionic liquids. *Ionic Liquids* 290: 185–212.

12 Dai, Z., Chen, Y., Liu, C. et al. (2021). Prediction and verification of heat capacities for pure ionic liquids. *Chinese Journal of Chemical Engineering* 31: 169–176.

13 Iqbal, M., Tao, Y., Xie, S. et al. (2016). Aqueous two-phase system (ATPS): an overview and advances in its applications. *Biological Procedures Online* 18 (1): 1–18.

14 Hatti-Kaul, R. (2001). Aqueous two-phase systems: a general overview. *Molecular Biotechnology* 19: 269–277.

15 Magalhães, F.F., Tavares, A.P., and Freire, M.G. (2021). Advances in aqueous biphasic systems for biotechnology applications. *Current Opinion in Green and Sustainable Chemistry* 27: 100417.

16 Goja, A.M., Yang, H., Cui, M., and Li, C. (2013). Aqueous two-phase extraction advances for bioseparation. *Journal of Bioprocessing & Biotechniques* 4 (1): 1–8.

17 Rito-Palomares, M. and Benavides, J. (2017). *Aqueous Two-phase Systems for Bioprocess Development for the Recovery of Biological Products*. Springer.

18 Anghel, C.I. and Cristea, M.V. (2010). Prediction of thermodynamic properties by artificial intelligence techniques. *Revista de Chimie* 61: 87–93.

19 Liu, X., Gao, J., Chen, Y. et al. (2023). Machine learning-assisted modeling study on the density and heat capacity of ionic liquid-organic solvent binary systems. *Journal of Molecular Liquids* 390: 122972.

20 Baskin, I., Epshtein, A., and Ein-Eli, Y. (2022). Benchmarking machine learning methods for modeling physical properties of ionic liquids. *Journal of Molecular Liquids* 351: 118616.

21 Yusuf, F., Olayiwola, T., and Afagwu, C. (2021). Application of artificial intelligence-based predictive methods in Ionic liquid studies: a review. *Fluid Phase Equilibria* 531: 112898.

22 Koutsoukos, S., Philippi, F., Malaret, F., and Welton, T. (2021). A review on machine learning algorithms for the ionic liquid chemical space. *Chemical Science* 12 (20): 6820–6843.

23 Dhakal, P. and Shah, J.K. (2021). Developing machine learning models for ionic conductivity of imidazolium-based ionic liquids. *Fluid Phase Equilibria* 549: 113208.

24 Datta, R., Ramprasad, R., and Venkatram, S. (2022). Conductivity prediction model for ionic liquids using machine learning. *The Journal of Chemical Physics* 156 (21): 214505.

25 Dhakal, P. and Shah, J.K. (2022). A generalized machine learning model for predicting ionic conductivity of ionic liquids. *Molecular Systems Design and Engineering* 7 (10): 1344–1353.

26 Karakasidis, T.E., Sofos, F., and Tsonos, C. (2022). The electrical conductivity of ionic liquids: numerical and analytical machine learning approaches. *Fluids* 7 (10): 321.

27 Yalcin, D., Le, T.C., Drummond, C.J., and Greaves, T.L. (2019). Machine learning approaches for further developing the understanding of the property trends observed in protic ionic liquid containing solvents. *The Journal of Physical Chemistry B* 123 (18): 4085–4097.

28 Chen, Y., Peng, B., Kontogeorgis, G.M., and Liang, X. (2022). Machine learning for the prediction of viscosity of ionic liquid-water mixtures. *Journal of Molecular Liquids* 350: 118546.

29 Chen Y, Ma S, Lei Y, et al. Ionic liquid binary mixtures: machine learning-assisted modeling, solvent tailoring, process design and optimization. *AIChE Journal* 2024; e18392. doi:https://doi.org/10.1002/aic.18392.

30 Fu, Y., Chen, Y., Zhang, C. et al. (2022). Prediction surface tension of ionic liquid–water mixtures using a hybrid group contribution and artificial neural network method. *Fluid Phase Equilibria* 563: 113571.

31 Lei, Y., Shu, Y., Liu, X. et al. (2023). Predictive modeling on the surface tension and viscosity of ionic liquid-organic solvent mixtures via machine learning. *Journal of the Taiwan Institute of Chemical Engineers* 151: 105140.

32 Pirdashti, M., Taheri, M., Dragoi, E.-N., and Curteanu, S. (2020). Machine learning approaches for prediction of phase equilibria in poly(ethylene glycol) + sodium phosphate aqueous two-phase systems. *Iranian Journal of Chemistry and Chemical Engineering* 39 (6): 185–197.

33 Chen, Y., Liang, X., Woodley, J.M., and Kontogeorgis, G.M. (2022). Modelling study on phase equilibria behavior of ionic liquid-based aqueous biphasic systems. *Chemical Engineering Science* 247: 116904.

34 Chen, Y., Meng, X., Cai, Y. et al. (2021). Optimal aqueous biphasic systems design for the recovery of ionic liquids. *Industrial and Engineering Chemistry Research* 60 (43): 15730–15740.

35 Pazuki, G., Taghikhani, V., and Vossoughi, M. (2010). Prediction of the partition coefficients of biomolecules in polymer–polymer aqueous two-phase systems using the artificial neural network model. *Particulate Science and Technology* 28 (1): 67–73.

36 Chen, Y., Liang, X., and Kontogeorgis, G.M. (2023). Artificial neural network modeling on the polymer-electrolyte aqueous two-phase systems involving biomolecules. *Separation and Purification Technology* 306: 122624.

37 Byrne, F.P., Jin, S., Paggiola, G. et al. (2016). Tools and techniques for solvent selection: green solvent selection guides. *Sustainable Chemical Processes* 4: 1–24.

38 Chen, Y. (2020). Computer-aided design methodology for separation processes with ionic liquids. Ph.D. thesis. Technical University of Denmark.

39 Clarke, C.J., Tu, W.-C., Levers, O. et al. (2018). Green and sustainable solvents in chemical processes. *Chemical Reviews* 118 (2): 747–800.

40 Chen, Y., Koumaditi, E., Gani, R. et al. (2019). Computer-aided design of ionic liquids for hybrid process schemes. *Computers and Chemical Engineering* 130: 106556.

41 Alshehri, A.S., Gani, R., and You, F. (2020). Deep learning and knowledge-based methods for computer-aided molecular design—toward a unified approach: state-of-the-art and future directions. *Computers and Chemical Engineering* 141: 107005.

42 Gani, R., Zhang, L., Kalakul, S., and Cignitti, S. (2016). Computer-aided molecular design and property prediction. In: *Computer Aided Chemical Engineering*, vol. 39 (ed. M. Martín, M.R. Eden, and N.G. Chemmangattuvalappil), 153–196. Elsevier.

43 Wen, H., Nan, S., Wu, D. et al. (2023). A systematic review on intensifications of artificial intelligence assisted green solvent development. *Industrial and Engineering Chemistry Research* 62 (48): 20473–20491.

44 Tripathi, N., Goshisht, M.K., Sahu, S.K., and Arora, C. (2021). Applications of artificial intelligence to drug design and discovery in the big data era: a comprehensive review. *Molecular Diversity* 25 (3): 1643–1664.

45 Liu, Q., Zhang, L., Tang, K. et al. (2021). Machine learning-based atom contribution method for the prediction of surface charge density profiles and solvent design. *AIChE Journal* 67 (2): e17110.

46 Wang, Z., Zhou, T., and Sundmacher, K. (2022). A novel machine learning-based optimization approach for the molecular design of solvents. *Computer Aided Chemical Engineering* 51: 1477–1482.

47 Sui, Y., Zhai, C., Wu, W., and Leung, M.K. (2022). Multi-scale computer-aided molecular design of ionic liquid for absorption heat transformer based on machine learning. *Energy Conversion and Management* 261: 115617.

48 Wang, K., Xu, H., Yang, C., and Qiu, T. (2021). Machine learning-based ionic liquids design and process simulation for CO_2 separation from flue gas. *Green Energy & Environment* 6 (3): 432–443.

49 Zhang, K., Wu, J., Yoo, H., and Lee, Y. (2021). Machine learning-based approach for tailor-made design of ionic liquids: application to CO_2 capture. *Separation and Purification Technology* 275: 119117.

50 Zhang, X., Wang, J., Song, Z., and Zhou, T. (2021). Data-driven ionic liquid design for CO_2 capture: molecular structure optimization and DFT verification. *Industrial and Engineering Chemistry Research* 60 (27): 9992–10000.

51 Liu, X., Chu, J., Huang, S. et al. (2023). Machine learning-based design of ionic liquids at the atomic scale for highly efficient CO_2 capture. *ACS Sustainable Chemistry & Engineering* 11 (24): 8978–8987.

52 Kuroki, N., Suzuki, Y., Kodama, D. et al. (2023). Machine learning-boosted design of ionic liquids for CO_2 absorption and experimental verification. *The Journal of Physical Chemistry B* 127 (9): 2022–2027.

53 Chen, Y., Liang, X., and Kontogeorgis, G.M. (2023). Optimal design of aqueous two-phase systems for biomolecule partitioning. *Industrial and Engineering Chemistry Research* 62 (28): 11165–11177.

54 Schweidtmann, A.M., Esche, E., Fischer, A. et al. (2021). Machine learning in chemical engineering: a perspective. *Chemie Ingenieur Technik* 93 (12): 2029–2039.

55 Jiang, J., Xiong, Y., Zhang, Z., and Rosen, D.W. (2022). Machine learning integrated design for additive manufacturing. *Journal of Intelligent Manufacturing* 33 (4): 1073–1086.

56 Chen, Y., Liu, X., Lei, Y. et al. (2023). A novel hybrid process design for efficient recovery of hydrophilic ionic liquids from dilute aqueous solutions. *AIChE Journal* 69: e18198.

2

Hunting for Better Aromatic Chemicals with AI Techniques

Qilei Liu, Haitao Mao, Lu Wang, and Lei Zhang

Dalian University of Technology, School of Chemical Engineering, Institute of Chemical Process Systems Engineering, Department of Pharmaceutical Sciences, Frontiers Science Center for Smart Materials Oriented Chemical Engineering, State Key Laboratory of Fine Chemicals, No. 2 Linggong Road, Ganjingzi District, Dalian, 116024, China

2.1 Introduction

Chemical-based products have always garnered significant interest in human society, as they are essential for our survival and contribute to improving our quality of life. Aromatic chemicals, in particular, are considered indispensable ingredients in various commercial products such as fragrances, shampoos, detergents, and food items. The incorporation of aromatic chemicals not only enhances the quality of these products but also sets them apart from their competitors. The global fragrance market has witnessed remarkable growth, more than doubling in size over the past 15 years, with profits soaring from US$9.6 billion in 1995 to US$22 billion in 2010 and reaching US$24 billion in 2016 [1]. The escalating demand for fragrances necessitates the development of more efficient design methods for aromatic chemicals.

However, the design of aromatic chemicals is difficult due to the lack of a reliable and efficient structure–odor relationship (SOR). The development of SOR can be classified as heuristic rule-based methods and model-based methods. Among the contributions of the heuristic rule-based methods, plenty of efforts have been put into the analysis of atom location and electronic effect to draw empirical/semiempirical rules [2]. With the help of expert knowledge, the established rules show promising applications in predicting several single odors [3]. However, heuristic rules often suffer from the limitation of application range. Besides, extra conditions could undermine the heuristic rules. In addition, the development of sophisticated rules demands a long period with enormous efforts and costs. One of the examples is that the "Tommy Girl" fragrance required close to 1100 iterations before the product was released to the market [4]. Recently, researchers have started turning to model-based methods, which need efficient and general SOR models. For the odor diffusion process, the macroscopic properties are considered initially, such as partition coefficient, gas chromatographic retention times, and solubility

Applied AI Techniques in the Process Industry: From Molecular Design to Process Design and Optimization, First Edition. Edited by Chang He and Jingzheng Ren.

[5]. Afterward, encouraged by the development of cheminformatics, studies of the integration of diverse categories of structural features/descriptors [6] and statistical models such as multiple linear regression are employed for the correlation of descriptors to odors [7]. One case is the development of the advanced data analysis and pattern recognition toolkit to generate seven categories of descriptors, which are further correlated to musk odor [8]. More recently, Shang et al. [9] have collected descriptors to predict gas chromatography/olfactometry of aromas, demonstrating the efficacy of model-based methods for the general prediction of various odors.

However, most of the developed model-based methods are still traditional ones, such as multiple linear regression models, which often intrinsically fail to accurately model the SOR. The understanding of the odor mechanism is so difficult that simple linear equations often fail to develop general SOR models [10]. In developing general SOR models, the selection of descriptors is necessary due to increasing descriptors causing loads to the development of SOR, but it often encounters the embarrassment where these selected descriptors are based on linear models, and it is hard to explain their responsibilities to the odor at a theoretical level [11].

One of the solutions to the above challenges is to select descriptors correlated tightly with odors to formulate a more sophisticated SOR model. Lavine et al. [12] have proposed an SOR model based on the electronic van der Waals surface descriptors, which has been validated by involving biological and nonbiological molecular behavior [13]. Nevertheless, the deficiency of this type of descriptor is that it only focuses on the process of odor perception but does not consider the information on macroscopic processes such as evaporation and diffusion, which are also essential factors affecting odors [14, 15].

After the selection of descriptors, the SOR model can be established. Recently, machine learning (ML) models have often been applied to overcome the deficiencies of finding appropriate mathematical correlations of SOR in traditional statistical methods. ML is the science of autonomously learning complex relationships from data and has experienced an immensely successful resurgence during the last decade [16]. ML offers the basis of a wealth of fascinating applications, including the estimation of chemical properties [17], the prediction of organic reaction outcomes [18], and facilitating the discovery/design of catalysts, drugs, aromas, and materials [19, 20]. There are also plenty of studies for the modeling of SOR using ML models such as support vector machines [21] and random forests [22].

With the ML-based SOR model developed, aromas can be designed using computer-aided molecular design methods. The design of chemical products involves the identification of molecular structures and/or the compositions of ingredients that satisfy a set of property targets representing a set of desired product needs or attributes. In computer-aided molecular design methods, the property prediction models play an important role, as they connect the molecular structure to the desired properties. The "generate-and-test" method is widely used in computer-aided molecular design, which was proposed by Gani and Brignole [23] and then Joback [24]. It generates feasible solutions and then uses the objective function to evaluate and rank the feasible solutions. However, combinatorial explosions are likely to be encountered, especially when the design problem involves a large number of atoms. Therefore, mathematical programming-based approaches have also been developed for computer-aided molecular design. In this approach, the molecular design problem is formulated as a mixed-integer linear/nonlinear programming (MILP/MINLP) problem. Odele and Macchietto [25] have solved the optimal solvent selection problems using a mathematical programming approach. Computer-aided molecular design methods were first applied to the design of small molecules [26] and mixtures [27]. Recently, the applications have been extended to more types of products, such as amino acids [28] and ionic liquids [29]. Methods and tools are often integrated into computer-aided molecular design to expand its scope to different product design problems. For example, Austin et al. [30] have used quantum mechanics to directly predict the reaction rate in different solvents for reaction solvent design, Kupgan et al. [31] have used molecular dynamics to design polymers for CO_2 capture and separations, and Guo et al. [32] have used ML methods for the discovery and optimization of additives in preparing Cu catalysts for CO_2 reduction. Review articles for computer-aided molecular design methods can be found in Zhang et al. [33, 34]. Although computer-aided molecular design methods have been studied for several decades, most of the research focuses on designing small molecules such as solvents, and the design of aromas with ML methods is limited [20, 35, 36].

Therefore, the aroma design framework using ML-based SOR models is proposed. The remainder of this chapter is structured as follows: In Section 2.2, the ML methods are used to develop the SOR models for predictions of odor properties for pure and mixture aromatic chemicals. In Section 2.3, a computer-aided aroma design (CAAD) framework is proposed for designing novel pure and mixture aromatic chemicals. In Section 2.4, four case studies for the designs of aromatic chemicals are carried out using the proposed ML-based SOR models and the CAAD framework.

2.2 Machine Learning-Based Odor Prediction Models

2.2.1 Odor Predictions for Pure Aromatic Chemicals Using Group-Based Machine Learning Method

2.2.1.1 Database Preparation

Here, the odor pleasantness (OP) and odor characters (OCs) are selected as the required key properties for a fragrance product, which are defined as follows. The odor pleasantness is a scale from the rating of people for a certain molecule, from 0 to 100; the odor characters are classified in terms of the following 20 categories based on people's perception [37], namely "edible," "bakery," "sweet," "fruit," "fish," "garlic," "spices," "cold," "sour," "burnt," "acid," "warm," "musky," "sweaty," "ammonia/urinous," "decayed," "wood," "grass," "flower," and "chemical." Therefore, a radar chart can be plotted for the representation of odor character for each molecule, as Figure 2.1 shows. These 20 categories of odor characters cover most of the odors for the design of fragrance products in the industry.

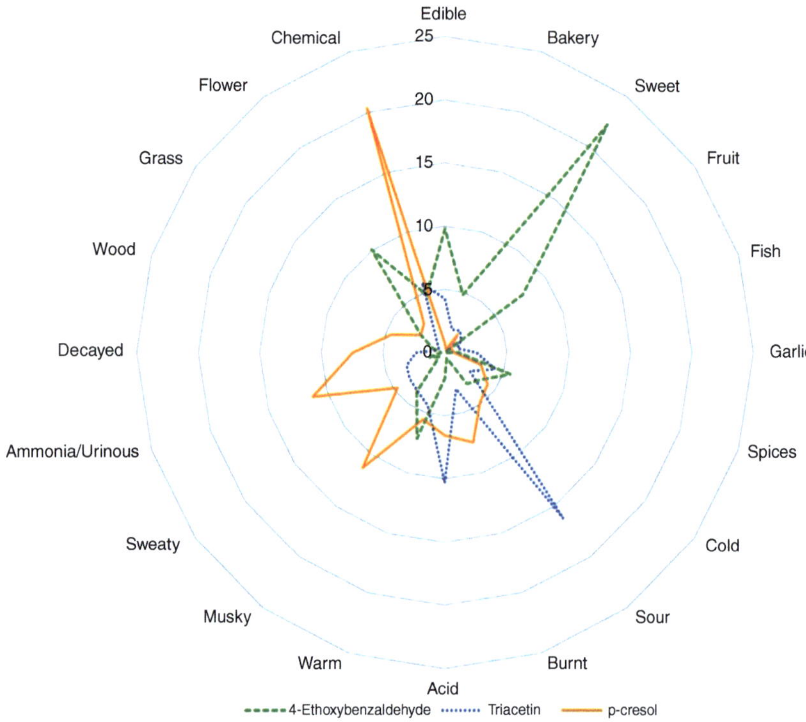

Figure 2.1 Radar chart for the representation of odor characters. Source: Zhang et al. [20]/with permission of Elsevier.

For each molecule in the radar chart, it can be found that there is one character among all the 20 that has the highest value, which is the key odor character. To simplify the problem, only the key odor character is reserved for each molecule. Therefore, we can say the odor of molecule "4-Ethoxybenzaldehyde" is sweet. Similarly, the odor of triacetin is sour, and the odor of *p*-cresol is chemical.

In this section, the database developed by Keller et al. [38] is used. This database has 480 molecules, which are listed in Table 2.A.1 in Appendix A. The molecules have between 1 and 28 nonhydrogen atoms, including 29 amines and 45 carboxylic acids. Two molecules contain halogen atoms, 53 have sulfur atoms, 73 have nitrogen atoms, and 420 have oxygen atoms. The molecules are structurally and chemically diverse, and many of them have unfamiliar smells; some have never been used in prior psychophysical experiments.

2.2.1.2 Molecular Representation

Next, the representation of molecular structures needs to be determined as the input of the ML model. Fragment (group)-based representation is commonly used in group contribution methods. It has been shown that the properties of a molecule can be determined with relatively high accuracy by summation of the contributions of the associated groups. This section uses group-based descriptors to develop ML models for predictions of the odor property. Here, 50 groups are selected, which are given in Table 2.A.2 in Appendix A.

2.2.1.3 Model Architecture

A convolutional neural network (CNN) is a class of deep, feed-forward artificial neural network (ANN), which consists of an input and an output layer, as well as multiple hidden layers, including convolutional, pooling, or fully connected layers. Python Keras [39] is used in this section for the development of the CNN-based SOR models. The input to the model is a 50×1 vector of groups, and the output is odor properties, including odor characters and odor pleasantness. The layer information is shown in Figure 2.2. In Keras, the embedding layers and the flatten layers are used for reshaping the data. The dropout layer is used to prevent overfitting. The dense layer is a fully connected layer, so the neurons in the layer are connected to those in the next layer.

As shown in Figure 2.2, the established CNN model consists of a 50×64 embedding layer, a 47×128 convolutional layer, a 44×128 convolutional layer, a 22×128 max-pooling layer, a 22×128 dropout layer, a 2816×1 flatten layer, a 128×1 dense layer, another 128×1 dropout layer, and a 20×1 dense layer. Finally, the properties of odor characters and odor pleasantness are predicted using this model architecture, with different trained parameters.

2.2.1.4 Results and Discussions

The 480 molecules in the database are tested using the established ML models. The training of the ML models is implemented on a desktop computer with Intel Core i7-7700 CPU and 16G memory. The training time is 696.8 seconds. The predicted results of odor characters and odor pleasantness are compared with the experimental data as shown in Figures 2.3 and 2.4. To account for the diverse psychophysical properties of odor pleasantness among individuals, it is not necessary to obtain continuous values. Instead, the odor pleasantness can be discretized into five levels based on the original odor pleasantness values. For example, level 1 can represent a range of 0–20, level 2 can represent a range of 20–40, and so on. This discretization approach enhances the representativeness and applicability of the model. Therefore, the prediction results shown in Figure 2.4 are also discretized into five levels.

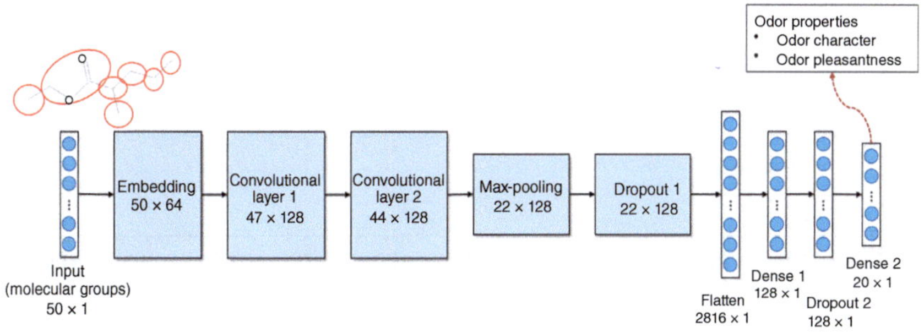

Figure 2.2 CNN layer information for SOR model. Source: Zhang et al. [20]/with permission of Elsevier.

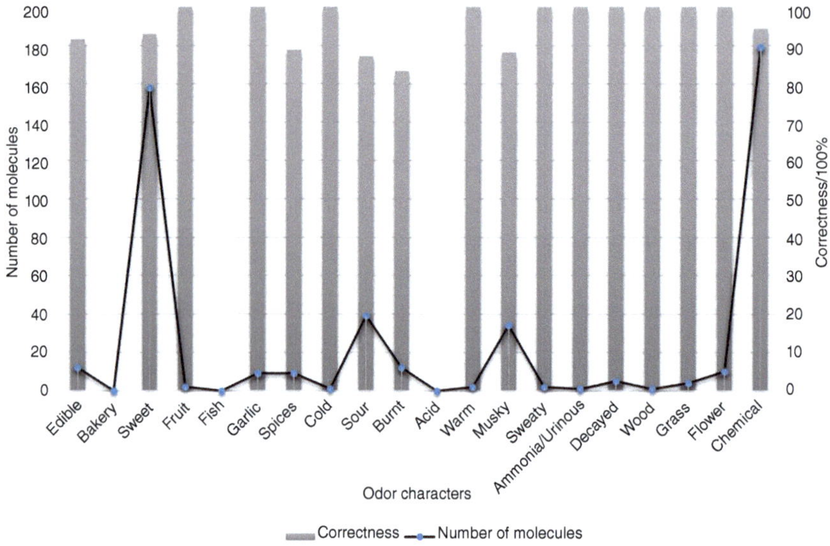

Figure 2.3 The predicted results of the 480 molecules in the database for odor characters. Source: Zhang et al. [20]/with permission of Elsevier.

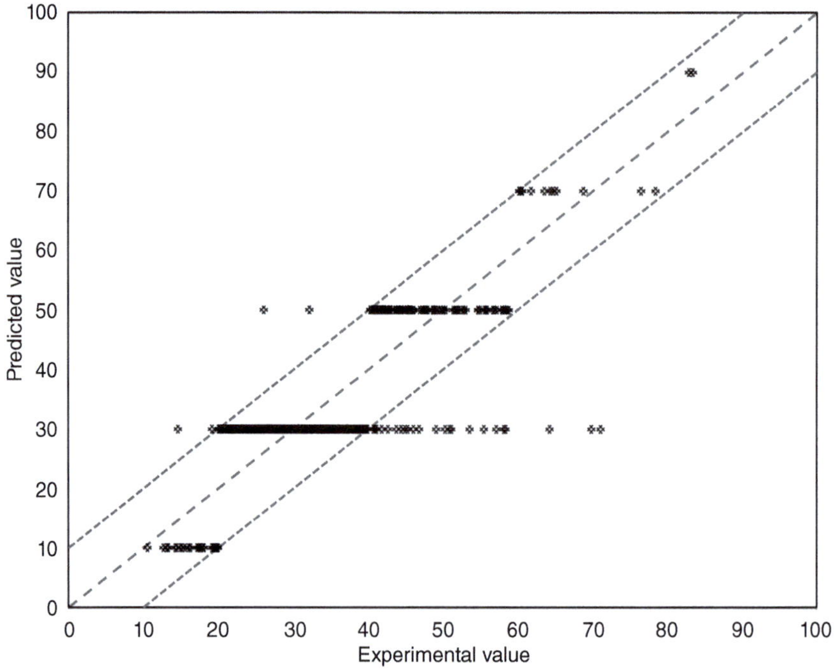

Figure 2.4 The comparison of experimental values and predicted results of odor pleasantness (scale from 0 to 100) for the 480 molecules in the database. Source: Zhang et al. [20]/with permission of Elsevier.

From the results in Figures 2.3 and 2.4, it can be seen that both predictions of odor characters and pleasantness are accurate using the developed ML model, which is tested using the 480 molecules in the database. The average correctness of odor characters is 92.9%, while the average prediction error of odor pleasantness is 18.4%. In Figure 2.3, the black line shows the number of molecules for a typical odor character in the database, while the bar shows the correctness of the model prediction. In the database, characters "sweet" and "chemical" possess the largest number of molecules, while other ones possess smaller but sufficient numbers of molecules. In Figure 2.4, the two dashed lines indicate the acceptable range for the predicted properties, that is, the predicted property (indicated by dots) must be inside the region covered by the dashed lines. From Figure 2.4, 27 molecules are out of the acceptable range. Since the odor pleasantness experimental values are obtained from the rating of people, the data may not be quite accurate. Therefore, although most of the characters and pleasantness have satisfactory correctness, the prediction of odor characters for molecules outside the database has to be re-evaluated. Several aromatic chemicals outside the database, which are commonly used in our daily lives, are evaluated using the ML model. The evaluation results show roughly 75% correctness for molecules outside the database using the developed ML model.

It is crucial to emphasize that each molecule is defined with only one odor character out of the 20 available. Consequently, molecules that possess multiple distinct smells are represented by a single odor character, which can impact the training of the CNN. This limitation is evident in the replacement of "floral" and "vanilla" with "sweet." This finding demonstrates the existence of correlations between different odor characters, which are quantified using mutual information as defined in Eq. (2.1). The mutual information between each pair of odor characters is illustrated in Figure 2.5.

$$I(X : Y) = \sum\nolimits_{x \in X} \sum\nolimits_{y \in Y} p(x, y) \log \frac{p(x, y)}{p(x)p(y)} \tag{2.1}$$

where $p(x)$ is the probability density of sample value x in a character vector X, $p(x, y)$ is the joint probability density of sample values x and y, and I is the mutual value of characters X and Y. From this definition, the greater the value I, the more correlated the two characters.

As shown in Figure 2.5, the mutual information value of "sweet" and "flower" is 0.78, while "sweet" and "fruity" is 0.85. This result shows that these characters are highly correlated. Therefore, the SOR model is acceptable for further aroma design problems. However, the prediction models can still be improved by introducing more molecules into the database, considering more detailed odor characters and more detailed molecular structures during the model development.

2.2.2 Odor Prediction for Mixture Aromatic Chemicals Using σ-Profiles-Based Machine Learning Method

2.2.2.1 Database Preparation

The odor property data is from Keller and Vosshall [37]. They have provided 480 different molecules at four dilutions ("1/10," "1/1000," "1/100 000," and "1/100 000")

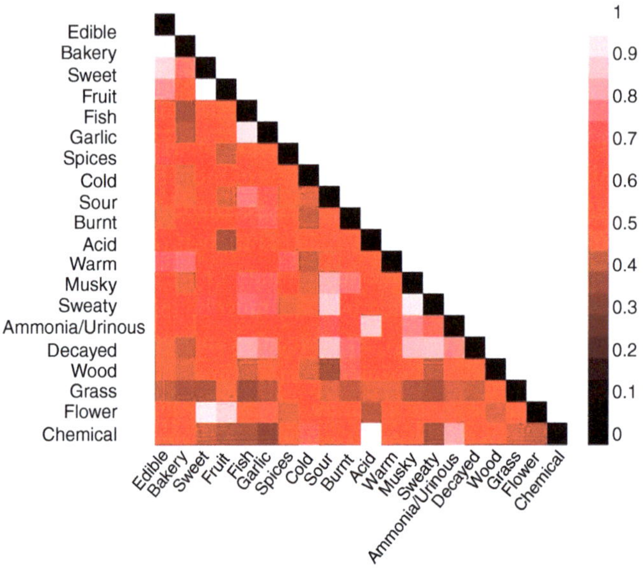

Figure 2.5 The heat map for mutual information of 20 odor characters. Source: Zhang et al. [20]/with permission of Elsevier.

to be tested by 49 healthy subjects. These subjects are required to evaluate molecules using a score from 0 to 100 for the perception of 22 specific odors, including 20 odor characters (edible, bakery, sweet, fruit, fish, garlic, spices, cold, sour, burnt, acid, warm, musky, sweaty, ammonia, decayed, wood, grass, flower, and chemical) and two odor perception characters (intensity and pleasantness). Their database of molecules includes not only familiar odor characters but also unfamiliar odor characters and even some odorless molecules like water. This odor database is applied in this section to establish the ML model.

Since different social habits, living environments, and individual differences exist when obtaining the data in the odor database, the data needs a preprocessing step to obtain standard, rescaled, and pruned data. First, the data of 1/1000 dilution is selected as it possesses the largest amount in the database. Since people have inherent diversities, someone might be insensitive to one type of odor character or even to some typical molecules, which means the score of odors made by the subject is blank. Therefore, scored data is preprocessed to eliminate the influence of individual differences. Here, two heuristic rules are applied for preprocessing scores from subjects who are not insensitive to the odor types:

Rule 1: If a subject is insensitive to one molecule, all scores of this subject to the molecule that are blanks will be eliminated.

Rule 2: If a subject is insensitive to one odor character of one molecule but is responsive to other molecules, the score of this odor character that is blank is set to zero.

With the above data selection steps, around 18 000 data are selected among 55 000 data points.

Second, to avoid the disturbance from individual differences, data averaging has proceeded for the scores of the odor characters (details can be found in Appendix B). Then, the odor property data is normalized to [0, 1] using the min–max normalization method.

To avoid overfitting, the preprocessed dataset is divided into training and validation sets. Here, k-folds cross-validation method is used [16], in which the whole dataset is divided into k ($k = 5$) subsets, and each subset is selected sequentially as the validation dataset, while the others are the training dataset. In this way, the model can be trained using the most representative dataset to avoid overfitting.

2.2.2.2 Molecular Representation

The molecular surface charge density profiles (σ-profiles) of a molecule characterize its electrostatic polarity and charge distribution, which is determined by its molecular structure. Therefore, the σ-profiles can be used as descriptors for property prediction due to their ability to represent the molecular structure. In this section, the σ-profiles are used as descriptors to establish the SOR model using ML methods. Here, the σ-profiles of a molecule are calculated from the conductor-like screening model-segment activity coefficient (COSMO-SAC) model [40]. Figure 2.6 presents the process of odor perception for aroma molecules as well as the workflow for obtaining the σ-profiles descriptors.

In Figure 2.6, an example molecule of Tyrosine (CAS No. 60-18-4) is shown in Figure 2.6a, and its odor is perceived when it binds to olfactory receptors in the human nose, which is shown in Figure 2.6d. The following two computation steps include geometry optimization and COSMO energy calculation. Then, the COSMO surface of the molecule is obtained, which is shown in Figure 2.6b. Afterward, the surface charge densities from the COSMO output are averaged, as shown in Figure 2.6c. These averaged charges are further projected into the two-dimensional spectrum (σ-profiles) (Figure 2.6e), which is used as descriptors for the prediction of physicochemical properties (Figure 2.6f).

After the above steps, all charges of the molecular COSMO surface are converted into a two-dimensional spectrum σ-profiles with the range from $-0.025\ \mathrm{e\,\mathring{A}^{-2}}$ ($-2.5\,\mathrm{e\,nm^{-2}}$) to $0.025\ \mathrm{e\,\mathring{A}^{-2}}$ ($2.5\,\mathrm{e\,nm^{-2}}$) with 51 data points in total. As shown in Figure 2.7, the σ-profiles of an aroma molecule are divided into 10 segments in the entire σ region. Thus, 10 areas (S_1–S_{10}) are obtained for each molecule by integrating those segments [42]. Afterward, the 10 area parameters are used as descriptors to quantify their effects on the odor properties. It is noted that min–max normalization is then employed to scale the σ-profiles descriptors (S_1–S_{10}) to [0, 1] to ensure all features (descriptors) have the same scale and for the purpose of obtaining the global optimum [16].

As the dilutions of the selected aroma data are 1/1000 from the database, which are dilute, the binary interaction of the ingredients can be neglected. Therefore, for aroma mixtures, a linear mixing rule (Eq. (2.2)) is applied for the σ-profiles descriptors, when the binary interaction between different ingredients of the mixture is neglected. A similar assumption has also been made from previous research [30, 43].

$$S_i^{\mathrm{M}} = \sum\nolimits_j x_j S_i^j \quad \forall i \in \{1, 2, \ldots, 10\} \tag{2.2}$$

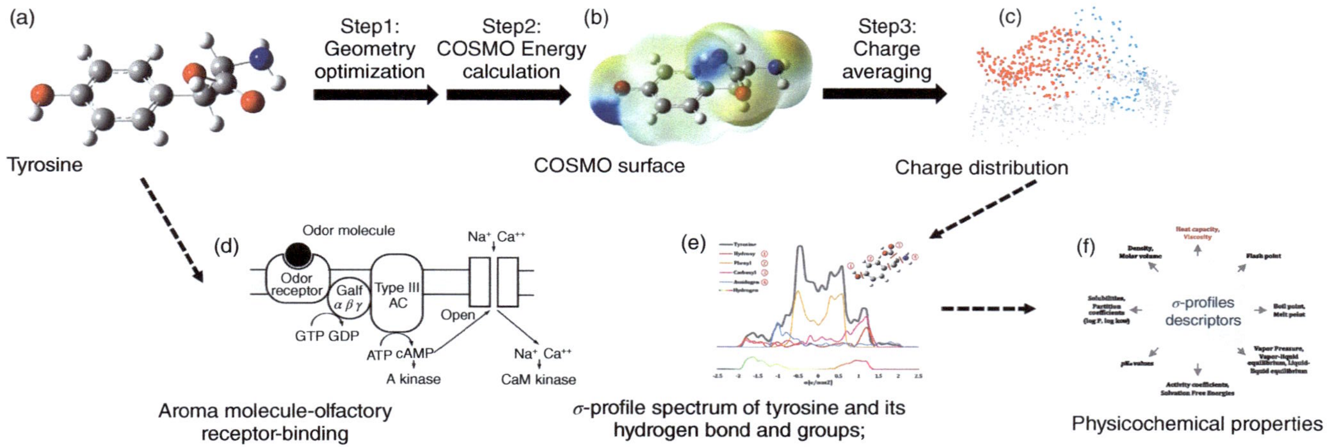

Figure 2.6 σ-Profiles descriptors for aroma molecules. Source: Zhang et al. [41]/with permission of Elsevier.

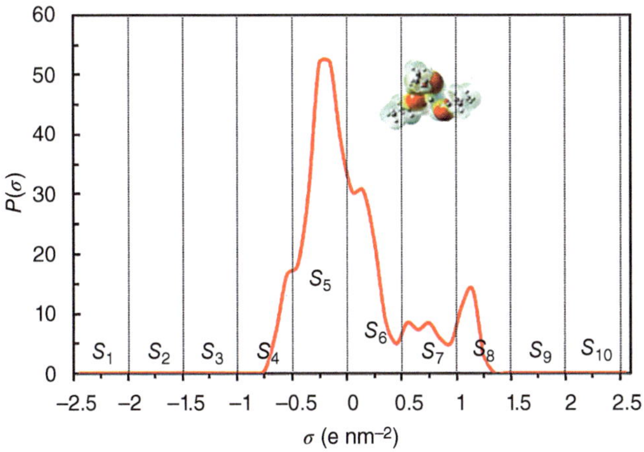

Figure 2.7 σ-Profiles descriptors of an example aroma. Source: Zhang et al. [41]/with permission of Elsevier.

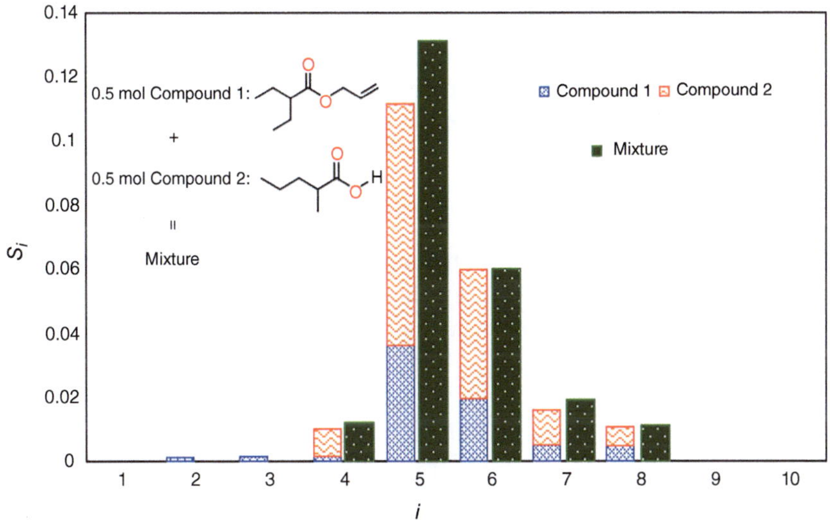

Figure 2.8 Linear mixing rule for aroma mixtures is reasonable. Source: Zhang et al. [41]/with permission of Elsevier.

In Eq. (2.2), i is the index of 10 area parameters of the descriptors, j is the index of compounds in the aroma mixture, S_i^{M} is the ith area parameter of the mixture, x_j is the volume fraction of compound j, and S_i^j is the ith area parameter of compound j.

Figure 2.8 shows the comparison of σ-profiles descriptors of two example aroma molecules and the mixture from the COSMO-SAC method and the linear mixing rule. From the comparison, it is seen that the error of the linear mixing rule for the calculation of σ-profiles descriptors for aroma mixtures is acceptable.

2.2.2.3 Model Architecture

After the selection of descriptors, the SOR model is established with the aroma mixture properties as the input and σ-profiles descriptors as the output. In this way, with a certain set of requirements for aroma design problems, the optimal aroma mixture can be designed directly with the ANN model. It should be noted that it is possible that the optimal aroma mixture from the established ML model is not unique. The design results can be further tested by experiments, which could greatly reduce the cost and manpower in the aroma design process.

ANN is employed for the modeling of SOR. As Figure 2.9 shows, the aroma mixture properties including seven odor properties (edible, bakery, sweet, fruit, sour, flower, and odor pleasantness) and two physical properties (vapor pressure and diffusion coefficient) are taken as the input, and the 10 area parameters ($S_1, S_2, \ldots, S_{10}$) of σ-profiles are taken as the output. Here, multiple-model strategy is used, where each ANN model uses all the input parameters to predict one output (S_i) separately.

In Figure 2.9, each ANN model utilizes a multilayer perceptron structure. The architecture of each ANN model comprises multiple layers, with each layer consisting of a set of nodes referred to as "neurons." The detailed architecture of each ANN model is depicted in Figure 2.10.

A single neuron contains two neuron layers: one consists of the input neurons $\boldsymbol{x}$ and the other consists of the output neurons $\boldsymbol{y}$. The connections between the inputs and outputs are called weights $\boldsymbol{w}$, the threshold values of outputs are called biases $\boldsymbol{b}$, and the activate function is f_{act}. The inputs and outputs of a single neuron can be represented as Eq. (2.3).

$$\boldsymbol{y} = f_{act}(\boldsymbol{w} \cdot \boldsymbol{x} + \boldsymbol{b}) \tag{2.3}$$

The established ANN model is then trained to obtain an optimized model using the training and validation dataset. The training process aims to minimize the cost function f_{cost} of model M by employing the learning algorithm F, which can be

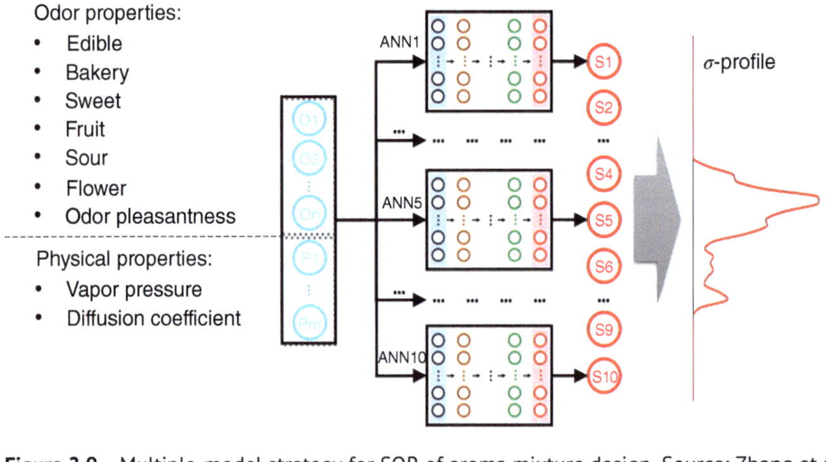

Figure 2.9 Multiple-model strategy for SOR of aroma mixture design. Source: Zhang et al. [41]/with permission of Elsevier.

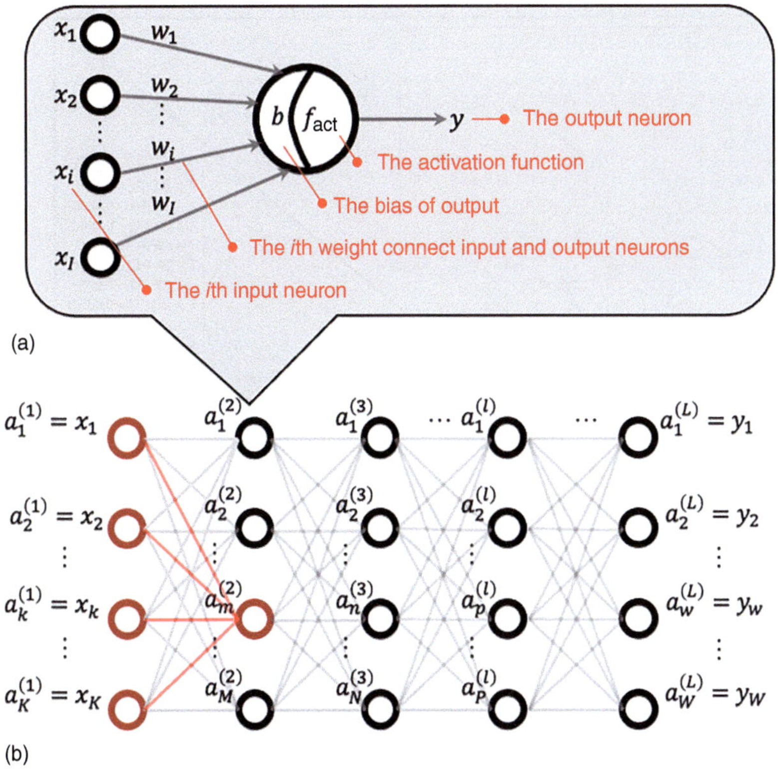

Figure 2.10 Model architecture of one ANN model. Source: Zhang et al. [41]/with permission of Elsevier. (a) A single neutron and (b) an ANN model representation.

represented as Eq. (2.4).

$$M = \mathcal{F}([\boldsymbol{x}', \boldsymbol{y}'], \boldsymbol{P}, f_{\text{cost}}(M(\boldsymbol{x}'), \boldsymbol{y}')) \quad \boldsymbol{P} \in \mathbb{R}^k \tag{2.4}$$

where $\boldsymbol{P}$ is the hyperparameters for the ML model, $\boldsymbol{x}'$ is the preprocessed input data (odor and physical properties), and $\boldsymbol{y}'$ is the preprocessed output data (σ-profiles descriptors). The performance of the ANN model is determined by the number of hyperparameters such as neuron layers [16]. When a new layer is added to the ANN model, the previously determined parameters have to be tuned again. The hyperparameters of the ANN model $\boldsymbol{P}$ are tuned by using the grid search method.

2.2.2.4 Results and Discussions

The ANN model is established and trained using Python Keras package [39] using a desktop with i7-7500 CPU, 8G memory. After determining the hyperparameters, each submodel is trained five times, and the average values are compared and analyzed, which are listed in Table 2.C.1 in Appendix C. Figure 2.11 shows the performances of 10 models in the testing dataset. The average R^2 for the full model is 0.88, which is acceptable considering the distribution of samples in the testing dataset may be inconsistent with the distribution of samples in the training dataset. Hence, the established ANN model can be used for further aroma mixture design.

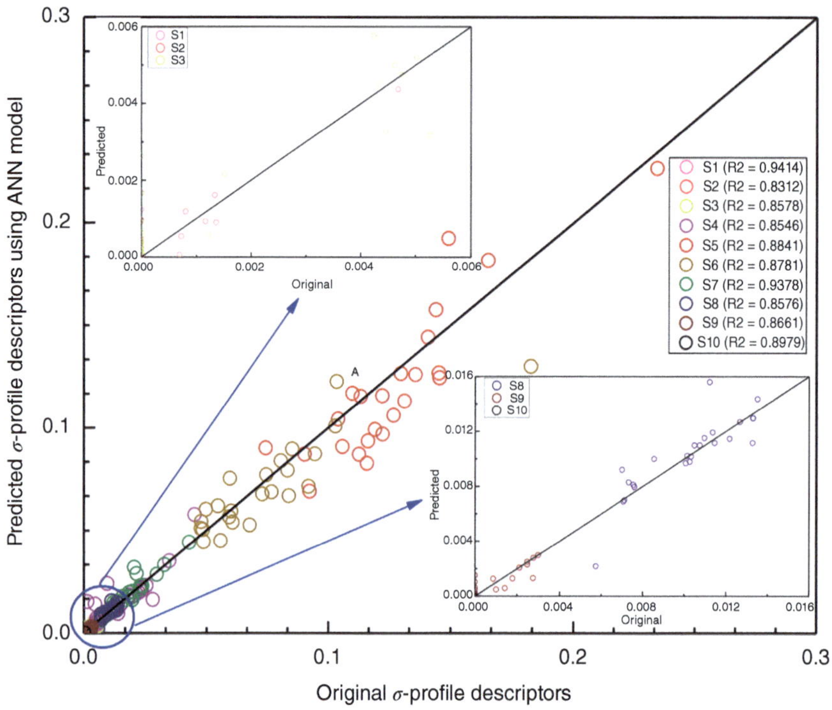

Figure 2.11 Performance of the established ANN model. Source: Zhang et al. [41]/with permission of Elsevier.

2.3 Computer-Aided Aroma Design (CAAD) Framework

2.3.1 CAAD for Pure Aromatic Chemicals

Aromatic chemicals need to be designed/screened to meet all their sensorial and technical requirements. CAAD uses mathematical modeling to determine the molecular structure that matches the given set of target properties with the given set of groups used in the property models. Here, a CAAD framework is established for the design of novel pure aromatic chemicals. In this CAAD framework, the odor of the molecules is predicted using the developed ML models, while group contribution-based models are included to predict the rest of the needed physical properties, such as vapor pressure, solubility parameter, and viscosity. The CAAD problem is formulated as an MINLP model for the design of aromatic chemicals. The decomposition-based solution approach [44] is used to obtain the optimal result. Figure 2.12 shows the steps for the development and solution of the MINLP model. The design steps of the aromatic chemicals are discussed in this section.

2.3.1.1 Identify Product Attributes

The first step is to identify the product attributes. The aromatic chemicals are the active ingredients or the additives of the fragrances. After the application of the

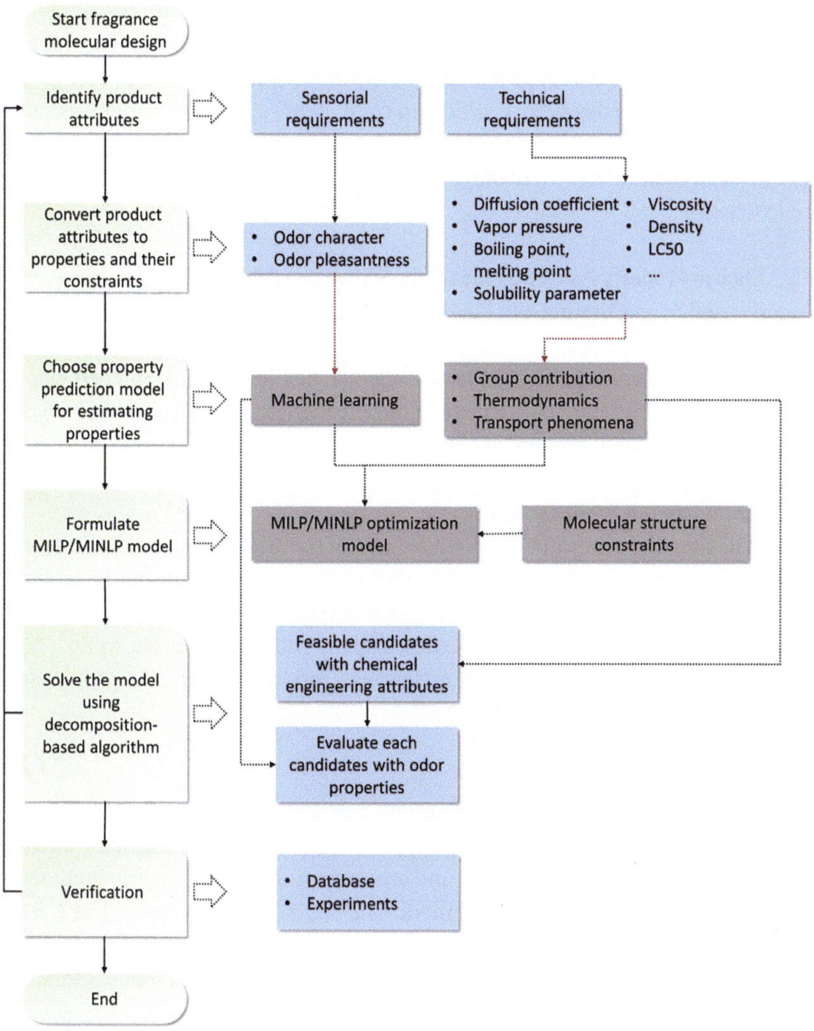

Figure 2.12 Steps for the development and solution of the MINLP model for the design of aromatic chemicals. Source: Zhang et al. [20]/with permission of Elsevier.

fragrance, the aromatic chemicals begin to evaporate into the headspace at different rates depending on their volatility, composition, and molecular interactions. Subsequently, the gas odorant molecules will diffuse through the surrounding air over time and distance, and finally, at a given time and distance, some of the aromatic chemicals will eventually reach the nose of the customer who perceives the odorants with a certain intensity and character [14, 15]. From this process, it can be seen that the odor properties, including odor character and odor pleasantness, and the physicochemical properties including diffusion, evaporation, and so on, are the required product attributes.

2.3.1.2 Convert Product Attributes to Properties and Their Constraints

In the second step, the product attributes are converted to properties and represented as constraints in the MINLP model. The product attributes are converted to physicochemical properties such as diffusion coefficient (diffusion), vapor pressure (evaporation), boiling and melting points (product form), solubility parameter (solubility), viscosity and density (rheology), and LC_{50} (environment and health), and their constraints.

2.3.1.3 Choose Property Prediction Model for Estimating Properties

These converted properties need to be estimated using appropriate predictive property models from the model library. The established ML model is used for odor characters and pleasantness, while for other properties (e.g., diffusion coefficient, vapor pressure, boiling point), group contribution method, theories for thermodynamics, and transport phenomena are used.

2.3.1.4 Formulate MILP/MINLP Model

The aroma design problem can be formulated as a MILP/MINLP optimization problem. The optimization model includes an objective function, molecular structure constraints, and property constraints. The objective function can be used to maximize/minimize one of the desired properties, such as odor pleasantness, as Eq. (2.5) shows. OP is predicted using the machine learning model, in which the groups are the input variables.

$$\max \text{OP} \tag{2.5}$$

The structure constraints define the molecular structure from the combination of a set of feasible groups. Therefore, the groups are selected first. For aromatic chemicals, groups containing oxygen, sulfur, and aromatic rings are commonly included, such as alcohols, ketones, esters, and ethers. The set of groups is defined as i or i' in the optimization model. The following equations define the property constraints. The odor character (OC) should be sweet, fruit, or flower from the 20 characters, as Eq. (2.6) shows.

$$\text{OC} = \text{sweet}, \quad \text{fruit}, \quad \text{or flower} \tag{2.6}$$

The constraint for odor pleasantness (OP) is defined in Eq. (2.7). The aromatic chemicals should have a pleasant odor; therefore, OP should be greater than a certain value (OP^L).

$$\text{OP} \geq \text{OP}^L \tag{2.7}$$

The detailed structure constraints and other property constraints can be found in Zhang et al. [20].

2.3.1.5 Solve the Model Using Decomposition-Based Algorithm

From the solution of the established MINLP model, the optimal aromatic chemicals are found. However, the model contains properties that have to be obtained from ML models, and at the same time, the property model equations for diffusion coefficient and vapor pressure are nonlinear equations. The existing MINLP solvers are not able to solve the design problems. Therefore, the decomposition-based algorithm [44] is used to solve the MINLP model. In the decomposition-based algorithm, the MILP/MINLP model is decomposed into an ordered set of subproblems. Each subproblem requires only the solution of a subset of the constraints from the original set. The final subproblem contains the objective function and the remaining constraints. In this way, the solution of the decomposed set of subproblems is equivalent to that of the original MILP/MINLP problem. Here, the structure constraints, and the property constraints, except the odor character, odor pleasantness, diffusion coefficient, and vapor pressure model, are considered in the first subproblem to generate feasible candidates. Then, each molecule in the set of candidates is tested using the equations of diffusion coefficient, vapor pressure, and the ML models of odor character and pleasantness. In this way, the size of the feasible candidates set becomes smaller, and finally, the optimal one (or several candidates) can be selected based on the evaluation of the objective function. Here, the model formulation and solution of the CADD problem has been implemented into a self-developed software "OptCAMD" [45]. Figure 2.13 shows the software architecture of OptCAMD.

The interface collects all the input data from the user, including structure information such as group selection, group numbers, and all the property constraints. Then the interface transforms and reformulates all the input data to an Excel file, which is needed in the GAMS code template. The GAMS code template calls the solver to solve the optimization problem and generates a list of feasible groups based on the ranking of the objective value. A molecular database (OptCAMD database) is integrated into the software, which contains around 10 000 commonly existing molecules. The generated groups are compared with the molecules in the database, and feasible molecules are returned. These molecules can be imported to ICAS ProPred [46] and PubChem database [47] for verification. Finally, the generated molecules are displayed in the interface. The user can also modify the GAMS code template to customize the optimization problem.

2.3.1.6 Verification

In this step, the obtained aromatic chemicals from the optimization model are verified through database search and/or experiments. If the designed molecules are known and exist in the database, then their measured property values can be verified from the database, or use of rigorous property prediction models, such as ICAS [46] and VPPD-LAB [48]. If the designed molecules cannot be found in any database, that means they are new molecules or their properties have not been measured yet, but they could be used in aroma products. Therefore, experiments are recommended to verify their properties and usability.

2.3.2 CAAD for Mixture Aromatic Chemicals

In this section, a CAAD framework is proposed for the design of a novel mixture of aromatic chemicals with the ML-based SOR models developed in Section 2.2.

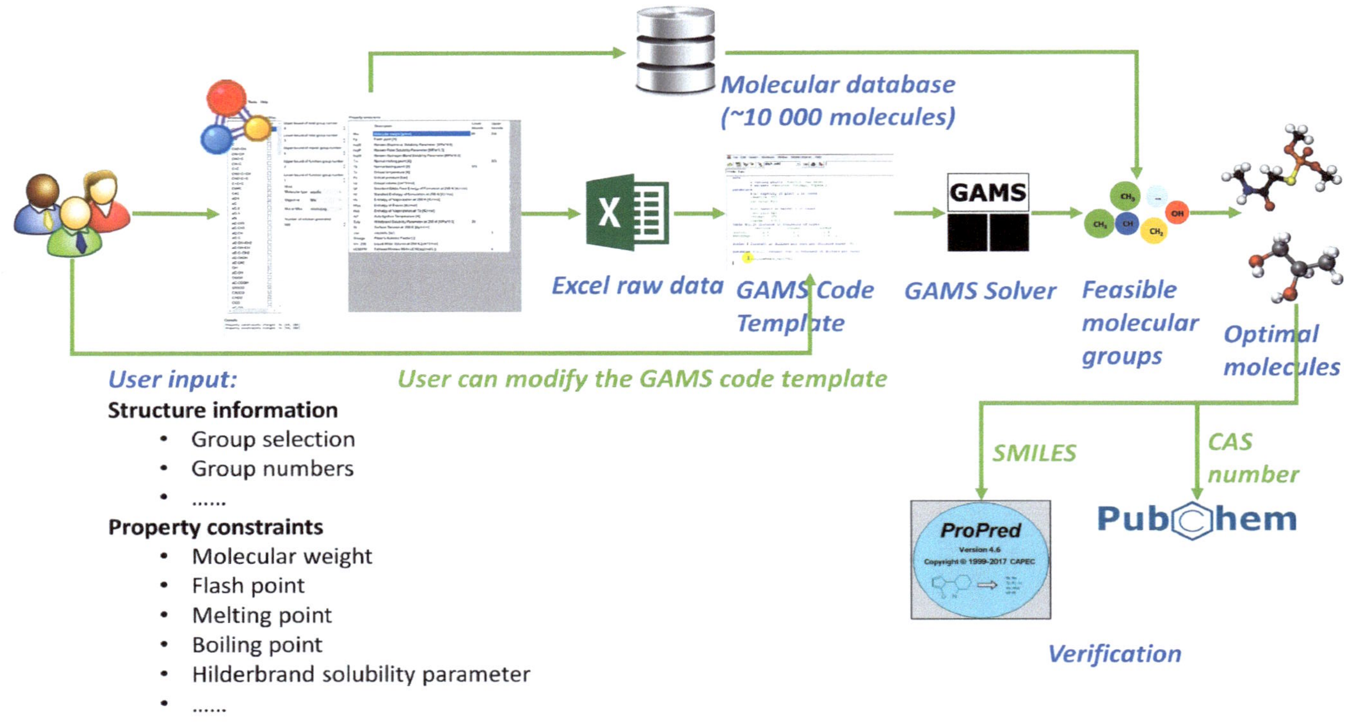

Figure 2.13 Software architecture of OptCAMD. Source: Zhang et al. [20]/with permission of Elsevier.

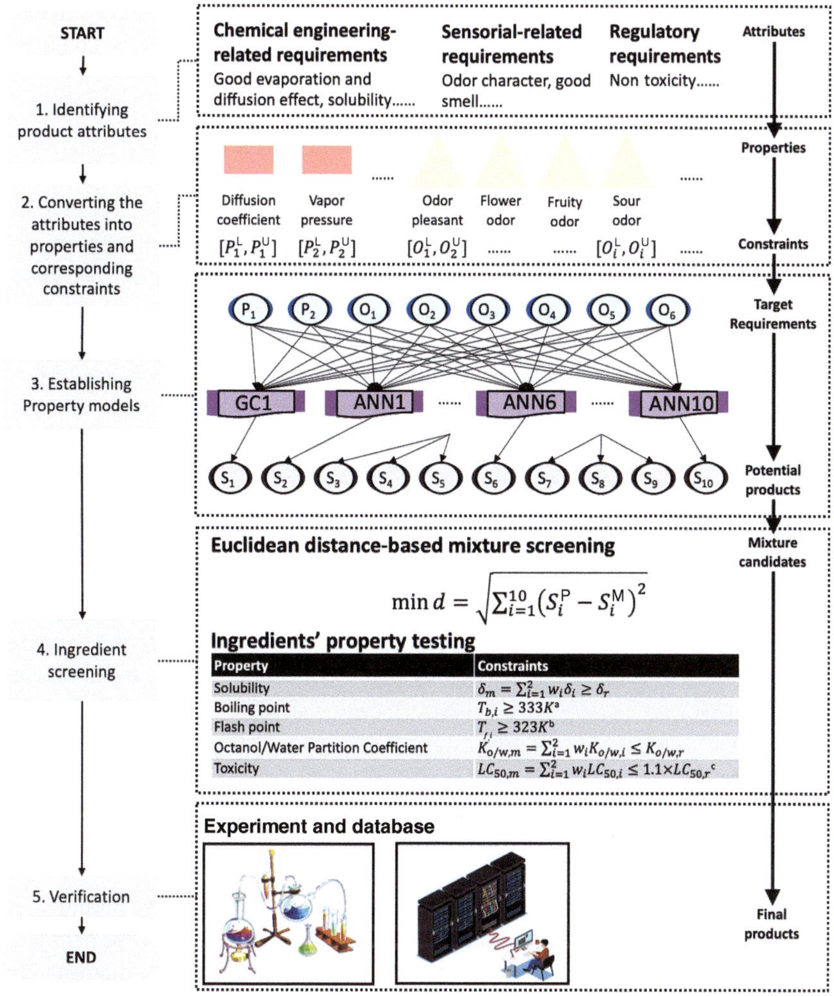

Figure 2.14 Computer-aided aroma design for novel mixture aromatic chemicals. Source: Zhang et al. [41]/with permission of Elsevier.

Figure 2.14 shows the proposed CAAD framework for mixture aroma design. The steps of the CAAD framework are discussed in detail in the following text.

2.3.2.1 Identify Product Attributes

The first step of the CAAD framework is identifying the product attributes. Here, a set of requirements are collected and determined according to the product application background. The requirements could be categorized into chemical engineering-related, sensorial-related, and regulatory attributes. For example, in candle fragrance products, the aroma is designed with the technical requirements of phase stability, evaporating and diffusing properties, duration, etc. The sensorial-related attributes include requirements of odor characters and pleasantness of the aroma, which determine the product performance. Furthermore, the regulatory requirements, including toxicity, safety, and so on, should be considered. These product attributes are usually collected by the market study of potential consumers.

2.3.2.2 Convert Product Attributes into Properties and Corresponding Constraints

After identifying the product attributes, they are converted into properties such as flash point, boiling and melting points, vapor pressure, diffusion coefficient, vapor pressure, LC_{50}, etc., which provide the translation from the marketing specialists to the R&D engineers. These product attributes are converted to properties, which can be predicted by theoretical property models, ML models, and/or obtained by experiments, with corresponding constraints. The boundaries of these constraints are obtained by satisfying the expectation of the product attributes, which are normally derived from consumer surveys or reference data from available products.

2.3.2.3 Establish Property Models

The property models are established in this step to correlate the product structure/mixture ingredient composition with the properties. According to different properties and their available models, different property prediction models are selected. For example, group contribution methods [26] are often applied to some common properties such as normal boiling point, critical properties, and so on, as Eq. (2.8) shows.

$$p = \sum_i f_i(n_i, p_i) \tag{2.8}$$

In Eq. (2.8), p is the properties, n_i is the number of group i, and p_i is the contribution of group i in property p, f_i is a certain function in group contribution methods, which is different in different properties.

For property models that are not available, but for which a large number of data exists, ML models can be established. Examples include the SOR model established in Section 2.2. If only limited experimental data exists, quantitative structure–property relationship models can be established by regressing the experimental data. For aroma mixture design problems, the σ-profiles descriptors ($S_1 \sim S_{10}$) are obtained from the input properties using the established ANN model.

2.3.2.4 Ingredient Screening

After the σ-profiles descriptors ($S_1 \sim S_{10}$) obtained, the ingredients of the mixture and their compositions need to be screened to find the corresponding aroma mixture that fits the σ-profiles descriptors. The ingredients are searched from the odor database [37], which has 480 molecules. The σ-profiles descriptors of the mixture are calculated using Eq. (2.2) if the ingredients and their compositions are known. In this step, Euclidean distance is used to evaluate the difference between the σ-profiles descriptors of the designed mixtures and the predicted values, which is shown in Eq. (2.9).

$$\min\ d = \sqrt{\sum_{i=1}^{10} \left(S_i^{\mathrm{P}} - S_i^{\mathrm{M}}\right)^2} \tag{2.9}$$

where d is the Euclidean distance between the σ-profiles descriptors of the designed mixtures and the predicted values, which should be minimized. S_i^{P} are the predicted values of σ-profiles area parameter i (i = 1, 2,…, 10) using the ANN-based SOR model. S_i^{M} are the values of σ-profiles area parameter i of the designed aroma mixture. S_i^{M} is calculated using Eq. (10).

$$S_i^{\mathrm{M}} = \sum\nolimits_{k=1}^{K} x_k S_i^k \quad \forall i = \{1, 2, \ldots, 10\} \tag{2.10}$$

where x_k is the volume fraction of ingredient k and K is the number of molecules in the odor database, which equals 480 in this section. S_i^k is the σ-profiles area parameter i of ingredient k, which is calculated using COSMO-SAC model and stored in the odor database.

The number of selected ingredients n should follow the constraints of Eqs. (2.11) and (2.12).

$$\frac{1}{M} y_k \leq x_k \leq y_k \quad \forall k = \{1, 2, \ldots, K\} \tag{2.11}$$

$$n = \sum\nolimits_{k=1}^{K} y_k \leq N \tag{2.12}$$

In Eq. (2.11), M is a big number and y_k is a binary variable that indicates whether compound k is selected as the ingredient of the aroma mixture. If $y_k = 1$, $\frac{1}{M} \leq x_k \leq 1$, otherwise, $x_k = 0$. In Eq. (2.12), n is the number of ingredients, and N is the upper bound of the ingredient number, which is set up by the users.

The property constraints such as solubility, boiling point, $\log(K_{\mathrm{o/w}})$, LC_{50}, flash point, and so on are also considered for the screening of aroma mixtures to guarantee the product performance, as shown in Eq. (2.13).

$$p^{\mathrm{L}} \leq p = f(x_k, p_k) \leq p^{\mathrm{U}} \tag{2.13}$$

where p^{L} and p^{U} are the lower and upper bounds of property p, respectively. p_k is the property value of ingredient k for property p and f is the function of the mixing rule.

An MINLP model is formulated for the ingredient screening of the optimal aroma mixture, in which Eq. (2.9) is the objective function, Eqs. (2.10)–(2.13) are the constraints. The decomposition-based solution approach [44] is applied to the solution of the optimization model. From the solution of the MINLP model, the ingredients and the volume fractions of its ingredients are obtained. It should be noted that the above mathematical model will always have feasible solutions, but the generated solution may be too far away from the required σ-profiles. In this situation, the odor database can be expanded to include a larger number of molecules.

2.3.2.5 Verification

Finally, the designed aroma mixtures are verified in the verification step. The verification of some properties can be performed using a database if the property data are available in the database. For odor properties, experimental verification is needed. An electronic nose is used to verify the odor properties. Besides, consumer testing studies can be applied to obtain more accurate verification of the design results after the product prototypes are made.

2.4 Case Studies

Based on the CAAD framework introduced above, four case studies for designing aromatic chemicals are presented in this section.

2.4.1 Pure Aroma Design for Shampoo Additives

Aromatic ingredients are commonly used in products such as shampoos, shower gels, shaving creams, and body lotions. Even some products labeled "unscented" may contain aromatic ingredients. This is because the manufacturer may add just enough aromas to mask the unpleasant smell of other ingredients, without giving the product a noticeable scent. On the other hand, the aromatic ingredient should meet other technical requirements. The objective of this case study is to find suitable aromatic chemicals as additives for shampoo using the developed MINLP model. According to the above analysis, the aromatic chemicals for shampoo additives should have the following properties. Table 2.1 shows the properties and constraints of the aromatic chemicals in this case study.

Groups CH_3, CH_2, CH, C, CH_2=CH, CH=CH, CH_2=C, CH=C, OH, CH_3CO, CH_2CO, CH_3COO, CH_2COO, CH_3O, and CH_2O are selected. The MINLP model for the case study is established based on their upper and lower bounds in Table 2.1. The decomposition-based algorithm is used for the solution of the MINLP model, as Figure 2.15 shows.

Table 2.1 Properties and constraints for the design of aromatic chemicals.

Properties	Constraints
Total group number	$4 \leq n \leq 10$
Repeat group number	$n_i \leq 4$
Functional group number	$1 \leq n_F \leq 3$
Odor character	OC = sweet, fruit or flower
Odor pleasantness	$OP \geq 40$
Diffusion coefficient ($m^2\ h^{-1}$)	$D \geq 0.15$
Vapor pressure (Pa)	$P^{sat} \geq 100$
Normal boiling point (K)	$T_b \geq 440$
Normal melting point (K)	$T_m \leq 293.15$
Solubility parameter ($MPa^{1/2}$)	$15 \leq S_p \leq 17$
Viscosity (cp)	$\eta \leq 2$
Density ($g\ cm^{-3}$)	$0.8 \leq \rho \leq 1$
$-\log(LC_{50})$ ($-\log(mol\ l^{-1})$)	$-\log(LC_{50}) \leq 4.2$

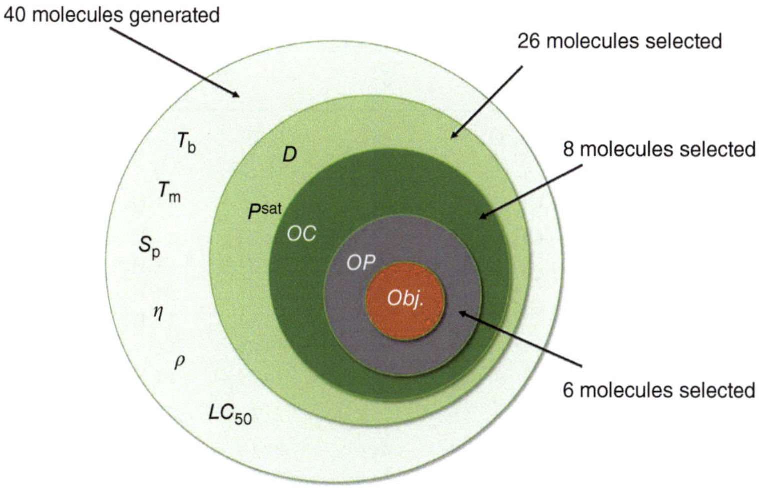

Figure 2.15 The decomposition-based algorithm for the aroma design case study. Source: Zhang et al. [20]/with permission of Elsevier.

First, feasible candidates are generated by matching constraints T_b, T_m, S_p, η, ρ and $-\log(LC_{50})$ as the model equations for these properties are linear. 40 feasible molecules are generated in this subproblem using the OptCAMD software. Then, constraints D and P^{sat} are added to evaluate each generated candidate to check if they satisfy these additional constraints. 26 molecules are selected in this subproblem. Then, the 26 molecules are tested using the ML model for odor character prediction to test if these molecules are "sweet," "fruit," or "flower" (as defined in Table 2.1), and 8 molecules are found to match these constraints. The odor pleasantness model is then used for the screening of these eight molecules, which finds six molecules matching this constraint. The final solution is the molecule that has the highest odor pleasantness within these six molecules. The six generated molecules satisfying all property constraints are listed in Table 2.2, together with their properties.

From the optimization result, molecule $C_9H_{18}O_2$ has the highest odor pleasantness. Therefore, it is selected as the best potential aroma in this case study. Database search has been made for all the six feasible molecules. The optimal molecule, however, is not found in any database as aromas and, therefore, it needs to be evaluated through experiments to verify if the odor properties are the same as predicted. The molecules $C_8H_{16}O$ (CAS No. 111-13-7) and $C_8H_{16}O_2$ (CAS No. 106-73-0) are found in the database as commonly used aromas for various purposes, which confirms the effectiveness of the CAAD framework.

2.4.2 Pure Aroma Design for the Ingredient in Insect Repellent Spray

An insect repellent spray usually constitutes an active ingredient (repel mosquitoes), a binary solvent mixture (deliver the active ingredient on the skin and vaporize after application), and additives (e.g., perfumes, moisturizing agents, etc.) [49]. It

Table 2.2 The generated feasible candidates.

No.	**1**	**2**	**3**	**4**	**5**	**6**
Formula	$C_8H_{16}O$	$C_8H_{16}O_2$	$C_7H_{12}O_2$	$C_7H_{12}O_3$	$C_8H_{14}O_2$	$C_9H_{18}O_2$
Groups	2 CH_3 4 CH_2 1 CH_2CO	1 CH_3 4 CH_2 1 CH_2CO 1 CH_3O	1 CH_3 1 CH 1 CH_2=C 1 CH_3CO 1 CH_3O	1 CH_3 1 CH 1 CH_2=CH 1 CH_2COO 1 CH_3O	1 CH_3 2 CH 1 CH_2=CH 1 CH_3CO 1 CH_3O	3 CH_3 3 CH_2 1 C 1 CH_3COO
T_m (K)	244	265	253	217	253	240
T_b (K)	443	469	443	442	459	458
S_p (Mpa$^{1/2}$)	16.47	16.88	16.55	16.49	16.32	15.23
η (cp)	1.08	0.91	0.21	0.2	0.15	0.89
ρ (g cm^{-3})	0.82	0.9	0.96	1	0.94	0.9
$-\log(LC_{50})$	3	2.58	2.83	3.53	3.58	3.13
P^{sat} (Pa)	1003.8	138.2	838.4	1298.3	318.4	501.7
D (m^2 h^{-1})	0.17	0.16	0.17	0.17	0.16	0.16
OC	Sweet	Sweet	Sweet	Sweet	Sweet	Sweet
OP	40	40	40	40	40	60
Available in database?	Y	Y	N	N	N	N
CAS No.	111-13-7	106-73-0	—	—	—	—
Molecular structure	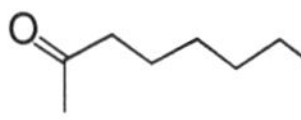	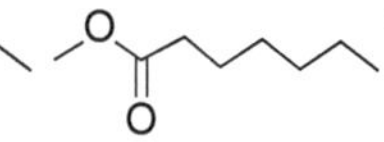	—	—	—	—
Odor in literature	Cheese-like, dairy nuances	Fruity	—	—	—	—

is designed with high effectiveness against mosquitoes, a water-based spray, pleasant scent, long durability, low toxicity, high stability, spray-ability, low price, and long shelf life. Picaridin is selected as the active ingredient for repelling mosquitoes by using a knowledge-based design method. However, the solubility of Picaridin in water is rather low. Therefore, a water–organic solvent mixture is an alternative to be used in order to dissolve Picaridin. A mixture of water and 2-propanol is designed as the solvent by Conte et al. [49] by using a computer-aided molecular design method for mixtures. Aromas should also be added to obtain a pleasant scent. The ingredient composition of the insect repellent spray is given in Table 2.3. Table 2.4 shows the properties and constraints of the aromatic chemicals in this case study. The objective of this case study is to design the aroma.

Table 2.3 Ingredient composition of the insect repellent spray.

Chemical	w_i (wt%)
Picaridin	9.74
2-Propanol	44.25
Water	42.01
Aroma	4

Table 2.4 Properties and constraints for the design of aromatic chemicals.

Properties	Constraints
Total group number	$7 \leq n \leq 10$
Repeat group number	$n_i \leq 3$
Functional group number	$2 \leq n_F \leq 3$
Odor character	OC = sweet, fruit or flower
Odor pleasantness	$OP \geq 40$
Diffusion coefficient ($m^2\ h^{-1}$)	$D \geq 0.154$
Vapor pressure (Pa)	$P^{sat} \geq 14$
Normal boiling point (K)	$T_b \geq 373.15$
Normal melting point (K)	$T_m \leq 300$
Solubility parameter ($MPa^{1/2}$)	$17 \leq S_p \leq 20$
Viscosity (cp)	$\eta \leq 2$
Molar volume ($cm^3\ mol^{-1}$)	$V_m \leq 200$
$-\log(LC_{50})$ ($-\log(mol\ l^{-1})$)	$-\log(LC_{50}) \leq 4.5$
Molecular weight ($g\ mol^{-1}$)	$100 \leq Mw \leq 200$
Flash point (K)	$F_p \geq 320$
Phase stability	Solubility in the solvent mixture

Groups CH_3, CH_2, CH, C, CH_2=CH, CH=CH, CH_2=C, CH=C, OH, CH_3COO, CH_2COO, and CHCOO are selected. According to the previous analysis, the aroma should have the following physical properties, as listed in Table 2.4. The MINLP model for the case study is established based on their upper and lower bounds in Table 2.4. The decomposition-based algorithm is used for the solution of the MINLP model. Constraints Mw, F_p, T_b, T_m, S_p, η, V_m, and $-\log(LC_{50})$ are first used to generate feasible candidates by using the OptCAMD software. A total of 132 feasible molecules are generated in this subproblem. Then, constraints D and P^{sat} are added to evaluate each generated candidate, and 44 molecules are found to satisfy these two additional constraints. Then, the 44 molecules are tested using the odor character prediction model to test if these molecules are "sweet," "fruit," or "flower," and 16 molecules are found to satisfy this requirement. The odor pleasantness model is then used for the screening of the remaining 16 molecules, which leads to 15 molecules. The final solution is the molecule that has the highest odor pleasantness among the 15 molecules. Details of the generated 15 molecules, together with their properties, are given in Table 2.D.1 in Appendix D. From the results in this table, the ninth molecule (Linalool, CAS No. 78-70-6) is found in the OptCAMD database, and the predicted physical properties as well as odor characters agree well with the available measured data. Therefore, Linalool is selected as the aroma in this product. The selection agrees with the heuristic selection reported by Conte et al. [49]. The experimental verification is conducted, and the properties and phase stability are tested by Conte et al. [49].

2.4.3 Mixture Aroma Design for Aroma Substitutes

With the increasing applications of aromas in chemical products such as fragrances, shampoos, cosmetics, etc., the design of aromas with consumer requirements becomes essential. The aroma *cis*-3-hexenyl propionate (CAS No. 33467-74-2) is a typical aroma applied widely, which smells fruity. In this case study, aroma mixtures with similar odor and physical properties with *cis*-3-hexenyl propionate are designed at a cheaper price. These product attributes are converted into properties that are listed in Table 2.5. Here, the property values are obtained from *cis*-3-hexenyl propionate.

Based on the property values of edible, bakery, sweet, fruity, flower, sour, pleasantness, vapor pressure, and diffusion coefficient, the established ANN-based SOR model is used to obtain the σ-profiles descriptors of the designed mixture. The result is shown in Figure 2.16.

Table 2.5 Properties of *cis*-3-hexenyl propionate.

Odor property	Value	Physical property	Value
Edible[a)]	33.42	Vapor pressure P_{sat} (Pa)[b)]	53.86
Bakery[a)]	43.00	Diffusion coefficient D_{AB} ($m^2\,h^{-1}$)[c)]	0.16
Sweet[a)]	34.37	Solubility δ ($mg\,l^{-1}$, 298 K)[b)]	158.9
Fruity[a)]	31.54	Boiling point T_b (K, 101.325 kPa)[b)]	454
Flower[a)]	29.95	Octanol–water partition coefficient $\log K_{o/w}$[b)]	2.909
Sour[a)]	25.93	Fathead minnow LC_{50} $-\log(LC_{50})$ ($\log mol\,l^{-1}$)[c)]	3.36
Pleasantness[a)]	60.19	Flash point F_p (K)[b)]	333

a) Data from the odor database [37].
b) Data from the Good Scents Company Information System (http://thegoodscentscompany.com).
c) Data from group contribution methods in ICAS (https://www.pseforspeed.com/icas).

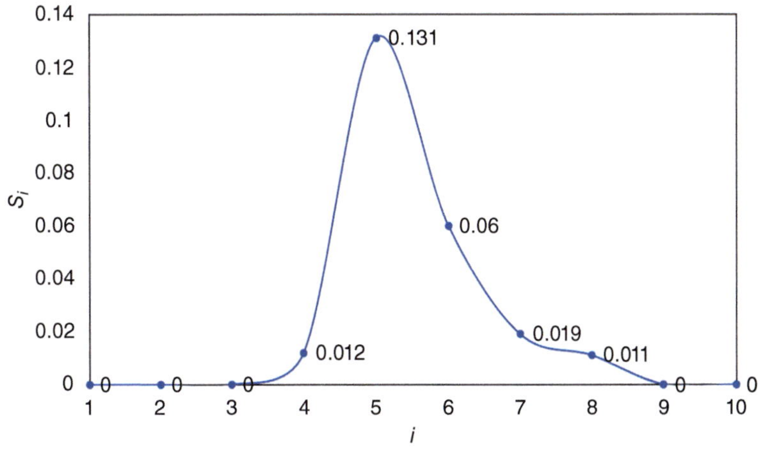

Figure 2.16 ANN-predicted results using the σ-profiles descriptors. Source: Zhang et al. [41]/with permission of Elsevier.

Then, the established MINLP model for ingredient screening is applied to search for feasible aroma mixture candidates, which considers the property constraints listed in Table 2.6. The compounds in the database for ingredient screening are listed in Table 2.E.1 in Appendix E. The number of ingredients is set to two. Finally, the rank of the feasible aroma mixtures based on the Euclidean distance is shown in Figure 2.17. The red line in Figure 2.17 is the tolerance of the Euclidean distance, which is set to 0.0035 in this case study.

Among the aroma mixture candidates, the ingredients of the second best one (CAS No. 97-61-0 and 7493-69-8) are all commonly used aromas, which are much cheaper than *cis*-3-hexenyl propionate. Therefore, the second best one is used for further verification. Table 2.7 lists the properties of the ingredients of the designed aroma mixture. From the results, it is seen that all property requirements are met, and the price of the designed aroma mixture is much cheaper than *cis*-3-hexenyl propionate.

Finally, the designed aroma mixture is verified by the electronic nose system (PEN3, Airsense Analytics GmbH, Germany) (E-nose). The samples of *cis*-3-hexenyl propionate and the designed aroma mixture are placed in two airtight 25 ml vials at $25 \pm 1\,°C$ for approximately 30 minutes, and odors are measured for 5 minutes at 1 second intervals. All samples are run with three repetitions and average values

Table 2.6 Property constraints of the case study.

Property	Constraint
δ (mg l^{-1}, 298 K)	$\delta \geq 158.6$
T_b (K, 101.325 kPa)	$T_b \geq 333$
F_p (K)	$F_p \geq 323$
$\log K_{o/w}$	$\log K_{o/w} \leq 2.91$
$-\log(LC_{50})$ (log mol l^{-1})	$-\log(LC_{50}) \leq 3.70$

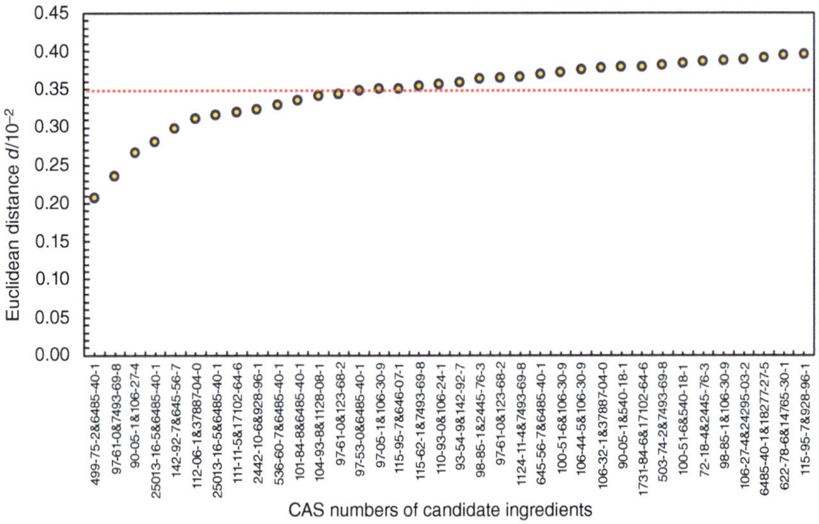

Figure 2.17 The rank of the feasible aroma mixtures based on the Euclidean distance. Source: Zhang et al. [41]/with permission of Elsevier.

Table 2.7 Ingredient properties of the designed aroma mixture.

		Designed aroma mixture	
	***cis*-3-Hexenyl propionate**	**Ingredient 1**	**Ingredient 2**
CAS No.	33467-74-2	97-61-0	7493-69-8
	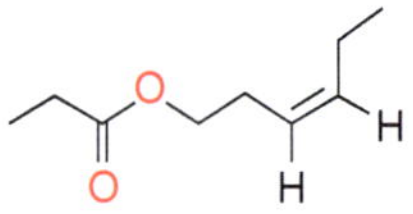	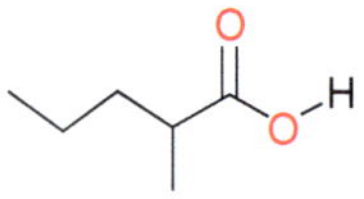	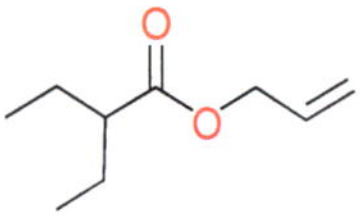
Odor[a)]	Green, grass, flower, fruity	Sweet, edible, fruity	Fruity, edible
Volume fraction	1	0.4	0.6
Vapor pressure (Pa, 298 K)[a)]	53.86	23.99	85.46
Diffusion coefficient[b)]	0.156	0.172	0.156
Solubility (mg l^{-1}, 298 K, water)[a)]	158.9	—	157.3
Boiling point (K, 101.325 kPa)[a)]	453–455	468.36	449.96
Flash point (K)[a)]	333	364.26	337.59
$\log K_{o/w}$[a)]	2.909	1.8	2.972
$-\log(LC_{50})$ (log mol l^{-1})[b)]	3.36	2.45	4.03
Price (CNY/kg)[c)]	6,059	2,377	574

a) The Good Scents Company Information System.
b) ICAS.
c) Sigma-Aldrich (https://www.sigmaaldrich.com).

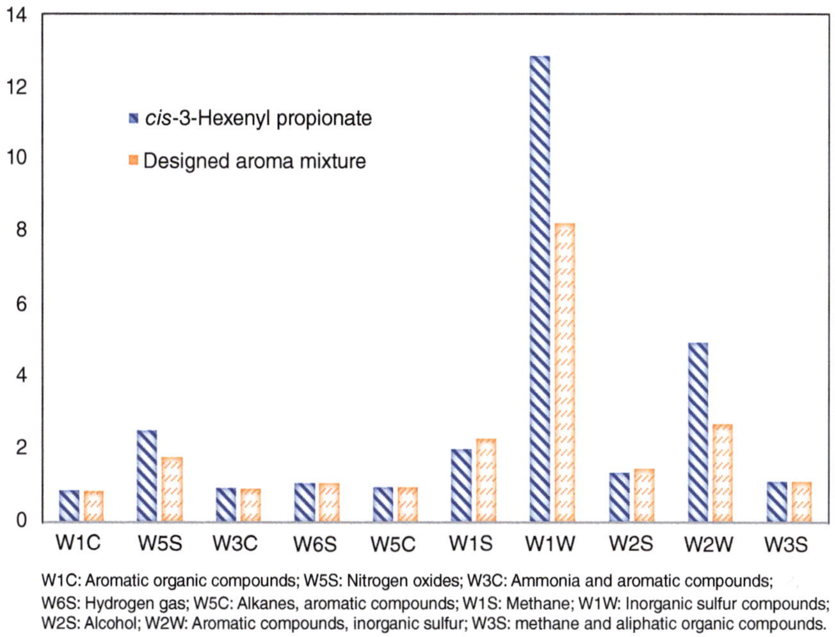

Figure 2.18 E-nose experimental results of *cis*-3-hexenyl propionate and the designed aroma mixture. Source: Zhang et al. [41]/with permission of Elsevier.

are obtained for the two samples. The E-nose results of the two samples are shown in Figure 2.18.

The results indicate the odor of the two samples is almost the same. Then, human tests are also conducted. From the tests, *cis*-3-hexenyl propionate smells like grass, and after evaporation for a certain time, the odor switches to fruity slightly. The designed aroma mixture smells like grass, which is quite similar to *cis*-3-hexenyl propionate. Therefore, from the experimental results, the effectiveness of the proposed CAAD framework and the ML model is verified.

2.4.4 Mixture Aroma Design for Odor Tuning

The tuning of the odor of aromas is also an interesting topic in aroma design. On the one hand, some types of pleasant odors can be enhanced to improve the quality of the product. On the other hand, some unpleasant odors can be covered through the tuning of odor. In this case study, the proposed CAAD framework and the ML model are applied for the odor tuning of aromas. Here, *cis*-4-heptenal (CAS No. 6728-31-0) is taken as an example, which is often used as a creamy fragrance. This aroma is tuned by the design of aroma mixtures with improved edible, sweet, and fruity odors to enhance its quality. Table 2.8 summarizes the properties and the tuning targets of *cis*-4-heptenal.

Table 2.8 Properties of *cis*-4-heptenal and the target values of the odor tuning.

Odor property	Value	Target value	Physical property	Value
Pleasantness	29.2	29.2	Vapor pressure P_{sat} (Pa)[a]	485.43
Edible	25.59	35.00	Diffusion coefficient D_{AB} ($m^2\,h^{-1}$)[b]	0.174
Bakery	38.11	38.11	Solubility δ ($mg\,l^{-1}$, 298 K)[a]	1810
Sweet	12.46	**30.00↑**	Boiling point T_b (K, 101.325 kPa)[a]	429.40
Fruit	11.55	**25.00↑**	Octanol–water partition coefficient $\log K_{o/w}$[a]	2.174
Sour	26.83	26.83	Fathead minnow LC_{50} $-\log(LC_{50})$ ($\log mol\,l^{-1}$)[b]	3.68
Flower	24.33	24.33	Flash point F_p (K)[a]	316.33

a) The Good Scents Company Information System.
b) ICAS.

Table 2.9 Ingredient properties of the designed aroma mixture.

		Designed aroma mixture	
	cis-4-Heptenal	Ingredient 1	Ingredient 2
CAS No.	6728-31-0	71-41-0	288-47-1
Volume fraction	1	0.9	0.1
Vapor pressure (Pa, 298 K)[a]	485.43	293.31	2881.50
Diffusion coefficient[b]	0.174	0.189	0.214
Solubility ($mg\,l^{-1}$, 298 K, water)[a]	1810	22000	53780
Boiling point (K, 101.325 kPa)[a]	429.40	406.75	387.84
$\log K_{o/w}$[a]	2.174	1.51	0.44
$-\log(LC_{50})$ ($\log mol\,l^{-1}$)[b]	3.68	2.83	2.94
Flash point (K)[a]	316.33	321.89	295.22
Price (CNY/kg)[c]	38 958.64	940.69	41179.3

a) The Good Scents Company Information System.
b) ICAS.
c) Sigma–Aldrich.

With the same design steps, the aroma mixture design results are listed in Table 2.9 and Figure 2.19. From the results of Table 2.8, it is seen that all properties meet the requirements, and from Figure 2.19, the odors of edible, sweet, and fruit of the designed mixture are enhanced, which means the requirements of the design target. The price of the designed mixture ($940.69 \times 0.9 + 41\,179.3 \times 0.1 = 4964.55$ CNY kg^{-1}) is also much cheaper than *cis*-4-heptenal (38 958.64 CNY kg^{-1}).

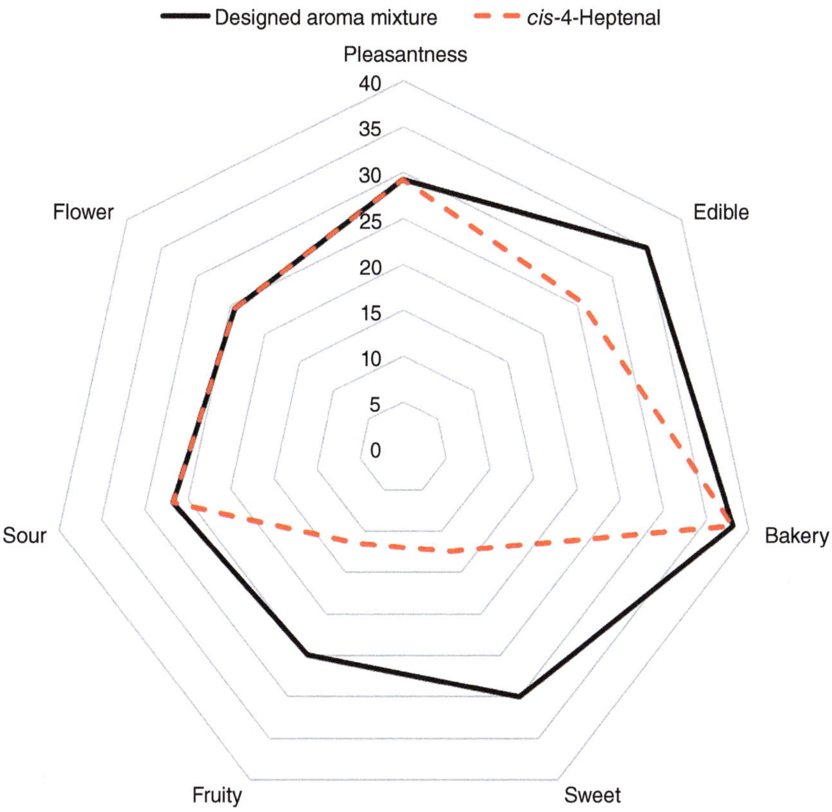

Figure 2.19 Radar plot of the odor properties of the designed aroma mixture and *cis*-4-heptenal. Source: Zhang et al. [41]/with permission of Elsevier.

2.5 Conclusions

Modern society has growing demands for aromas. Since odor properties are essential for aroma design, the SOR has been proposed for better understanding the process of odor perception and further discovering potential aromas. The model-based aroma design method is utilized in this chapter because of its efficiency in SOR development and general application. The descriptors of groups and σ-profiles are used to develop SOR models for predictions of odor properties of pure and mixture aromatic chemicals using ML algorithms. The accuracy of the SOR model for pure aromatic chemicals is 92.9%, while the R^2 of the SOR model for mixture aromatic chemicals is 0.88, showing the high accuracy of the established SOR models in predicting odor properties. Subsequently, a CAAD framework for designing novel pure and mixture aromatic chemicals is proposed. Four case studies are presented and verified by experiments and literature data, the results of which demonstrate the feasibility and effectiveness of the CAAD framework as well as the established ML models.

2.A The CAS Number of Molecules and the Selected Groups

Table 2.A.1 CAS registry number of the fragrance molecules in the database.

100-06-1	111-14-8	13708-12-8	34413-35-9	592-88-1	7553-56-2
1003-04-9	111-27-3	137-32-6	34451-19-9	5989-27-5	7554-12-3
10031-82-0	111-47-7	13925-00-3	350-03-8	60-01-5	7558-79-4
100-41-4	111-61-5	140-10-3	35250-53-4	60047-17-8	75-65-0
100-51-6	111-65-9	140-11-4	352-93-2	60-12-8	76-22-2
100-52-7	111-66-0	140-39-6	35836-72-7	60-18-4	76-49-3
100-66-3	111-70-6	140-67-0	3658-80-8	6032-29-7	765-70-8
101-41-7	1117-55-1	141-12-8	36653-82-4	60-33-3	7732-18-5
101-84-8	111-79-5	141-78-6	3681-71-8	606-45-1	7764-50-3
102-04-5	111-80-8	142-19-8	37887-04-0	611-13-2	7779-41-1
102-13-6	111-87-5	142-62-1	3796-70-1	613-70-7	7779-50-2
102-16-9	1120-21-4	142-92-7	39212-23-2	614-99-3	7779-65-9
102-20-5	112-06-1	143-07-7	39255-32-8	616-25-1	77-83-8
102-76-1	112-12-9	143-08-8	40018-26-6	61-90-5	7785-70-8
103-05-9	1122-62-9	143-13-5	4075-07-4	61931-81-5	7786-58-5
103-09-3	112-30-1	1438-94-4	4077-47-8	620-02-0	7787-20-4
103-36-6	112-31-2	14765-30-1	4112-89-4	622-78-6	77-90-7
103-45-7	112-37-8	147-85-3	41453-56-9	623-37-0	77-92-9
103-50-4	1124-11-4	149-57-5	41519-23-7	623-42-7	77-93-0
103-60-6	112-42-5	150-30-1	4180-23-8	624-24-8	78-59-1
104-21-2	112-44-7	150-60-7	4208-49-5	625-84-3	78-70-6
104-54-1	1125-21-9	150-78-7	4221-99-2	626-93-7	78761-38-3
104-76-7	112-53-8	151-05-3	42436-07-7	627-90-7	78-83-1
104-93-8	1125-88-8	151-10-0	431-03-8	628-63-7	78-84-2
10519-33-2	112-66-3	1516-17-2	4411-89-6	629-19-6	78-93-3
10521-91-2	1128-08-1	15356-60-2	464-49-3	629-33-4	79-09-4
105-37-3	1131-62-0	15679-12-6	470-82-6	63012-97-5	79-20-9
105-53-3	115-95-7	15679-19-3	4864-61-3	6309-51-9	79-31-2
105-54-4	118-58-1	15707-23-0	491-35-0	63-68-3	79-77-6
105-57-7	118-93-4	15707-24-1	4938-52-7	637-64-9	80-62-6
105-60-2	1191-43-1	1618-26-4	498-02-2	6378-65-0	81-14-1
105-66-8	119-36-8	16409-46-4	499-75-2	638-25-5	81925-81-7
105-87-3	1193-79-9	16630-66-3	502-42-1	63-91-2	821-41-0
10599-70-9	119-61-9	1679-07-8	503-74-2	64-04-0	821-55-6
106-02-5	119-84-6	17102-64-6	505-10-2	64-17-5	823-22-3

Table 2.A.1 (Continued)

106-21-8	120-14-9	1731-84-6	505-79-3	64-19-7	825-51-4
106-24-1	120-24-1	1759-28-0	50-69-1	64275-73-6	85-91-6
106-25-2	120-50-3	1797-74-6	50-70-4	645-56-7	87-22-9
106-27-4	120-51-4	18172-67-3	513-85-9	646-07-1	88-15-3
106-30-9	120-57-0	18277-27-5	5146-66-7	6485-40-1	881-68-5
106-32-1	120-58-1	18368-91-7	51-67-2	65405-77-8	882-33-7
106-36-5	120-92-3	18409-17-1	5271-38-5	656-53-1	88-69-7
106-44-5	121-32-4	18486-69-6	52-90-4	65-85-0	89-79-2
106-65-0	121-33-5	18640-74-9	529-20-4	66-25-1	89-83-8
106-68-3	121-34-6	1866-31-5	5292-21-7	67-03-8	90-00-6
106-70-7	122-03-2	2035-99-6	531-59-9	6728-26-3	90-02-8
106-73-0	122-40-7	2051-78-7	534-22-5	6728-31-0	90-05-1
1076-56-8	122-43-0	21145-77-7	536-60-7	67-56-1	91-22-5
107-85-7	122-63-4	2153-26-6	539-82-2	67-63-0	91-62-3
107-92-6	122-67-8	2257-09-2	540-07-8	67-64-1	91-64-5
107-95-9	122-70-3	2305-25-1	540-18-1	67-68-5	92-48-8
108-21-4	122-74-7	2396-84-1	540-42-1	67-71-0	92-52-4
108-22-5	122-97-4	24295-03-2	5405-41-4	67715-80-4	927-49-1
108-29-2	123-08-0	2432-51-1	541-31-1	67883-79-8	928-96-1
108-48-5	123-11-5	2442-10-6	54300-08-2	6789-88-4	93-04-9
108-94-1	123-19-3	2445-76-3	5461-08-5	6790-58-5	93-08-3
108-98-5	123-25-1	24683-00-9	5471-51-2	693-54-9	93-15-2
109-15-9	123-32-0	2497-18-9	54947-74-9	693-95-8	93-16-3
109-19-0	123-51-3	25013-16-5	551-93-9	698-10-2	93-28-7
109-21-7	123-66-0	25152-85-6	554-12-1	705-86-2	93-29-8
109-42-2	123-68-2	2530-10-1	56-40-6	706-14-9	93-51-6
109-52-4	123-72-8	2568-25-4	56-41-7	71-00-1	93-54-9
109-60-4	123-75-1	2623-23-6	56-81-5	710-04-3	93-58-3
109-73-9	123-86-4	2639-63-6	56-84-8	71-23-8	93-60-7
109-94-4	123-92-2	2705-87-5	56-85-9	71-36-3	93-89-0
110-02-1	124-04-9	2721-22-4	56-86-0	71-41-0	93905-03-4
110-15-6	124-06-1	2785-89-9	56-87-1	71-43-2	94278-27-0
110-17-8	124-07-2	27939-60-2	57074-37-0	7149-32-8	94-47-3
110-27-0	124-13-0	2847-30-5	57-55-6	7217-59-6	94-62-2
110-38-3	127-17-3	28664-35-9	583-60-8	72-18-4	96-22-0
110-40-7	128-37-0	288-47-1	584-02-1	7452-79-1	96-48-0

(Continued)

Table 2.A.1 (Continued)

110-43-0	133-18-6	290-37-9	589-92-4	74-79-3	97-42-7
110-62-3	133-37-9	3149-28-8	589-98-0	7493-63-2	97-53-0
110-81-6	134-20-3	31704-80-0	590-01-2	7493-69-8	97-54-1
110-82-7	135-02-4	3208-16-0	59020-90-5	7493-71-2	97-61-0
110-86-1	13532-18-8	3268-49-3	591-24-2	75-07-0	97-99-4
110-89-4	13623-11-5	32974-92-8	591-31-1	75-18-3	98-00-0
110-93-0	13678-67-6	334-48-5	591-78-6	7540-51-4	98-85-1
111-11-5	137-00-8	33467-74-2	591-80-0	7541-49-3	98-86-2
111-13-7	137-06-4	3391-86-4	592-84-7	75-47-8	99-87-6

2.B The Calculation Formula of Odor Score

The scores of all odor characters for one molecule are averaged as shown in Eq. (2.B.1):

$$S'_{p,d,i} = \frac{\sum_{d=1}^{n} S_{p,d,i} \times y_{p,d,i}}{\sum_{d=1}^{n} y_{p,d,i}} \tag{2.B.1}$$

where $S_{p,\,d,\,i}$ is the score of property p for the molecule i from a subject d, $y_{p,\,d,\,i}$ is a binary value, which is assigned to 0 or 1 for whether it is eliminated.

Table 2.A.2 Selected groups for the representation of aromatic molecules.

CH_3	AC	CH_3COO	CH_2N	$HCON(CH_2)_2$
CH_2	$ACCH_3$	CH_2COO	$ACNH_2$	COO
CH	$ACCH_2$	HCOO	C_5H_5N	$CONH_2$
C	ACCH	CH_3O	C_5H_4N	$CONHCH_2$
CH_2=CH	OH	CH_2O	C_5H_3N	$C_2H_4O_2$
CH=CH	CH_3OH	CHO	COOH	CH_3S
CH_2=C	H_2O	CH_2NH_2	CH_2SH	CH_2S
CH=C	ACOH	$CHNH_2$	I	CHS
C=C	CH_3CO	CH_3NH	C#C	C_4H_4S
ACH	CH_2CO	CH_2NH	DMSO	C_4H_3S

2.C The Parameters and Results of the ANN Model

Table 2.C.1 ANN model parameters and result analysis.

Model structure	Model parameters	Model output	R^2 of training set	R^2 of testing set
9-24-(Tanh)-20-(Tanh)-16-(Sigmoid)-12-(Tanh)-1	Weight initialization = Xavier; Bias initialization = Zero; Optimizer = Adam; Epoch = 7000;	S_1	0.9999	0.9905
9-24-(Tanh)-20-(Tanh)-16-(Sigmoid)-1	Weight initialization = Normal; Bias initialization = Zero; Optimizer = RMSprop; Epoch = 6000;	S_2	0.9783	0.9188
9-24-(Tanh)-20-(Sigmoid)-1	Weight initialization = Normal; Bias initialization = Zero; Optimizer = Levenberg–Marquardt; Epoch = 4000;	S_3	0.9850	0.9563
9-24-(Tanh)-20-(Sigmoid)-1	Weight initialization = Uniform; Bias initialization = Zero; Optimizer = Levenberg–Marquardt; Epoch = 3000;	S_4	0.9805	0.9272
9-18-(Tanh)-1	Weight initialization = Uniform; Bias initialization = Zero; Optimizer = Levenberg–Marquardt; Epoch = 1000;	S_5	0.9101	0.7764

(Continued)

Table 2.C.1 (Continued)

Model structure	Model parameters	Model output	R^2 of training set	R^2 of testing set
9-18-(Tanh)-1	Weight initialization = Uniform; Bias initialization = Zero; Optimizer = Levenberg–Marquardt; Epoch=1000;	S_6	0.9007	0.8289
9-24-(Tanh)-20-(Tanh)-1	Weight initialization = Uniform; Bias initialization = Zero; Optimizer = Levenberg–Marquardt; Epoch = 3000;	S_7	0.9931	0.9018
9-24-(Tanh)-20-(Sigmoid)-1	Weight initialization = Normal; Bias initialization = Zero; Optimizer = Levenberg–Marquardt; Epoch = 4000;	S_8	0.9874	0.8735
9-24-(Tanh)-20-(Tanh)-16-(Sigmoid)-1	Weight initialization = Normal; Bias initialization = Zero; Optimizer = RMSprop; Epoch = 6000;	S_9	0.9773	0.8747
9-24-(Tanh)-20-(Tanh)-16-(Sigmoid)-12-(Tanh)-1	Weight initialization = Xavier; Bias initialization = Zero; Optimizer = Adam; Epoch = 7000;	S_{10}	0.9999	0.9998

2.D The Designed Results of Molecules for Case Study 2

Table 2.D.1 The generated feasible candidates for case study 2.

No.	Groups	OC	OP	Available in database?	CAS number	Molecular structure	Odor in literature
1	3 CH_3 1 CH 1 C 1 CH_2=CH 1 OH 1 CH_2COO	Sweet	40	N	—	—	—
2	3 CH_3 2 CH_2 1 C 1 OH 1 CH_2COO	Sweet	40	N	—	—	—
3	2 CH_3 1 CH_2 1 C 1 CH_2=CH 1 OH 1 CH_2COO	Sweet	40	N	—	—	—
4	3 CH_3 1 C 1 CH_2=C 1 OH 1 CH_2COO	Sweet	40	N	—	—	—
5	3 CH_3 1 C 1 CH=CH 1 OH 1 CH_3COO	Sweet	40	N	—	—	—
6	2 CH_3 1 CH_2 1 C 1 CH_2=C 1 OH 1 CH_3COO	Sweet	40	N	—	—	—
7	2 CH_3 2 CH 1 CH_2=CH 1 OH 1 CH_2COO	Sweet	40	N	—	—	—

(Continued)

Table 2.D.1 (Continued)

No.	Groups	OC	OP	Available in database?	CAS number	Molecular structure	Odor in literature
8	2 CH_3 1 CH_2 1 CH 1 $CH_2{=}C$ 1 OH 1 CH_2COO	Sweet	40	N	—	—	—
9	3 CH_3 2 CH_2 1 C 1 $CH_2{=}CH$ 1 CH=C 1 OH	Sweet	40	Y	78-70-6	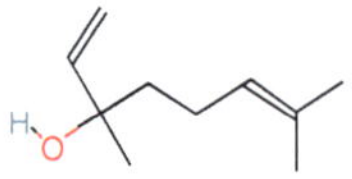	Sweet, floral, petitgrain-like
10	3 CH_3 1 CH_2 1 C 1 OH 1 CH_2COO	Sweet	40	N	—	—	—
11	2 CH_3 2 CH_2 1 CH 1 OH 1 CH_2COO	Sweet	40	N	—	—	—
12	3 CH_3 2 CH 1 OH 1 CH_2COO	Sweet	40	N	—	—	—
13	1 CH_3 3 CH_2 1 CH 1 OH 1 CH_3COO	Sweet	40	N	—	—	—
14	3 CH_3 3 CH_2 1 CH 1 CH=C 1 OH	Sweet	40	N	—	—	—
15	2 CH_3 2 CH_2 1 CH 1 CH=CH 1 $CH_2{=}C$ 1 OH	Sweet	40	N	—	—	—

2.E Aroma Compounds for Ingredient Screening

Table 2.E.1 Aroma compounds for ingredient screening.

CAS No.	Name	Smiles
100-06-1	Acetanisole	COc1ccc(cc1)C(C)=O
1003-04-9	45-Dihydro-3(2*H*)-thiophenone	O=C1CCSC1
10031-82-0	4-Ethoxybenzaldehyde	CCOc1ccc(C=O)cc1
100-41-4	Ethylbenzene	CCc1ccccc1
100-51-6	Benzyl alcohol	OCc1ccccc1
100-52-7	Benzaldehyde	O=Cc1ccccc1
100-66-3	Anisole	COc1ccccc1
101-41-7	Methyl phenylacetate	COC(=O)Cc1ccccc1
101-84-8	Diphenyl ether	O(c1ccccc1)c2ccccc2
102-04-5	13-Diphenyl-2-propanone	O=C(Cc1ccccc1)Cc2ccccc2
102-13-6	Isobutyl phenylacetate	CC(C)COC(=O)Cc1ccccc1
102-16-9	Benzyl phenylacetate	O=C(Cc1ccccc1)OCc2ccccc2
102-20-5	Phenethyl phenylacetate	O=C(Cc1ccccc1)OCCc2ccccc2
102-76-1	Triacetin	CC(=O)OCC(COC(C)=O)OC(C)=O
103-05-9	Dimethyl benzene propanol	CC(C)(O)CCc1ccccc1
103-09-3	2-Ethylhexyl acetate	CCCCC(CC)COC(C)=O
103-36-6	Ethyl cinnamate	CCOC(=O)\C=C\c1ccccc1
103-45-7	Phenethyl acetate	CC(=O)OCCc1ccccc1
103-50-4	Benzyl ether	C(OCc1ccccc1)c2ccccc2
103-60-6	2-Phenoxyethyl isobutyrate	CC(C)C(=O)OCCOc1ccccc1
104-21-2	Anisyl acetate	COc1ccc(COC(C)=O)cc1
104-54-1	Cinnamyl alcohol	OC\C=C/c1ccccc1
104-76-7	2-Ethyl-1-hexanol	CCCCC(CC)CO
104-93-8	4-Methylanisole	COc1ccc(C)cc1
10519-33-2	3-Decen-2-one	CCCCCC/C=C/C(C)=O
10521-91-2	5-Phenyl-1-pentanol	OCCCCCc1ccccc1
105-37-3	Ethyl propionate	CCOC(=O)CC
105-53-3	Diethyl malonate	CCOC(=O)CC(=O)OCC
105-54-4	Ethyl butyrate	CCCC(=O)OCC
105-57-7	Acetal	CCOC(C)OCC
105-60-2	ε-Caprolactam	O=C1CCCCCN1
105-66-8	Propyl butyrate	CCCOC(=O)CCC
105-87-3	Geranyl acetate	CC(=O)OC/C=C(C)/CCC=C(C)C
10599-70-9	3-Acetyl-25-dimethylfuran	CC(=O)c1cc(C)oc1C

(Continued)

Table 2.E.1 (Continued)

CAS No.	Name	Smiles
106-02-5	Ω-Pentadecalactone	O=C1CCCCCCCCCCCCCCO1
106-21-8	37-Dimethyl-1-octanol	CC(C)CCCC(C)CCO
106-24-1	Geraniol	CC(C)=CCC\C(C)=C\CO
106-25-2	Nerol	CC(C)=CCC\C(C)=C/CO
106-27-4	Isoamyl butyrate	CCCC(=O)OCCC(C)C
106-30-9	Ethyl heptanoate	CCCCCCC(=O)OCC
106-32-1	Ethyl octanoate	CCCCCCCC(=O)OCC
106-36-5	Propyl propionate	CCCOC(=O)CC
106-44-5	*p*-Cresol	Cc1ccc(O)cc1
106-65-0	Dimethyl succinate	COC(=O)CCC(=O)OC
106-68-3	3-Octanone	CCCCCC(=O)CC
106-70-7	Methyl caproate	CCCCCC(=O)OC
106-73-0	Methyl heptanoate	CCCCCCC(=O)OC
1076-56-8	1-Methyl-3-methoxy-4-isopropylbenzene	COc1cc(C)ccc1C(C)C
107-85-7	Isoamylamine	NCCC(C)C
107-92-6	Butyric acid	CCCC(O)=O
107-95-9	β-Alanine	NCCC(O)=O
108-21-4	Isopropyl acetate (replicate)	CC(C)OC(C)=O
108-22-5	Isopropenyl acetate	CC(=C)OC(C)=O
108-29-2	γ-Valerolactone	CC1CCC(=O)O1
108-48-5	26-Dimethylpyridine	Cc1cccc(C)n1
108-94-1	Cyclohexanone	O=C1CCCCC1
108-98-5	Benzenethiol	Sc1ccccc1
109-15-9	Octyl isobutyrate	CCCCCCCCOC(=O)C(C)C
109-19-0	Isobutyl acetate	O=C(C)OCC(C)C
109-21-7	Butyl butyrate	CCCCOC(=O)CCC
109-42-2	Butyl 10-undecenoate	CCCCOC(=O)CCCCCCCCC=C
109-52-4	Valeric acid	CCCCC(O)=O
109-60-4	Propyl acetate	CCCOC(C)=O
109-73-9	Butylamine	CCCCN
109-94-4	Ethyl formate	CCOC=O
110-02-1	Thiophene (replicate)	s1cccc1
110-15-6	Succinic acid	OC(=O)CCC(O)=O
110-17-8	Fumaric acid	OC(=O)\C=C\C(O)=O
110-27-0	Isopropyl myristate	CCCCCCCCCCCCCC(=O)OC(C)C

Table 2.E.1 (Continued)

CAS No.	Name	Smiles
110-38-3	Ethyl decanoate	CCCCCCCCCC(=O)OCC
110-40-7	Diethyl sebacate	CCOC(=O)CCCCCCCCC(=O)OCC
110-43-0	2-Heptanone	CCCCCC(C)=O
110-62-3	Valeraldehyde	CCCCC=O
110-81-6	Diethyl disulfide	CCSSCC
110-82-7	Cyclohexane	C1CCCCC1
110-86-1	Pyridine	c1ccncc1
110-89-4	Piperidine	C1CCNCC1
110-93-0	6-Methyl-5-hepten-2-one	CC(C)=CCCC(C)=O
111-11-5	Methyl caprylate	CCCCCCCC(=O)OC
111-13-7	2-Octanone	CCCCCCC(C)=O
111-14-8	Heptanoic acid	CCCCCCC(O)=O
111-27-3	Hexanol	CCCCCCO
111-47-7	Propyl sulfide	CCCSCCC
111-61-5	Ethyl stearate	CCCCCCCCCCCCCCCCCC(=O)OCC
111-65-9	Octane	CCCCCCCC
111-66-0	1-Octene	CCCCCCC=C
111-70-6	Heptanol	CCCCCCCO
1117-55-1	Hexyl octanoate	CCCCCCCC(=O)OCCCCCC
111-79-5	Methyl *trans*-2-nonenoate	CCCCCC\C=C/C(=O)OC
111-80-8	Methyl 2-nonynoate	CCCCCCC#CC(=O)OC
111-87-5	Octanol	CCCCCCCCO
1120-21-4	Undecane	CCCCCCCCCCC
112-06-1	Heptyl acetate	CCCCCCCOC(C)=O
112-12-9	2-Undecanone	CCCCCCCCCC(C)=O
1122-62-9	2-Acetyl pyridine	CC(=O)c1ccccn1
112-30-1	Decanol	CCCCCCCCCCO
112-31-2	Decanal	CCCCCCCCCC=O
112-37-8	Undecanoic acid	CCCCCCCCCCC(O)=O
1124-11-4	2356-Tetramethylpyrazine	Cc1nc(C)c(C)nc1C
112-42-5	1-Undecanol	CCCCCCCCCCCO
112-44-7	Undecanal	CCCCCCCCCCC=O
1125-21-9	4-Oxoisophorone	CC1=CC(=O)CC(C)(C)C1=O
112-53-8	Lauryl alcohol	CCCCCCCCCCCCO
1125-88-8	Benzaldehyde dimethyl acetal	COC(OC)c1ccccc1
112-66-3	Lauryl acetate	CCCCCCCCCCCCOC(C)=O

(Continued)

Table 2.E.1 (Continued)

CAS No.	Name	Smiles
1128-08-1	Dihydrojasmone	CCCCCC1=C(C)CCC1=O
1131-62-0	34-Dimethoxyacetophenone	COc1ccc(cc1OC)C(C)=O
115-95-7	Linalyl acetate	CC(C)=CCCC(C)(OC(C)=O)C=C
118-58-1	Benzyl salicylate	Oc1ccccc1C(=O)OCc2ccccc2
118-93-4	2-Hydroxyacetophenone	CC(=O)c1ccccc1O
1191-43-1	16-Hexanedithiol	SCCCCCCS
119-36-8	Methyl salicylate	COC(=O)c1ccccc1O
1193-79-9	2-Acetyl-5-methylfuran	CC(=O)c1oc(C)cc1
119-61-9	Benzophenone	O=C(c1ccccc1)c2ccccc2
119-84-6	Dihydrocoumarin	O=C1CCc2ccccc2O1
120-14-9	Veratraldehyde	COc1ccc(C=O)cc1OC
120-24-1	Isoeugenyl phenylacetate	COc1cc(\C=C\C)ccc1OC(=O)Cc2ccccc2
120-50-3	Isobutyl benzoate	CC(C)COC(=O)c1ccccc1
120-51-4	Benzyl benzoate	O=C(OCc1ccccc1)c2ccccc2
120-57-0	Piperonal	O=Cc1ccc2OCOc2c1
120-58-1	Isosafrole	C/C=C/c1ccc2OCOc2c1
120-92-3	Cyclopentanone	O=C1CCCC1
121-32-4	Ethyl Vanillin	CCOc1cc(C=O)ccc1O
121-33-5	Vanillin	COc1cc(C=O)ccc1O
121-34-6	4-Hydroxy-3-methoxybenzoic acid	COc1cc(ccc1O)C(O)=O
122-03-2	Cuminaldehyde	CC(C)c1ccc(C=O)cc1
122-40-7	α-Amylcinnamaldehyde	CCCCC/C(C=O)=C/c1ccccc1
122-43-0	Butyl phenylacetate	CCCCOC(=O)Cc1ccccc1
122-63-4	Benzyl propionate	CCC(=O)OCc1ccccc1
122-67-8	Isobutyl cinnamate	CC(C)COC(=O)\C=C\c1ccccc1
122-70-3	Phenethyl propionate	CCC(=O)OCCc1ccccc1
122-74-7	3-Phenylpropyl propionate	CCC(=O)OCCCc1ccccc1
122-97-4	3-Phenyl-1-propanol	OCCCc1ccccc1
123-08-0	4-Hydroxybenzaldehyde	Oc1ccc(C=O)cc1
123-11-5	*p*-Anisaldehyde	COc1ccc(C=O)cc1
123-19-3	4-Heptanone	CCCC(=O)CCC
123-25-1	Diethyl succinate	CCOC(=O)CCC(=O)OCC
123-32-0	25-Dimethyl pyrazine	Cc1cnc(C)cn1
123-51-3	Isoamyl alcohol	CC(C)CCO
123-66-0	Ethyl hexanoate	CCCCCC(=O)OCC

Table 2.E.1 (Continued)

CAS No.	Name	Smiles
123-68-2	Allyl hexanoate	CCCCCC(=O)OCC=C
123-72-8	Butyraldehyde	CCCC=O
123-75-1	Pyrrolidine	N1CCCC1
123-86-4	Butyl acetate	CCCCOC(C)=O
123-92-2	Isopentyl acetate	CC(C)CCOC(C)=O
124-04-9	Adipic acid	OC(=O)CCCCC(O)=O
124-06-1	Ethyl myristate	CCCCCCCCCCCCCC(=O)OCC
124-07-2	Octanoic acid	CCCCCCCC(O)=O
124-13-0	Octanal	CCCCCCCC=O
127-17-3	Pyruvic acid	CC(=O)C(O)=O
128-37-0	Butylated hydroxytoluene	Cc1cc(c(O)c(c1)C(C)(C)C)C(C)(C)C
133-18-6	Phenethyl anthranilate	Nc1ccccc1C(=O)OCCc2ccccc2
133-37-9	DL-Tartaric acid	OC(C(O)C(O)=O)C(O)=O
134-20-3	Methyl anthranilate	COC(=O)c1ccccc1N
135-02-4	*o*-Anisaldehyde	COc1ccccc1C=O
13532-18-8	Methyl 3-(methylthio)propionate	COC(=O)CCSC
13623-11-5	245-Trimethylthiazole	Cc1sc(C)c(C)n1
13678-67-6	22′-(Thiodimethylene)difuran	C(SCc1occc1)c2occc2
137-00-8	4-Methyl-5-thiazoleethanol	Cc1ncsc1CCO
137-06-4	*o*-Toluenethiol	Cc1ccccc1S
13708-12-8	5-Methylquinoxaline	Cc1cccc2nccnc12
137-32-6	2-Methyl-1-butanol (replicate)	CCC(C)CO
13925-00-3	2-Ethyl pyrazine	CCc1cnccn1
140-10-3	*trans*-Cinnamic acid	OC(=O)\C=C\c1ccccc1
140-11-4	Benzyl acetate	CC(=O)OCc1ccccc1
140-39-6	*p*-Tolyl acetate	CC(=O)Oc1ccc(C)cc1
140-67-0	4-Allylanisole	COc1ccc(CC=C)cc1
141-12-8	Neryl acetate	CC(=O)OC\C=C(C)/CCC=C(C)C
141-78-6	Ethyl acetate	CCOC(C)=O
142-19-8	Allyl heptanoate	CCCCCCC(=O)OCC=C
142-62-1	Caproic acid	CCCCCC(O)=O
142-92-7	Hexyl acetate	CCCCCCOC(C)=O
143-07-7	Lauric acid (replicate)	CCCCCCCCCCCC(O)=O
143-08-8	Nonanol	CCCCCCCCCO
143-13-5	Nonyl acetate	CCCCCCCCCOC(C)=O
1438-94-4	1-Furfurylpyrrole	C(n1cccc1)c2occc2

(Continued)

Table 2.E.1 (Continued)

CAS No.	Name	Smiles
14765-30-1	2-*sec*-Butylcyclohexanone	CCC(C)C1CCCCC1=O
147-85-3	L-Proline	OC(=O)C1CCCN1
149-57-5	2-Ethylhexanoic acid	CCCCC(CC)C(O)=O
150-30-1	DL-Phenylalanine (replicate)	NC(Cc1ccccc1)C(O)=O
150-60-7	Benzyl disulfide	C(SSCc1ccccc1)c2ccccc2
150-78-7	Dimethoxy benzene	COc1ccc(OC)cc1
151-05-3	α,α-Dimethylphenethyl acetate	CC(=O)OC(C)(C)Cc1ccccc1
151-10-0	13-Dimethoxybenzene	COc1cccc(OC)c1
1516-17-2	*transtrans*-24-Hexadienyl acetate (replicate)	C/C=C/C=C/COC(C)=O
15356-60-2	(+)-Menthol	CC(C)[C@@H]1CC[C@@H](C)C[C@@H]1O
15679-12-6	2-Ethyl-4-methylthiazole	CCc1scc(C)n1
15679-19-3	2-Ethoxythiazole	CCOc1sccn1
15707-23-0	2-Ethyl-3-methylpyrazine	CCc1nccnc1C
15707-24-1	23-Diethylpyrazine	CCc1nccnc1CC
1618-26-4	Bis(methylthio)methane	CSCSC
16409-46-4	Menthyl isovalerate	CCC(C)C(=O)OC1CC(C)CCC1C(C)C
16630-66-3	Methyl (methylthio)acetate	COC(=O)CSC
1679-07-8	Cyclopentanethiol	SC1CCCC1
17102-64-6	*transtrans*-24-Hexadien-1-ol	C/C=C/C=C/CO
1731-84-6	Methyl nonanoate	CCCCCCCCC(=O)OC
1759-28-0	4-Methyl-5-vinylthiazole	Cc1ncsc1C=C
1797-74-6	Allyl phenylacetate	C=CCOC(=O)Cc1ccccc1
18172-67-3	(−)-β-Pinene	CC1(C)[C@H]2CCC(=C)[C@@H]1C2
18277-27-5	2-(1-Methylpropyl)thiazole	CCC(C)c1sccn1
18368-91-7	2-Ethylfenchol	CCC1(O)C(C)(C)C2CCC1(C)C2
18409-17-1	*trans*-2-Octen-1-ol	CCCCC\C=C\CO
18486-69-6	(1*R*)-(−)-Myrtenal	O=C(C)C1=CCC2CC1C2(C)C
18640-74-9	2-Isobutylthiazole	CC(C)Cc1sccn1
1866-31-5	Allyl cinnamate	C=CCOC(=O)\C=C\c1ccccc1
2035-99-6	Isoamyl octanoate	CCCCCCCC(=O)OCCC(C)C
2051-78-7	Allyl butyrate	CCCC(=O)OCC=C
21145-77-7	6-Acetyl-112447-Hexamethyltetralin	CC1CC(C)(C)c2cc(C(C)=O)c(C)cc2C1(C)C
2153-26-6	Terpinyl formate	CC1=CCC(CC1)C(C)(C)OC=O

Table 2.E.1 (Continued)

CAS No.	Name	Smiles
2257-09-2	2-Phenylethyl isothiocyanate	S=C=NCCc1ccccc1
2305-25-1	Ethyl-3-hydroxyhexanoate	CCCC(O)CC(=O)OCC
2396-84-1	Ethyl sorbate	CCOC(=O)\C=C\C=C\C
24295-03-2	2-Acetylthiazole	CC(=O)c1sccn1
2432-51-1	Methyl thiobutyrate	CCCC(=O)SC
2442-10-6	1-Octen-3-yl acetate	CCCCCC(OC(C)=O)C=C
2445-76-3	Hexyl propionate	CCCCCCOC(=O)CC
24683-00-9	2-Isobutyl-3-methoxypyrazine	COc1nccnc1CC(C)C
2497-18-9	*trans*-2-Hexenyl acetate	CCC\C=C/COC(C)=O
25013-16-5	Butylated hydroxyanisole	COc1ccc(O)c(c1)C(C)(C)C
25152-85-6	*cis*-3-Hexenyl benzoate	CC\C=C/CCOC(=O)c1ccccc1
2530-10-1	3-Acetyl-25-dimethylthiophene	CC(=O)c1cc(C)sc1C
2568-25-4	Benzaldehyde propylene glycol acetal	CC1COC(O1)c2ccccc2
2623-23-6	L-Menthyl acetate (replicate)	CC(C)[C@@H]1CC[C@@H](C)C[C@H]1OC(C)=O
2639-63-6	Hexyl butyrate	CCCCCCOC(=O)CCC
2705-87-5	Allyl cyclohexanepropionate	C=CCOC(=O)CCC1CCCCC1
2721-22-4	δ-Tetradecalactone (replicate)	CCCCCCCCCC1CCCC(=O)O1
2785-89-9	4-Ethylguaiacol	CCc1ccc(O)c(OC)c1
27939-60-2	Trivertal	CC1C=CCCC1(C)C=O
2847-30-5	2-Methoxy-3-methylpyrazine	COc1nccnc1C
28664-35-9	45-Dimethyl-3-hydroxy-25-dihydrofuran-2-one	CC1OC(=O)C(=C1C)O
288-47-1	Thiazole	s1ccnc1
290-37-9	Pyrazine	c1cnccn1
3149-28-8	2-Methoxypyrazine	COc1cnccn1
31704-80-0	3-(5-Methyl-2-furyl)butanal	CCC(C=O)c1oc(C)cc1
3208-16-0	2-Ethylfuran	CCc1occc1
3268-49-3	3-(Methylthio)propionaldehyde	CSCCC=O
32974-92-8	2-Acetyl-3-ethylpyrazine	CCc1nccnc1C(C)=O
334-48-5	Decanoic acid	CCCCCCCCCC(O)=O
33467-74-2	*cis*-3-Hexenyl propionate	CC\C=C/CCOC(=O)CC
3391-86-4	1-Octen-3-ol	CCCCCC(O)C=C
34413-35-9	5678-Tetrahydroquinoxaline	C1CCc2nccnc2C1
34451-19-9	Butyl (*S*)-(−)-lactate	CCCCOC(=O)C(C)O
350-03-8	3-Acetylpyridine	CC(=O)c1cccnc1

(Continued)

Table 2.E.1 (Continued)

CAS No.	Name	Smiles
35250-53-4	Pyrazineethanethiol	SCCc1cnccn1
352-93-2	Diethyl sulfide	CCSCC
35836-72-7	Nopol acetate	O=C(C)OCCC1CCC2CC1C2(C)C
3658-80-8	Dimethyl trisulfide (replicate)	CSSSC
36653-82-4	1-Hexadecanol	CCCCCCCCCCCCCCCCO
3681-71-8	*cis*-3-Hexenyl acetate	CC\C=C/CCOC(C)=O
37887-04-0	2-Mercapto-3-butanol	CC(O)C(C)S
3796-70-1	610-Dimethyl-59-undecadien-2-one	CC(C)=CCC\C(C)=C\CCC(C)=O
39212-23-2	Whiskey lactone	CCCCC1OC(=O)CC1C
39255-32-8	Ethyl 2-methylpentanoate (replicate)	CCCC(C)C(=O)OCC
40018-26-6	25-Dihydroxy-14-dithiane	OC1CSC(O)CS1
4075-07-4	Androstadienone	CC12CCC3C(CCC4=CC(=O)CCC34C)C1CC=C2
4077-47-8	25-Dimethyl-4-methoxy-3(2*H*)-furanone	COC1=C(C)OC(C)C1=O
4112-89-4	Guaiacyl phenylacetate	COc1ccccc1OC(=O)Cc2ccccc2
41453-56-9	*cis*-2-Nonen-1-ol	CCCCCC\C=C/CO
41519-23-7	*cis*-3-Hexenyl isobutyrate	CC\C=C/CCOC(=O)C(C)C
4180-23-8	*trans*-Anethole	COc1ccc(\C=C\C)cc1
4208-49-5	Allyl 2-furoate	C=CCOC(=O)c1occc1
4221-99-2	(*S*)-(+)-2-butanol	CCC(C)O
42436-07-7	*cis*-3-Hexenyl phenylacetate	CC\C=C/CCOC(=O)Cc1ccccc1
431-03-8	23-Butanedione	CC(=O)C(C)=O
4411-89-6	2-Phenyl-2-butenal	C\C=C(C=O)/c1ccccc1
464-49-3	D-Camphor	CC1(C)[C@H]2CC[C@@]1(C)C(=O)C2
470-82-6	18-Cineole	CC12CCC(CC1)C(C)(C)O2
4864-61-3	3-Octyl acetate	CCCCCC(CC)OC(=O)C
491-35-0	4-Methylquinoline	Cc1ccnc2ccccc12
4938-52-7	1-Hepten-3-ol	CCCCC(O)C=C
498-02-2	Acetovanillone	COc1cc(ccc1O)C(C)=O
499-75-2	Carvacrol	CC(C)c1ccc(C)c(O)c1
502-42-1	Cycloheptanone	O=C1CCCCCC1
503-74-2	Isovaleric acid	CC(C)CC(O)=O
505-10-2	3-(Methylthio)-1-propanol	CSCCCO
505-79-3	3-(Methylthio)propyl isothiocyanate	CSCCCN=C=S

Table 2.E.1 (Continued)

CAS No.	Name	Smiles
50-69-1	D-(−)-Ribose	OCC1OC(O)C(O)C1O
50-70-4	D-Sorbitol	OCC(O)C(O)C(O)C(O)CO
513-85-9	Butanediol	C[C@@H](O)[C@H](C)O
5146-66-7	37-Dimethyl-26-octadienenitrile	CC(C)=CCC\C(C)=C\C#N
51-67-2	Tyramine	NCCc1ccc(O)cc1
5271-38-5	2-(Methylthio)ethanol	CSCCO
52-90-4	L-Cysteine	NC(CS)C(O)=O
529-20-4	*o*-Tolualdehyde	Cc1ccccc1C=O
5292-21-7	Cyclohexaneacetic acid	OC(=O)CC1CCCCC1
531-59-9	7-Methoxycoumarin	COc1ccc2C=CC(=O)Oc2c1
534-22-5	2-Methylfuran	Cc1occc1
536-60-7	4-Isopropylbenzyl alcohol	CC(C)c1ccc(CO)cc1
539-82-2	Ethyl valerate	CCCCC(=O)OCC
540-07-8	Amyl hexanoate	CCCCCOC(=O)CCCCC
540-18-1	Amyl butyrate	CCCCCOC(=O)CCC
540-42-1	Isobutyl propionate	CCC(=O)OCC(C)C
5405-41-4	Ethyl-3-hydroxybutyrate	CCOC(=O)CC(C)O
541-31-1	3-Methyl-1-butanethiol	CC(C)CCS
54300-08-2	2-Acetyl-35(6)-dimethylpyrazine	CC(=O)c1ncc(C)nc1C
5461-08-5	Piperonyl isobutyrate	CC(C)C(=O)OCc1ccc2OCOc2c1
5471-51-2	4-(4-Hydroxyphenyl)-2-butanone	CC(=O)CCc1ccc(O)cc1
54947-74-9	(±)-4-Methyloctanoic acid	CCCC[C@@H](C)CCC([O-])=O
551-93-9	2-Aminoacetophenone	CC(=O)c1ccccc1N
554-12-1	Methyl propionate	CCC(=O)OC
56-40-6	Glycine	NCC(O)=O
56-41-7	L-Alanine	CC(N)C(O)=O
56-81-5	Glycerol	OCC(O)CO
56-84-8	Aspartic acid	NC(CC(O)=O)C(O)=O
56-85-9	L-Glutamine	N[C@@H](CCC(N)=O)C(O)=O
56-86-0	L-Glutamic acid	NC(CCC(O)=O)C(O)=O
56-87-1	L-Lysine	NCCCCC(N)C(O)=O
57074-37-0	*cis*-4-Decen-1-ol	CCCCC\C=C\CCCO
57-55-6	Propylene glycol	CC(O)CO
583-60-8	2-Methylcyclohexanone	CC1CCCCC1=O
584-02-1	3-Pentanol	CCC(O)CC
589-92-4	4-Methylcyclohexanone	CC1CCC(=O)CC1

(Continued)

Table 2.E.1 (Continued)

CAS No.	Name	Smiles
589-98-0	3-Octanol	CCCCCC(O)CC
590-01-2	Butyl propionate	CCCCOC(=O)CC
59020-90-5	2-Furanmethanethiol formate	O=CSCc1occc1
591-24-2	3-Methylcyclohexanone	CC1CCCC(=O)C1
591-31-1	*m*-Anisaldehyde	COc1cccc(C=O)c1
591-78-6	2-Hexanone	CCCCC(C)=O
591-80-0	4-Pentenoic acid	OC(=O)CCC=C
592-84-7	Butyl formate	CCCCOC=O
592-88-1	Allyl sulfide	C=CCSCC=C
5989-27-5	D-Limonene	CC(=C)[C@@H]1CCC(=CC1)C
60-01-5	Tributyrin	CCCC(=O)OCC(COC(=O)CCC)OC(=O)CCC
60047-17-8	Linalool oxide	CC(C)(O)C1CCC(C)(O1)C=C
60-12-8	2-Phenylethanol	OCCc1ccccc1
60-18-4	L-Tyrosine	NC(Cc1ccc(O)cc1)C(O)=O
6032-29-7	2-Pentanol	CCCC(C)O
60-33-3	Linoleic acid	CCCCC/C=C\C\C=C/CCCCCCCC(O)=O
606-45-1	Methyl 2-methoxybenzoate	COC(=O)c1ccccc1OC
611-13-2	Methyl 2-furoate	COC(=O)c1occc1
613-70-7	2-Methoxyphenyl acetate	COc1ccccc1OC(C)=O
614-99-3	Ethyl 2-furoate	CCOC(=O)c1occc1
616-25-1	1-Penten-3-ol	CCC(O)C=C
61-90-5	L-Leucine	CC(C)CC(N)C(O)=O
61931-81-5	*cis*-3-Hexenyl lactate	CC\C=C/CCOC(=O)C(C)O
620-02-0	5-Methylfurfural	Cc1oc(C=O)cc1
622-78-6	Benzyl isothiocyanate	#VALUE!
623-37-0	3-Hexanol	CCCC(O)CC
623-42-7	Methyl butyrate	CCCC(=O)OC
624-24-8	Methyl valerate	CCCCC(=O)OC
625-84-3	25-Dimethylpyrrole	CC1=CC=C(C)N1
626-93-7	2-Hexanol	CCCCC(C)O
627-90-7	Ethyl undecanoate	CCCCCCCCCCC(=O)OCC
628-63-7	Pentyl acetate	CCCCCOC(C)=O
629-19-6	Dipropyl disulfide	CCCSSCCC
629-33-4	Hexyl formate	CCCCCCOC=O
63012-97-5	2-Methyl-3-methylthiofuran	CSC1=C(C)OC=C1

Table 2.E.1 (Continued)

CAS No.	Name	Smiles
6309-51-9	Isoamyl laurate	CCCCCCCCCCCC(=O)OCCC(C)C
63-68-3	L-Methionine	CSCCC(N)C(O)=O
637-64-9	Tetrahydrofurfuryl acetate	CC(=O)OCC1CCCO1
6378-65-0	Hexyl hexanoate	CCCCCCOC(=O)CCCCC
638-25-5	*N*-Amyl octanoate	CCCCCCCC(=O)OCCCCC
63-91-2	L-Phenylalanine	NC(Cc1ccccc1)C(O)=O
64-04-0	Phenethylamine	NCCc1ccccc1
64-17-5	Ethanol	CCO
64-19-7	Acetic acid	CC(O)=O
64275-73-6	*cis*-5-Octen-1-ol	CC\C=C/CCCCO
645-56-7	4-Propylphenol	CCCc1ccc(O)cc1
646-07-1	4-Methylvaleric acid	CC(C)CCC(O)=O
6485-40-1	(−)-Carvone	CC(=C)[C@@H]1CC=C(C)C(=O)C1
65405-77-8	*cis*-3-Hexenyl salicylate	CC/C=C/CCOC(=O)c1ccccc1O
656-53-1	4-Methyl-5-thiazoleethanol acetate	CC(=O)OCCc1scnc1C
65-85-0	Benzoic acid	OC(=O)c1ccccc1
66-25-1	Hexanal	CCCCCC=O
67-03-8	Thiamine hydrochloride	[H+].[Cl−].[Cl−].Cc1ncc(C[n+]2csc(CCO)c2C)c(N)n1
6728-26-3	2-Hexenal	CCC/C=C/C=O
6728-31-0	*cis*-4-Heptenal	CC/C=C/CCC=O
67-56-1	Methanol	CO
67-63-0	Isopropyl alcohol	CC(C)O
67-64-1	Acetone	CC(C)=O
67-68-5	Methyl sulfoxide	C[S](C)=O
67-71-0	Methyl sulfone	C[S](C)(=O)=O
67715-80-4	2-Methyl-4-propyl-13-oxathiane	CCCC1CCOC(C)S1
67883-79-8	*cis*-3-Hexenyl tiglate	CC/C=C/CCOC(=O)C(/C)=C\C
6789-88-4	Hexyl benzoate	CCCCCCOC(=O)c1ccccc1
6790-58-5	Ambrox	CC1(C)CCCC2C1CCC3(C)OCCC23
693-54-9	2-Decanone	CCCCCCCCC(C)=O
693-95-8	4-Methylthiazole	Cc1cscn1
698-10-2	5-Ethyl-3-hydroxy-4-methyl-2(5*H*)-furanone (replicate)	CCC1OC(=O)C(=C1C)O

(Continued)

Table 2.E.1 (Continued)

CAS No.	Name	Smiles
705-86-2	δ-Decalactone	CCCCCC1CCCC(=O)O1
706-14-9	γ-Decalactone	CCCCCCC1CCC(=O)O1
71-00-1	L-Histidine	O=C(O)C(N)CC1=CNC=N1
710-04-3	δ-Undecalactone	CCCCCCC1CCCC(=O)O1
71-23-8	Propanol	CCCO
71-36-3	Butanol	CCCCO
71-41-0	Pentanol	CCCCCO
71-43-2	Benzene	c1ccccc1
7149-32-8	Phenethyl 2-furoate	O=C(OCCc1ccccc1)c2occc2
7217-59-6	2-Methoxythiophenol	COc1ccccc1S
72-18-4	L-Valine	CC(C)C(N)C(O)=O
7452-79-1	Ethyl 2-methylbutyrate	CCOC(=O)C(C)CC
74-79-3	L-Arginine	NC(CCCN=C(N)N)C(O)=O
7493-63-2	Allyl anthranilate	Nc1ccccc1C(=O)OCC=C
7493-69-8	Allyl 2-ethylbutyrate	CCC(CC)C(=O)OCC=C
7493-71-2	Allyl tiglate	C\C=C(/C)C(=O)OCC=C
75-07-0	Acetaldehyde	CC=O
75-18-3	Methyl sulfide	CSC
7540-51-4	(−)-Citronellol	C[C@H](CCO)CCC=C(C)C
7541-49-3	Phytol	CC(C)CCC[C@@H](C)CCC [C@@H](C)CCC/C(C)=C/CO
75-47-8	Iodoform	IC(I)I
7553-56-2	Iodine	II
7554-12-3	Diethyl malate	CCOC(=O)CC(O)C(=O)OCC
7558-79-4	Sodium phosphate dibasic	[Na+].[Na+].O[P]([O−])([O−])=O
75-65-0	*tert*-Butanol	CC(C)(C)O
76-22-2	Camphor	CC1(C)C2CCC1(C)C(=O)C2
76-49-3	Bornyl acetate	CC(=O)OC1CC2CCC1(C)C2(C)C
765-70-8	Methylcyclopentenolone	CC1CCC(=O)C1=O
7732-18-5	Water	O
7764-50-3	D-Dihydrocarvone	CC1CCC(CC1=O)C(C)=C
7779-41-1	Decanal dimethyl acetal	CCCCCCCCCC(OC)OC
7779-50-2	Ω-6-Hexadecenlactone	O=C1OCCCCCCCCCC=CCCCC1
7779-65-9	Isoamyl cinnamate	CC(C)CCOC(=O)/C=C/c1ccccc1
77-83-8	Ethyl 3-methyl-3-phenylglycidate	CCOC(=O)C1OC1(C)c2ccccc2
7785-70-8	α-Pinene	CC1=CC[C@@H]2C[C@H]1C2(C)C

Table 2.E.1 (Continued)

CAS No.	Name	Smiles
7786-58-5	Octyl isovalerate	CCCCCCCCOC(=O)CC(C)C
7787-20-4	L-Fenchone	CC1(C)[C@H]2CC[C@](C)(C2)C1=O
77-90-7	Tributyl-2-Acetylcitrate	CCCCOC(=O)CC(CC(=O)OCCCC)(OC(C)=O)C(=O)OCCCC
77-92-9	Citric acid	OC(=O)CC(O)(CC(O)=O)C(O)=O
77-93-0	Triethyl citrate	CCOC(=O)CC(O)(CC(=O)OCC)C(=O)OCC
78-59-1	Isophorone	CC1=CC(=O)CC(C)(C)C1
78-70-6	Linalool	CC(C)=CCCC(C)(O)C=C
78761-38-3	*trans*-Cinnamyl propionate	CCC(=O)OC=Cc1ccccc1
78-83-1	2-Methyl-1-propanol	CC(C)CO
78-84-2	Isobutyraldehyde	CC(C)C=O
78-93-3	2-Butanone	CCC(C)=O
79-09-4	Propionic acid	CCC(O)=O
79-20-9	Methyl acetate	COC(C)=O
79-31-2	Isobutyric acid	CC(C)C(O)=O
79-77-6	β-Ionone	CC(=O)/C=C/C1=C(C)CCCC1(C)C
80-62-6	Methyl methacrylate	COC(=O)C(C)=C
81-14-1	Musk ketone	CC(=O)c1c(C)c(c(c(c1C)[N+]([O−])=O)C(C)(C)C)[N+]([O-])=O
81925-81-7	5-Methyl-2-hepten-4-one	CCC(C)C(=O)\C=C\C
821-41-0	5-Hexen-1-ol	OCCCCC=C
821-55-6	2-Nonanone	CCCCCCCC(C)=O
823-22-3	δ-Hexalactone	CC1CCCC(=O)O1
825-51-4	Decahydro-2-naphthol	OC1CCC2CCCCC2C1
85-91-6	Dimethyl anthranilate	CNc1ccccc1C(=O)OC
87-22-9	Phenethyl salicylate	Oc1ccccc1C(=O)OCCc2ccccc2
88-15-3	2-Acetylthiophene	CC(=O)c1sccc1
881-68-5	Vanillin acetate	COc1cc(C=O)ccc1OC(C)=O
882-33-7	Phenyl disulfide	S(Sc1ccccc1)c2ccccc2
88-69-7	2-Isopropylphenol	CC(C)c1ccccc1O
89-79-2	(−)-Isopulegol	CC1CCC(C(O)C1)C(C)=C
89-83-8	Thymol	CC(C)c1ccc(C)cc1O
90-00-6	2-Ethylphenol	CCc1ccccc1O
90-02-8	Salicylaldehyde	Oc1ccccc1C=O

(Continued)

Table 2.E.1 (Continued)

CAS No.	Name	Smiles
90-05-1	Guaiacol	COc1ccccc1O
91-22-5	Quinoline	c1ccc2ncccc2c1
91-62-3	6-Methyl quinoline	Cc1ccc2ncccc2c1
91-64-5	Coumarin	O=C1Oc2ccccc2C=C1
92-48-8	6-Methylcoumarin	Cc1ccc2OC(=O)C=Cc2c1
92-52-4	Biphenyl	c1ccc(cc1)c2ccccc2
927-49-1	6-Undecanone	CCCCCC(=O)CCCCC
928-96-1	*cis*-3-Hexen-1-ol	CC\C=C/CCO
93-04-9	Nerolin Yara Yara	COc1ccc2ccccc2c1
93-08-3	Methyl β-naphthyl ketone	CC(=O)c1ccc2ccccc2c1
93-15-2	4-Allyl-12-dimethoxy benzene	COc1ccc(CC=C)cc1OC
93-16-3	Methyl isoeugenol	COc1ccc(\C=C\C)cc1OC
93-28-7	Eugenyl acetate	COc1cc(CC=C)ccc1OC(C)=O
93-29-8	Isoeugenol acetate	COc1cc(/C=C/C)ccc1OC(C)=O
93-51-6	2-Methoxy-4-methylphenol	COc1cc(C)ccc1O
93-54-9	1-Phenyl-1-propanol	CCC(O)c1ccccc1
93-58-3	Methyl benzoate	COC(=O)c1ccccc1
93-60-7	Methyl nicotinate	COC(=O)c1cccnc1
93-89-0	Ethyl benzoate	CCOC(=O)c1ccccc1
93905-03-4	2-Methoxy-3(5 or 6)-isopropylpyrazine	COc1cncc(n1)C(C)C
94278-27-0	Ethyl 3-(furfurylthio)propionate	CCOC(=O)CCSCc1occc1
94-47-3	Phenethyl benzoate	O=C(OCCc1ccccc1)c2ccccc2
94-62-2	Piperine	O=C(/C=C/C=C/c1ccc2OCOc2c1)N3CCCCC3
96-22-0	3-Pentanone	CCC(=O)CC
96-48-0	4-Hydroxybutanoic acid lactone	O=C1CCCO1
97-42-7	Carvyl acetate	CC(=C)C1CC=C(C)C(C1)OC(C)=O
97-53-0	Eugenol	COc1cc(CC=C)ccc1O
97-54-1	Isoeugenol	COc1cc(\C=C\C)ccc1O
97-61-0	2-Methylpentanoic acid	CCCC(C)C(O)=O
97-99-4	Tetrahydrofurfuryl alcohol	OCC1CCCO1
98-00-0	Furfuryl alcohol	OCc1occc1
98-85-1	α-Methylbenzyl alcohol	CC(O)c1ccccc1
98-86-2	Acetophenone	CC(=O)c1ccccc1
99-87-6	*para*-Cymene	CC(C)c1ccc(C)cc1

Acknowledgments

The authors are grateful for the financial support of NSFC (22078041, 22278053), and "the Fundamental Research Funds for the Central Universities" (DUT22YG218).

References

1 Leffingwell & Associates (2018). Flavor & Fragrance Industry Leaders.

2 Ohloff, G. (1971). *Gustation and Olfaction*, 178–183. London, New York: Academic Press.

3 Boelens, H. (1976). Molecular structure and olfactive properties. In: *Structure–Activity Relationship in Chemoreception* (ed. G. Benz), 197–210. London: Information Retrieval Ltd.

4 Wolfson, W. (2005). In the fragrance business, the right molecule smells like money. *Chemistry & Biology* 12 (8): 857–858.

5 Chastrette, M. (1981). An approach to a classification of odours using physicochemical parameters. *Chemical Senses* 6 (3): 157–163.

6 Jurs, P.C., Ham, C.L., and Bruegger, W.E. (1981). Computer-assisted studies of chemical structure and olfactory quality using pattern recognition techniques. In: *Odor Quality and Chemical Structure*, vol. 148, 143–160. Washington, DC: American Chemical Society.

7 Eminet, B.P. and Chastrette, M. (1983). Discrimination of camphoraceous substances using physicochemical parameters. *Chemical Senses* 7 (3–4): 293–300.

8 Ham, C.L. and Jurs, P.C. (1985). Structure-activity studies of musk odorants using pattern recognition: monocyclic nitrobenzenes. *Chemical Senses* 10: 491–505.

9 Shang, L., Liu, C., and Tomiura, Y. (2017). Machine-learning-based olfactometer: prediction of odor perception from physicochemical features of odorant molecules. *Analytical Chemistry* 89 (22): 11999–12005.

10 Lin, D.Y., Zhang, S., Block, E., and Katz, L.C. (2005). Encoding social signals in the mouse main olfactory bulb. *Nature* 434: 470–477.

11 Rossiter, K.J. (1996). Structure-odor relationships. *Chemical Reviews* 96: 3201–3240.

12 Lavine, B.K., Davidson, C.E., Breneman, C., and Katt, W. (2003). Electronic van der Waals surface property descriptors and genetic algorithms for developing structure-activity correlations in olfactory databases. *Journal of Chemical Information and Computer Sciences* 43: 1890–1905.

13 Breneman, C.M. and Rhem, M. (1997). QSPR analysis of HPLC column capacity factors for a set of high-energy materials using electronic van der Waals surface property descriptors computed by transferable atom equivalent method. *Journal of Computational Chemistry* 18 (2): 182–197.

14 Teixeira, M.A., Rodríguez, O., Rodrigues, A.E. et al. (2013a). Prediction model for the odor intensity of fragrance mixtures: a valuable tool for perfumed product design. *Industrial and Engineering Chemistry Research* 52 (2): 963–971.

15 Teixeira, M.A., Rodríguez, O., Gomes, P. et al. (2013b). Chapter 1: A product engineering approach in the perfume industry. In: *Perfume Engineering*, 1–13. Oxford: Butterworth-Heinemann.

16 Goodfellow, I., Bengio, Y., and Courville, A. (2016). *Deep Learning*. MIT Press.

17 Chinta, S. and Rengaswamy, R. (2019). Machine learning derived Quantitative Structure Property Relationship (QSPR) to predict drug solubility in binary solvent systems. *Industrial and Engineering Chemistry Research* 58 (8): 3082–3092.

18 Gao, H., Struble, T.J., Coley, C.W. et al. (2018). Using machine learning to predict suitable conditions for organic reactions. *ACS Central Science* 4 (11): 1465–1476.

19 Butler, T.K., Davies, W.D., Cartwright, H. et al. (2018). Machine learning for molecular and materials science. *Nature* 559: 547–555.

20 Zhang, L., Mao, H., Liu, L. et al. (2018). A machine learning based computer-aided molecular design/screening methodology for fragrance molecules. *Computers and Chemical Engineering* 115 (12): 295–308.

21 Ivanciuc, O. (2002). Structure–odor relationships for pyrazines with support vector machines. *Internet Electronic Journal of Molecular Design* 1 (5): 269–284.

22 Li, H., Panwar, B., Omenn, G.S., and Guan, Y. (2017). Accurate prediction of personalized olfactory perception from large-scale chemoinformatic features. *GigaScience* 7 (2): gix127.

23 Gani, R. and Brignole, E.A. (1983). Molecular design of solvents for liquid extraction based on UNIFAC. *Fluid Phase Equilibria* 13: 331–340.

24 Joback, K.G. (1989). Designing molecules possessing desired physical property values.

25 Odele, O. and Macchietto, S. (1993). Computer aided molecular design: a novel method for optimal solvent selection. *Fluid Phase Equilibria* 82: 47–54.

26 Marrero, J. and Gani, R. (2001). Group-contribution based estimation of pure component properties. *Fluid Phase Equilibria* 183: 183–208.

27 Gmehling, J., Constantinescu, D., and Schmid, B. (2015). Group contribution methods for phase equilibrium calculations. *Annual Review of Chemical and Biomolecular Engineering* 6: 267–292.

28 Jhamb, S., Liang, X., Gani, R., and Kontogeorgis, G.M. (2019). Systematic model-based methodology for substitution of hazardous chemicals. *ACS Sustainable Chemistry & Engineering* 7: 7652–7666.

29 Dong, Y., Huang, S., Guo, Y., and Lei, Z. (2019). COSMO-UNIFAC model for ionic liquids. *AIChE Journal* 66 (1): e16787.

30 Austin, N.D., Sahinidis, N.V., Konstantinov, I.A., and Trahan, D.W. (2018). COSMO-based computer-aided molecular/mixture design: a focus on reaction solvents. *AIChE Journal* 64: 104–122.

31 Kupgan, G., Abbott, L.J., Hart, K.E., and Colina, C.M. (2018). Modeling amorphous microporous polymers for CO_2 capture and separations. *Chemical Reviews* 118: 5488–5538.

32 Guo, Y., He, X., Su, Y. et al. (2021). Machine-learning-guided discovery and optimization of additives in preparing Cu catalysts for CO_2 reduction. *Journal of the American Chemical Society* 143: 5755–5762.

33 Zhang, L., Babi, D.K., and Gani, R. (2016). New vistas in chemical product and process design. *Annual Review of Chemical and Biomolecular Engineering* 7: 557–582.

34 Zhang, L., Mao, H., Liu, Q., and Gani, R. (2019). Chemical product design – recent advances and perspectives. *Current Opinion in Chemical Engineering* 27: 22–34.

35 Ahn, Y.Y., Ahnert, S.E., Bagrow, J.P., and Barabasi, A.L. (2011). Flavor network and the principles of food pairing. *Scientific Reports* 1: 196.

36 Goodwin, R., Maria, J., Das, P., Horesh, R., Segal, R., Fu, J., Harris, C., 2017. AI for fragrance design. *31st Conferece on Neural Information Processing Systems (NIPS 2017)* (4–9 December 2017), Long Beach, CA, USA.

37 Keller, A. and Vosshall, L.B. (2016). Olfactory perception of chemically diverse molecules. *BMC Neuroscience* 17: 55.

38 Keller, A., Gerkin, R.C., Guan, Y. et al. (2017). Predicting human olfactory perception from chemical features of odor molecules. *Science* 355: 820.

39 Chollet, F. (2015). Keras. GitHub. https://github.com/fchollet/keras (accessed 19 January 2018).

40 Lin, S.T. and Sandler, S.I. (2002). A priori phase equilibrium prediction from a segment contribution solvation model. *Industrial and Engineering Chemistry Research* 41 (5): 899–913.

41 Zhang, L., Mao, H., Zhuang, Y. et al. (2021). Odor prediction and aroma mixture design using machine learning model and molecular surface charge density profiles. *Chemical Engineering Science* 245: 116947.

42 Zhou, T., Mcbride, K., Zhang, X. et al. (2015). Integrated solvent and process design exemplified for a Diels–Alder reaction. *AIChE Journal* 61 (1): 147–158.

43 Mullins, E., Liu, Y.A., Ghaderi, A., and Fast, S.D. (2008). Sigma profile database for predicting solid solubility in pure and mixed solvent mixtures for organic pharmacological compounds with COSMO-based thermodynamic methods. *Industrial and Engineering Chemistry Research* 47 (5): 1707–1725.

44 Karunanithi, A.T., Achenie, L.E.K., and Gani, R. (2005). A new decomposition-based computer-aided molecular/mixture design methodology for the design of optimal solvents and solvent mixtures. *Industrial and Engineering Chemistry Research* 44: 4785–4797.

45 Liu, Q., Zhang, L., Liu, L. et al. (2019). OptCAMD: an optimization-based framework and tool for molecular and mixture product design. *Computers and Chemical Engineering* 124: 285–301.

46 Gani, R., Hukkerikar, A.S., Sin, G., Abildskov, J., and Sarup, B. (2013). Development of pure component property models for chemical product-process design and analysis. CAPEC Internal Report. Technical University of Denmark.

47 Kim, S., Thiessen, P.A., Bolton, E.E. et al. (2016). PubChem substance and compound databases. *Nucleic Acids Research* 44: D1202–D1213.

48 Kalakul, S., Cignitti, S., Zhang, L., and Gani, R. (2015). VPPD-Lab: the chemical product simulator. *Computer Aided Chemical Engineering* 37: 1415–1420.

49 Conte, E., Gani, R., and Ng, K.M. (2011). Design of formulated products: a systematic methodology. *AIChE Journal* 57: 2431–2449.

3

Machine Learning-Aided Rational Screening of Task-Specific Ionic Liquids

Ruofan Gu and Zhen Song

State Key Laboratory of Chemical Engineering, School of Chemical Engineering, East China University of Science and Technology, 130 Meilong Road, Shanghai 200237, China

3.1 Introduction

Ionic liquids (ILs) are commonly acknowledged as prospective substitutes for organic solvents because of their distinctive physicochemical features such as negligible vapor pressure, broad liquidus window, nonflammability, high thermal and chemical stability, high ionic conductivity, selectively strong solvating ability toward many different chemicals, etc. [1]. The properties of ILs are widely customizable to fit task-specific requirements through deliberate combinations of cation, anion, and substituents [2–4]. They have been extensively explored in many chemical separation processes like extractive distillation (ED), liquid–liquid extraction, and gas absorption [5–9]. For a particular separation task, it is crucial to know the phase equilibrium behavior and identify an appropriate IL among numerous different ILs with unique properties. However, such a selection of ILs, as the key decision for the design of a specific IL-based process, still relies heavily on experimental testing of a small number of candidates in most cases. Considering the wide variety of possible IL structures and potential applications, experimental methods for screening ILs are labor-intensive, expensive, time-consuming, and not even realistic to screen the optimal IL candidate. Hence, proper theoretical methods to model the thermodynamic properties of ILs and IL-based separation systems are highly required to aid in selecting the best ILs for particular tasks.

In contrast to conventional organic molecules, ILs are a unique class of compounds composed entirely of anions and cations (in most cases, large asymmetric organic cations and organic/inorganic anions). Consequently, a great deal of effort has been invested in IL structure–property modeling and molecular design methodology. As the combinations of cations, anions, and substituents are virtually unlimited, it is particularly suitable to apply the computer-aided molecular design (CAMD) approach to IL design, i.e. computer-aided ionic liquid design (CAILD). In recent years, CAILD-related studies have increased considerably in literature. On the one hand, IL structure–property models including physical and thermodynamic

Applied AI Techniques in the Process Industry: From Molecular Design to Process Design and Optimization, First Edition. Edited by Chang He and Jingzheng Ren.

properties as well as some properties related to environment, health, and safety (EHS) or sustainability have exploded due to the increase in the amount of available data and/or the growing popularity of advanced modeling methods (e.g. machine learning, ML). In addition, the CAMD approach has also been extended to cover IL mixtures (also known as double-salt ionic liquids [DSILs]).

Our group has also conducted substantial research on the above aspects, and this chapter will briefly summarize the main direction of ML-boosted rational screening of task-specific ILs. The whole chapter is organized as follows. Section 3.2 presents molecular representations of ILs, and following that, Section 3.3 describes the research on ML-based structure–property models. Section 3.4 describes some representative applications of ML-assisted rational screening of ILs for specific tasks. Finally, Section 3.5 briefly summarizes this chapter and makes some future perspectives in the field.

3.2 Molecule Representation of ILs

In general, CAILD aims to efficiently select IL from a large number of potential chemical structures to meet specific task requirements. Similar to the well-known concept of CAMD, CAILD can also be divided into two subproblems (see Figure 3.1). The forward subproblem involves linking the molecular structure of IL with its properties through a structure–property model. In this case, a specific term, quantitative structure–property relationship (QSPR), can be widely referred to. These structural property models can be based on diverse molecular representations such as groups or fragments, molecular descriptors or fingerprints, and machine-learned representations. The reverse subproblem is to determine the IL structure that can provide the optimal value of target properties (such as physical, thermodynamic, and process) based on the established structure–property models.

When conducting CAILD, if a known IL database is prespecified, the term "screening" can be used, meaning to search for the best IL candidate from the existing structural space. However, new cation–anion combinations that have not been experimentally synthesized or reported can still be identified and can be considered as new IL. The term "design" can be used to potentially identify IL with novel cationic and/or anionic structures based on the virtual combinations of smaller IL substructures such as groups or fragments.

As in the CAILD scheme above, the scope of application of IL structure–property modeling directly determines the molecular space that can be explored by the

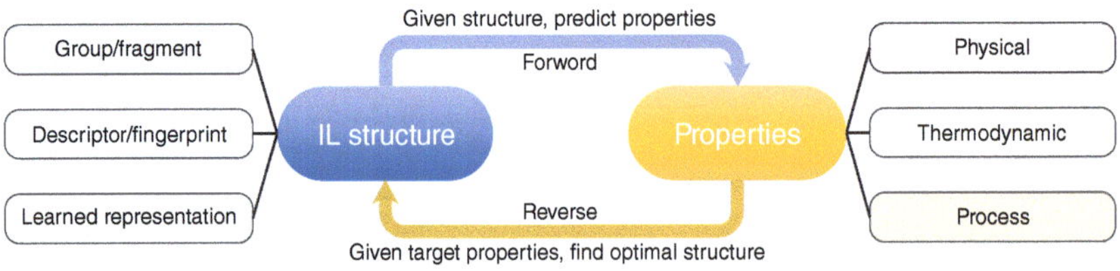

Figure 3.1 Scheme of the forward and reverse subproblems in CAILD.

CAILD method. Thus, a reliable IL structure–property model is a prerequisite for enabling computer-aided IL screening or design. Two key pillars in structure–property modeling are the proper molecular representation of ILs and the modeling algorithm. Different molecular representations of ILs can be employed, such as groups or fragments, molecular descriptors or fingerprints derived from cheminformatics or quantum chemical calculations, as well as learned representations. Accordingly, various modeling algorithms can be considered for pairing the IL representations, ranging from classical linear algorithms to nonlinear algorithms and deep learning (DL) algorithms. It is worth mentioning that the development or extension of some physically based models, such as UNIFAC (universal quasichemical functional-group activity coefficients) and COSMO (conductor-like screening model) types of activity coefficient models follows the corresponding theoretical foundations and does not involve different modeling algorithms. However, from the molecular representations point of view, UNIFAC models belong to molecular group-based structure–property models, whereas COSMO-type activity coefficient models essentially belong to structure–property models based on quantum chemical molecular fingerprints. Therefore, the foundations of different molecular representations will be briefly introduced in this section.

3.2.1 Groups or Fragment-Based Representation

Just like the property modeling of conventional molecules, the use of groups or fragments as molecular representations is a fundamental and commonly employed approach for modeling the properties of ILs and IL-related systems. Among the models that rely on such molecular representations, the group contribution (GC) method is a well-established framework. This method adheres to the additivity rule, postulating that the properties of a given molecule can be approximated by summing the contributions of its constituent groups, typically composed of a few atoms and bonds. Within the context of group decomposition for ILs, three distinct approaches are generally employed, as illustrated in Figure 3.2: (a) decomposing into one cation and one

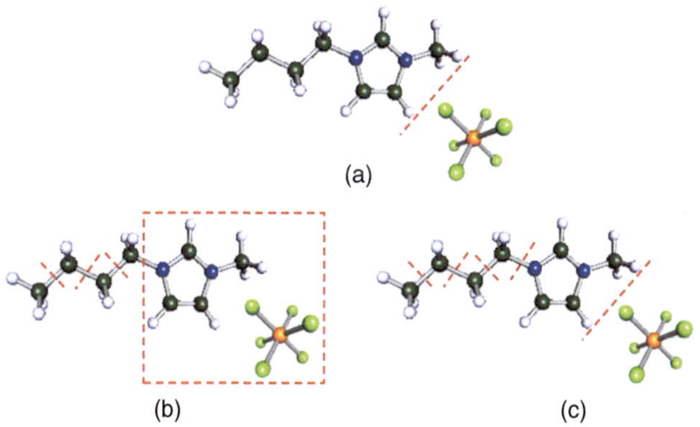

Figure 3.2 Three different approaches to decompose ILs into UNIFAC functional groups (illustrated with $[C_4MIm][PF_6]$). Source: Song et al. [10]/with permission of John Wiley & Sons.

anion, (b) decomposing into multiple groups while treating the cation skeleton and the anion as unified entities, and (c) decomposing into several groups with separate treatment of the cation skeleton.

Among the three approaches, the first method necessitates the inclusion of parameters associated with each potential cation and anion, limiting its coverage to specific cation–anion combinations. The second approach can accommodate the influence of cation substituents and avoid the use of the charged groups, which is sometimes considered theoretically advantageous, as adopted in the UNIFAC model extension proposed by Lei et al. [11]. The third approach is the most widely adopted for modeling IL properties due to its superior adaptability in capturing structural variations in cations, anions, and substituents. This approach offers the broadest applicability and flexibility during the CAILD stage. For instance, (i) Chen et al. [12] employed the third group decomposition to develop GC models for IL properties, namely density, heat capacity, viscosity, surface tension, and melting point. (ii) Song et al. [10, 13] extended the original UNIFAC models to encompass IL-solute systems, resulting in an extensive parameter matrix for the UNIFAC–IL model.

In the domain of ILs property modeling, it is observed that the group or fragment-based modeling approach is, to an extent, influenced by the experience and choices of researchers. Such variability in GC schemes presents challenges in the implementation of CAILD. Specifically, researchers may need to simultaneously navigate multiple group decomposition strategies to achieve effective molecular encoding. In this context, ambiguities in group decomposition can sometimes lead to misconceptions among model users, subsequently affecting the predictive accuracy of the models [14]. Moreover, it is imperative to strike a balance between simplicity and predictive accuracy when defining the groups or fragments for IL property modeling. Opting for a smaller number of functional groups may make molecular representation simpler, but it often undermines the accuracy of predictions. Conversely, incorporating a larger number of groups, reaching the complexity of molecular levels, amplifies the dimensional breadth of the molecular representation. This enhanced dimensionality reflects the proximal effects and interactions between groups more accurately, thereby augmenting the model's predictive or at least correlative performance. However, it is crucial to note that the development of such methodologies typically depends on extensive property datasets. Currently, only a few IL properties, such as density, viscosity, and CO_2 solubility, have been extensively studied experimentally, allowing for the application of these high-accuracy predictive models.

Despite the relative simplicity of implementing GC models in the CAMD of ILs, they also exhibit certain limitations. First, not all properties adhere to the simple additive rule. Second, these models' fundamental assumptions overlook the proximal effects of groups within molecules and are unable to differentiate between structures with varying connectivity. To address these limitations, enhancements in the property modeling of conventional molecules have been introduced, including the incorporation of multiple-order groups (first-, second-, and/or third-order groups, also known as the GC^+ approach) or additional interaction terms [15, 16].

Nevertheless, the advancement of such models for IL properties remains limited to date, partly attributable to the relatively small size of the datasets available [17, 18].

3.2.2 COSMO-Derived Descriptors or Fingerprints

The use of fingerprints based on the COSMO theory, particularly those utilizing the screening charge density (SCD or σ) distribution, has demonstrated notable potential in the area of molecular property modeling. This type of molecular fingerprint can quantitatively represent information such as polarity and hydrogen bond accepting/donating capabilities. Specifically, the COSMO calculations based on quantum chemical provide a discrete surface for a molecule embedded in a virtual conductor, where each surface segment is characterized by its area a_i and σ_i, taking into account the electrostatic screening around the molecule and the back polarization of the molecule [19]. The SCD distribution, commonly referred to as the σ-profile, represents the probability distribution of σ_i on each segment i [20, 21]. Although initially designed to link microscopic surface-interaction energies to macroscopic thermodynamic properties (i.e. as the input for COSMO–RS (conductor-like screening model for real solvent) thermodynamic calculations), σ-profiles have also been proven to be an effective fingerprint for property modeling [22–26].

In the realm of COSMO-derived fingerprints and descriptors, they have been successfully applied to the prediction of IL properties. For instance, (i) Palomar et al. [27] calculated the $S_{\sigma\text{-profile}}$ descriptors for 45 imidazolium-based ILs, establishing a strong correlation with molecular volume and density ($R^2 > 0.99$), demonstrating that the $S_{\sigma\text{-profile}}$ is an effective representation method for ILs. (ii) Zhao et al. [28] utilized the $S_{\sigma\text{-profile}}$ to predict the heat capacity of 46 ILs, encompassing a dataset of 2416 data points. Combined with the extreme learning machine algorithm, the resulting model showed reasonable performance in predicting heat capacity, with an average absolute relative deviation (AARD) of 0.60%. It is noteworthy that COSMO-derived molecular fingerprints or descriptors are contingent on the calculations of σ-profile and σ-potential; in their absence, quantum chemical calculations are required initially.

3.2.3 Machine-Learned Representations

Many of the above molecular representations have easily interpretable meanings that can support the interpretation of the models developed from them. However, there are also limitations of such molecular representations: (i) the descriptors are limited by the available knowledge space and (ii) the fingerprints are generally of large lengths when the dataset has a large variety, which leads to a large number of sparse vectors and affects model training.

Learned molecular representations are gaining popularity in the field of predicting molecular properties. One approach is based on sequence modeling, where chemical structures are efficiently transformed into strings using simplified molecular-input line-entry system (SMILES) or the international chemical identifiers (InChIs). Chen et al. [29] recently extended the application of the SMILES transformer to enable the prediction of σ-profile as well as the V_{COSMO} of conventional molecules. First, the transformer model was pretrained on the ChEMBL25 database with 1 768 081 molecules using an encoder–decoder architecture that can effectively capture molecular fingerprints from the canonical SMILES of molecules. A convolutional neural network (CNN) model was trained and tested for predicting the σ-profile and V_{COSMO} on the VT-2005 database by employing the pretrained molecular fingerprints. The transformer–CNN model demonstrates significantly improved performance compared to the GC–COSMO approach. Recently, a similar framework called ILTransR [30] has been successfully developed for predicting 11 IL properties. It was pretrained on 10 243 410 IL-like SMILES and fine-tuned corresponding to each small, labeled IL property dataset.

Another type of learned representation for molecules is based on the neural network entity embeddings [31]. Entity embedding can place similar entities closer together in the embedding space by mapping high-dimensional categorical variables to a low-dimensional learned representation. Research shows that when data are sparse and statistical information is unknown, entity embedding helps neural networks better generalize. Recently, Chen et al. [32] employed the neural network entity embeddings as the molecule representation method to develop the neural recommender system (RS) for predicting activity coefficients of IL-involved systems. Specifically, the neural network entity embedding was used to map each solute and each IL to a 128-number latent vector. The inner product or the cosine of the angle between latent vectors can assess the latent similarity of each IL and each solute. If one-hot-encoding is used for ILs and solutes, there would be a vector with the same dimensions as the number of ILs and solutes, where a “1” denotes a specific IL or solute and all other elements in the vector are “0.” In this way, vectors representing similar ILs or solutes are not “closer.” In contrast, it is possible to represent all ILs and solutes in smaller dimensions while mapping similar ILs and solutes closer together in the embedding space by training a neural network to learn entity embeddings. Entity embeddings can be learned as part of a neural network that predicts γ^{∞}, with the embedding layer weights adjusted during training to minimize losses on the prediction problem.

3.3 Machine Learning-Based Structure–Property Models

3.3.1 Machine Learning for COSMO-Based Models

Based on COSMO, two models of activity coefficient have emerged: one for realistic solvents known as COSMO–RS [33] and the other for COSMO segment activity coefficient known as COSMO–SAC [34, 35]. These two models share a similar methodological framework but differ in the approach to handling the combinatorial and residual contributions in chemical property calculations. As both models require only the σ-profile and V_{COSMO} derived from quantum chemistry as input, they have been widely evaluated and used to predict the thermodynamic properties of IL-involved systems. Coutinho's research group [36, 37] extensively evaluated the predictive performance of COSMO–RS for the liquid–liquid equilibrium (LLE) of over 150 binary and 180 ternary {IL + aromatic hydrocarbon + aliphatic hydrocarbon} systems, concluding that COSMO–RS mainly provides qualitative and semiquantitative trends in the phase behavior dependence of ILs and hydrocarbon molecules.

The predictive performance of COSMO–RS has also been extensively evaluated for vapor–liquid equilibrium (VLE) and LLE of binary {IL + molecular compound} systems, as well as the mutual solubility of water and ILs. For example, to develop efficient separation processes (such as ED) based on ILs, it is necessary to understand the VLE behavior related to ILs. As experimental measurements are often time-consuming and expensive, reliable VLE prediction models are highly desired. Qin et al. [38] systematically evaluated the predictive performance of COSMO–RS for ternary and binary VLE systems containing ILs, with a large experimental database established, including 6531 ternary and 2265 binary systems. The VLE systems were divided into five ternary system subsets and three binary system subsets based on the type of solute molecule. The performance of COSMO–RS in capturing the effects of IL structure, solute structure, IL concentration, and temperature on VLE was also analyzed. COSMO–RS provides an overall qualitative description of VLE behavior and, in some cases, a quantitative description, making it a valuable tool for IL screening and process analysis related to the separation processes. The solubility of water and environmentally friendly solvent ILs is crucial for proper IL selection in chemical reactors and separation processes. Zhou et al. [39] predicted the mutual solubility of water and 1500 ILs (50 cations and 30 anions) at 298.15 K, using the COSMO–RS as a thermodynamic model. Based on the COSMO–RS calculations, the influence of cation and anion types, side chains, and substituent groups on water solubility was extensively analyzed, providing a pre-screening of candidate solvent ILs. Additionally, to gain a deeper understanding of their inherent solubility behavior, multiple water–IL interaction energies were calculated based on COSMO–RS calculations, comparing different types of molecular interactions between ILs and water in solution. It was confirmed that hydrogen bonding interactions between anions and water molecules have a major impact on solubility. Finally, to rapidly estimate solubility and select solvents, molecular descriptors were calculated for typical anions and anion families derived from COSMO–RS, indicating the strength of hydrogen bond acceptors on anions.

Considering the quantum chemistry calculation of σ-profile is the most computationally expensive part of COSMO-type activity coefficient calculations; there have been attempts to generate σ-profiles without quantum chemical calculations, e.g. developing GC or ML methods based on existing σ-profile databases. For instance, (i) Peng et al. [40] proposed a CAILD method for designing IL solvents for specific separation processes, wherein they established a GC–COSMO method for estimating the σ-profile and cavity volume of cations of ILs. (ii) Zhang et al. [41] proposed a modified GC method, the COSMO–GC–IL method, for the quick prediction of IL σ-profiles and V_{COSMO} based on the σ-profile of the IL groups in their reference state as well as the two modification parameters. The effect of the molecular environment on the target group can be well described, allowing for the differentiation of the IL isomers. They validated the method by calculating the activity coefficients of 28 644 binary systems and 3410 ternary mixtures in LLE, using the σ-profiles and V_{COSMO} generated by the present method and the Dmol^3 database, demonstrating the reliability of the COSMO–GC–IL method. (iii) As for the combination with the ML method, Chen et al. [29] have developed a DL method for quickly predicting the σ-profiles and V_{COSMO} of molecules. Molecular fingerprints can be obtained from the encoder state of the pretrained transformer model on the ChEMBL database, allowing for transfer learning from large-scale unlabeled data and improving generalization performance. They trained and tested a CNN model for the σ-profiles and V_{COSMO} on the VT-2005 database using the pretrained molecular fingerprints. The obtained transformer–CNN model notably outperforms the GC–COSMO method, predicting σ-profiles and V_{COSMO} of millions of molecules in just a few minutes. Of course, this promising method can be extended for the prediction of σ-profiles and V_{COSMO} of cations and anions of ILs, provided that a considerable database of σ-profiles and V_{COSMO} of ILs is available.

3.3.2 Machine Learning for UNIFAC Extensions

The UNIFAC model is an efficient and practical thermodynamic prediction model based on GC methods. By knowing the groups in the mixture and the related group parameters, it can accurately calculate the activity coefficient of molecules in the system. Currently, the UNIFAC models for IL-involved systems (also known as the UNIFAC–IL model) have successfully fitted a large number of interaction parameters in the literature and have been widely used in IL screening or design for many typical separation processes.

By adopting the group decomposition method in Figure 3.2c, Song et al. [10] successfully extended the UNIFAC model to the extraction of desulfurization systems. They used thiophene, alkanes (hexane, heptane, octane), aromatics (benzene, toluene, ethylbenzene), and cycloalkanes (cyclopentane, cyclohexane, cycloheptane) to regress UNIFAC interaction parameters between solute groups and IL groups based on experimental γ^{∞}. This included 7 cation skeleton groups, 13 anion groups, and 7 conventional groups. They explicitly provided the AARD between experimental γ^{∞} and model-calculated γ^{∞} for various solutes, followed by external validation by predicting the LLE behavior of 171 systems. The low AARD values during γ^{∞} calculations and small root-mean-square deviation (RMSD) during LLE predictions fully demonstrate the reliability of the obtained model. Inspired by this encouraging result, they subsequently proposed a larger extension. They compiled a dataset of experimental γ^{∞} data for 83 molecular solutes in 214 ILs, totaling 39 358 data points. These 83 solutes included various types of conventional molecular species such as alkanes, cycloalkanes, aromatics, olefins, ketones, alcohols, ethers, ketones, sulfur compounds, nitrogen compounds, and water, while the 214 ILs were from 9 cation families and 29 anion families. Therefore, the UNIFAC–IL extension results covered 21 conventional main groups, 9 cation skeleton groups, and 29 anion groups. Due to the fact that ILs were divided into independent cation skeletons, anions, and substituent groups, the UNIFAC–IL model could cover most ILs, and molecular solutes involved in relevant studies. Meanwhile, they also provided the AARD between experimental and calculated γ^∞ for various solutes and predicted a total of 860 ternary LLE systems and 99 ternary VLE systems for verification. These analyses demonstrate the predictive performance of the UNIFAC–IL model for different types of systems.

From the above extension of UNIFAC based on experimental data, it can be seen that due to the shortage and uneven distribution of available data, there are still a large number of missing parameters in the parameter matrix. To overcome this limitation, Song et al. [13] systematically evaluated the accuracy of COSMO–RS model in predicting γ^∞ of different solute families in various ILs based on a large experimental database. It was found that the AARDs between uncalibrated COSMO–RS predicted and experimental γ^∞ differ significantly, ranging from 22.78% for thiophene to 207.93% for CCl_4. To reduce systematic errors associated with this model, they calibrated the solute families, but the AARD still ranged from 19.23% for carboxylic acids to 56.80% for alkanes. Therefore, it is not suitable to use COSMO–RS predicted data generally in the UNIFAC–IL expansion process, especially for solute series with large AARDs. In view of this, they only generated a calculated pseudo-experimental database based on calibrated COSMO–RS predicted values with AARDs less than 30% for 7 solute families (i.e. alcohols, esters, aldehydes, nitroalkanes, thiophene, pyridine, and H_2O) to specifically fill in the missing 72 parameters in the UNIFAC–IL parameter matrix.

In addition to the activity coefficient method of COSMO-type, Chen et al. [32] also proposed to use ML-based models to generate pseudo-experimental data. They took advantage of the highly sparse feature of the currently available solute–IL γ^∞ database and carefully developed a deep neural network-based RS to predict γ^∞. The results showed that the obtained RS has high accuracy in predicting γ^∞ for different solutes and solute families, outperforming calibrated COSMO–RS models and UNIFAC–IL models. Therefore, the establishment of a sparse experimental γ^∞ database enabled the extended UNIFAC–IL model to cover all interaction parameters involved (the completed UNIFAC–IL model parameter matrix is shown in Figure 3.3). The application of γ^∞ data obtained from the recommendation system to UNIFAC–IL extension is a good example of combining the advantages of ML and classical thermodynamic methods.

It should be emphasized that the above UNIFAC–IL extensions all follow the framework of the original UNIFAC model. However, despite the widespread research and successful applications of the original UNIFAC model, it still has some limitations. For example, for certain strongly asymmetric systems, its predictions for γ^∞, excess enthalpy (H^E), and other aspects are not satisfactory. To overcome these weaknesses of the original UNIFAC model, Weidlich and Gmehling [43] proposed a modified UNIFAC model (also known as the Dortmund model) in 1987. The difference between the modified UNIFAC model and the original UNIFAC model lies in its unique combinatorial term and the introduction of temperature-dependent group interaction parameters and different van der Waals group parameters. These improvements make the Dortmund model more accurate in predicting the

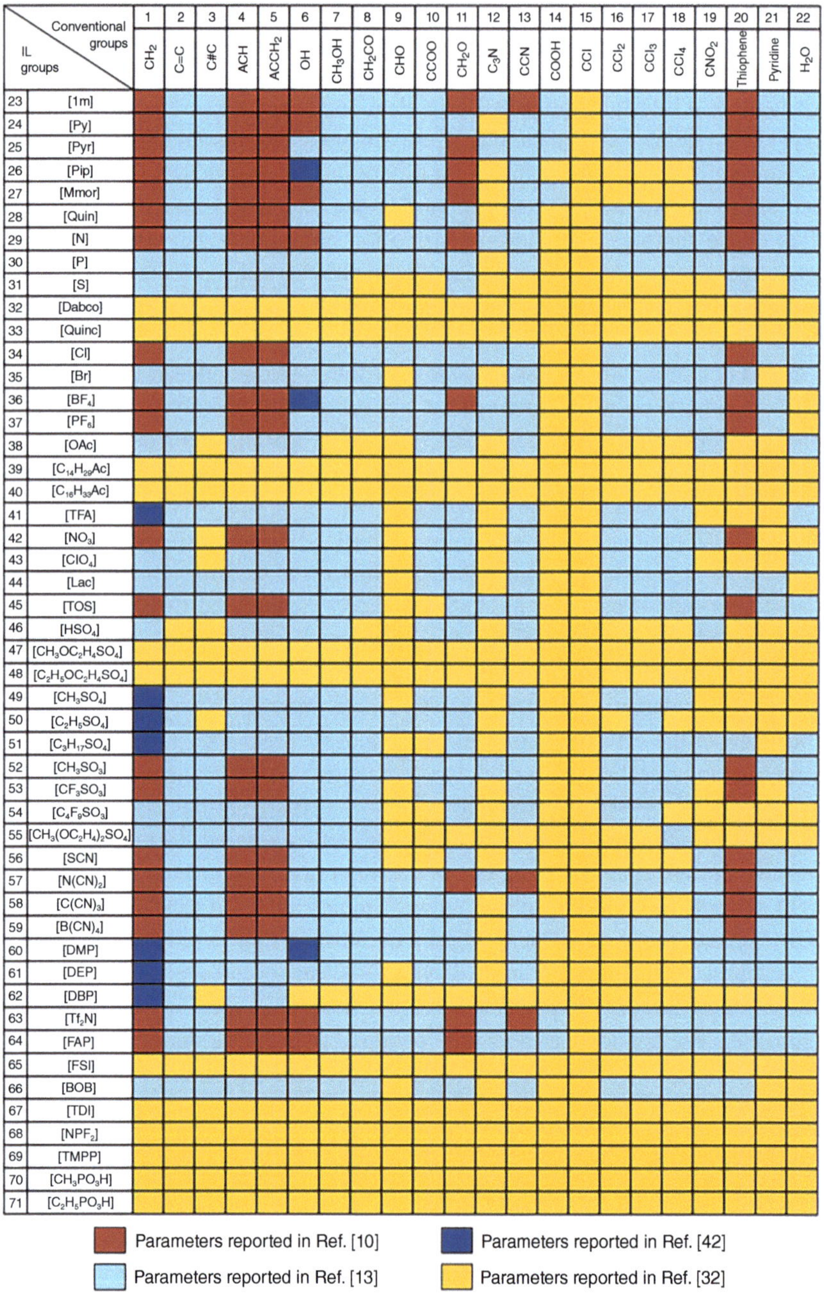

Figure 3.3 Extended UNIFAC-IL parameter matrix by combining experimental and neural recommender system predicted γ^{∞} data. Source: Chen et al. [32]/with permission of John Wiley & Sons.

thermodynamic properties of certain specific systems. The modified UNIFAC model has also been extended to systems involving ILs.

Kato and Gmehling [44] reported the extension of the modified UNIFAC model to IL-involved systems, especially the first extension to [RR′IM][BTI]. They introduced two main groups, imidazolium ([im]) and bis(trifluoromethanesulfonyl)imide ($[BTI]^-$), and obtained their group interaction parameters with main groups such as alkanes, cycloalkanes, olefins, and alcohols by simultaneously fitting experimental VLE, γ^∞, and H^E data of related systems composed of different kinds of imidazolium cations and $[BTI]^-$ anions. In cases where the parameters were temperature-dependent, the modified UNIFAC model showed better consistency with experimental data than the original UNIFAC model in γ^∞ and H^E, while the deviation of the two models was similar in VLE. After this work, the same research group published a series of similar extensions covering 81 group interaction parameters in total [45–48]. Compared with the corresponding work of the original UNIFAC model, the modified UNIFAC model's extension to IL-involved systems still covers a very limited number of groups. Therefore, obtaining more experimental data, especially temperature-related data such as VLE and H^E, and further expanding the improved UNIFAC–IL model will be of great significance. This will enable us to predict the thermodynamic properties of complex systems involving ILs more accurately and promote the development of ILs and related fields.

3.3.3 Pure Machine Learning Models

Another type of ML-based structure–property model is the ones built in a purely data-driven method. For instance, for one widely studied thermodynamic property of IL-involved systems, i.e. the solubility of carbon dioxide in ILs, Song et al. [49] reported two models by combining ML algorithms (ANN and SVM) with GC concepts. The model development employed a comprehensive dataset containing 10 116 solubility data of carbon dioxide in various ILs at different temperatures and pressures, with IL group, temperature, and pressure as model inputs. Based on an artificial-random hybrid strategy, the entire dataset was divided into a training set consisting of 8093 points and a test set consisting of 2023 points. Specifically, the least commonly used group's data points were evenly divided into five folders, while all other data points were randomly distributed among the five folders. After training the model using four of the five folders, the estimated mean absolute error (MAE) and R^2 for the ANN–GC model were 0.0202 and 0.9836, respectively, while the estimated MAE and R^2 for the SVM–GC model were 0.0240 and 0.9783, respectively.

Recently, DL algorithms have also been employed to predict the solubility of carbon dioxide in ILs. Chen et al. [30] extended the DL-based transformer–CNN framework to predict the solubility of CO_2 in ILs (1 of the 11 benchmarks for validating the DL strategy). As only the SMILES representation of ILs, temperature, and pressure are required as inputs, and a rigorous 10-fold cross-validation procedure employing IL-based random dataset splitting strategies is utilized, the obtained model has a testing MAE of 0.022 and can thus be considered a valuable method for predicting CO_2 solubility in ILs.

3.4 Applications

The evolution and advancement of IL-related structure–property models, coupled with the assistance of ML methods, have sparked numerous initiatives in the computer-aided screening and design of ILs. These endeavors vary in their scope and approach, encompassing a range of properties and their interconnections, and can be broadly categorized into five groups: single-objective property screening, multilevel hierarchical screening, IL molecular design, sequential molecular design with process evaluation, and integrated molecular and process design. This spectrum covers diverse separation tasks, including LLE, ED, and gas absorption, as well as reaction separation processes like reactive extraction. Moreover, the realm of energy storage and conversion materials, such as heat transfer fluids, materials for solar energy storage, and electrolytes, has also been explored. In the following Sections 3.4.1, 3.4.2, 3.4.3, we will delve into exemplary applications of ML-facilitated IL screening and design, highlighting their varied uses in different application scenarios.

3.4.1 Computer-Aided IL Screening

The early availability of predictive thermodynamic methods such as COSMO–RS and COSMO–SAC has made it possible to assess various thermodynamic properties and has led to the screening of task-specific ILs for different separation tasks.

The extraction of nitrogen and sulfur compounds from fuel oil, which is typically carried out to remove the final traces of these compounds, is an example of a solvent-sensitive system where the use of ILs as solvents in extraction systems can result in significant downstream issues or a significant decrease in fuel product quality. This means that even a small potential solubility of IL in the extract phase can result in significant downstream issues or a significant decrease in the fuel product quality. Therefore, special attention should be paid to this issue. A similar threat exists when ILs are used to separate organic compounds and metal ions from water mixtures, as well as to extract biopolymers or drugs. For these solvent-sensitive systems, Song et al. [50] proposed that, in addition to extraction capacity and selectivity, a strict criterion should be established to select suitable IL solvents for such systems, taking into account their solubility in the extract phase. To illustrate this point, they carried out a stepwise analysis of the influence of different anions, alkyl chain length, and the number of alkyl chains on the three thermodynamic parameters to screen ILs for desulfurization. Through COSMO–RS screening, a potential IL [Bmim][H_2PO_4] was successfully selected, which was then verified for its solubility in the extract phase and its extraction capacity through IL solubility, LLE, and multistage extraction experiments. The proposed solvent screening method was verified in the desulfurization process, and [Bmim][H_2PO_4] was selected through this screening method with its high extraction performance verified experimentally.

Deodorizer distillate is a byproduct of vegetable oil refining and an important source of natural vitamin E. Qin et al. [51] used a combination of COSMO–RS prediction and experimental evaluation to screen for an IL that could be used as a carrier agent to extract natural vitamin E from deodorizer distillate (using a mixture of α-tocopherol and methyl linoleate as a model). First, COSMO–RS was used to study 47 different ILs with different anions based on the [Bmim] cation to identify the most promising candidate. The influence of alkyl chain length on the imidazole ring was also studied experimentally. It was found that the selectivity for vitamin E and the solubility of IL in methyl linoleate were directly proportional to the HB_acc3 descriptor factor of the anion. Taking into account the influence of the alkyl side chain length and number, [Hmim][OAc] was ultimately identified as the most promising IL, with a selectivity of 108.23 and a distribution coefficient of 8.18. Corn oil deodorization distillate was used as a model to verify its performance, which showed excellent performance in processing low concentrations of vitamin E.

Following the establishment of these initial solvent selection frameworks tailored for specific process applications, numerous researchers have been actively adapting them. This includes the incorporation of various computational methods and models, along with adjustments to target characteristics to suit particular case studies. The separation of 1,3-butadiene (1,3-C_4H_6) and 1-butene (n-C_4H_8) presents a significant challenge due to their closely aligned boiling points and analogous molecular structures. ED is widely acknowledged as a viable method for this separation. In the context of ED processes, selecting an appropriate entrainer is of paramount importance. ILs have emerged as potential substitutes for traditional organic solvents, offering a pathway to reduce the energy consumption of these processes. Qin et al. [52] used COSMO–RS for high throughput computational screening of ILs, calculating the infinite dilution extraction capacity and selectivity of 36 260 ILs to find suitable entrainers. After screening, ILs that met preset thermodynamic criteria and physical property limitations were identified and subsequently sent for rigorous process simulation and design. It was found that 1,2,3,4,5-pentamethylimidazolium methyl carbonate was the best IL solvent. Compared to a benchmark ED process using the organic solvent *N*-methyl-2-pyrrolidone, energy consumption was reduced by 26%.

To select suitable ILs as extraction solvents for industrial applications, Song et al. [53] proposed a systematic approach that combines COSMO–RS phase equilibrium calculations, GC model-based physical property predictions, and process simulation based on Aspen Plus to screen for practical ILs suitable for extraction. COSMO–RS can be used to predict the LLE of systems composed of the target mixture to be separated and different ILs, allowing for the pre-screening of ILs with higher mass distribution coefficients, stronger selectivity, and lower solvent losses. Then, the GC methods can be used to estimate the key physical properties of the pre-screened ion exchange resins, and further recommend candidate substances that meet certain physical property restrictions. Subsequently, the performance of the best candidate IL in a continuous process can be analyzed using Aspen Plus, and the optimal solvent based on the process can be ultimately determined.

Taking the separation task as an example, a comprehensive hierarchical solvent screening framework can be developed by considering all important aspects from different studies, as shown in Figure 3.4. First, specific situations should be evaluated based on expert knowledge or database searches, such as potential solvent residues

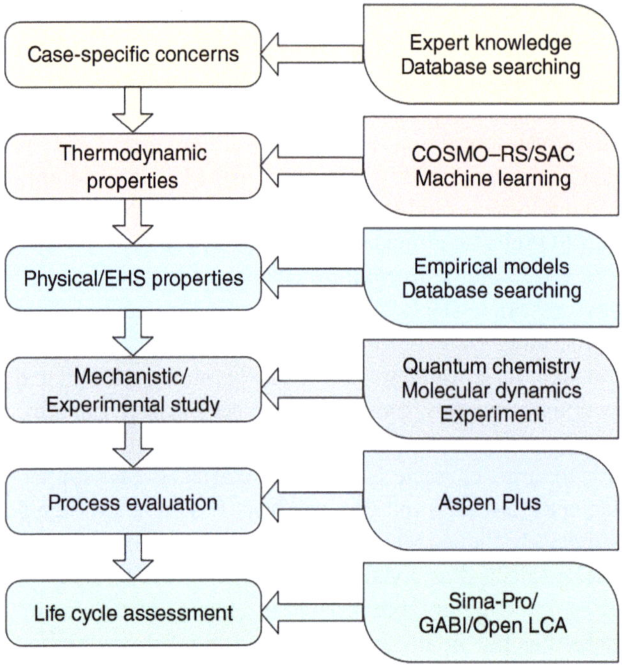

Figure 3.4 A multilevel hierarchical IL screening framework for separation tasks.

(losses), availability or synthetic accessibility, and reactivity with components in the system [54]. Subsequently, thermodynamic calculations based on COSMO models can be used to screen for IL candidates with satisfactory separation performance, such as phase equilibrium calculations, activity coefficient predictions, and Henry's law constant calculations. In addition, important physical and EHS-related properties (e.g. melting point, viscosity, toxicity, and biodegradability) can be evaluated using empirical models or database searches to eliminate IL candidates that are unsuitable for practical applications, resulting in a shortlist of promising candidates. For these candidates, mechanism studies such as quantum chemical-based molecular interaction analysis or molecular dynamics simulations can be conducted to gain insights into their potential functions, while experiments can also be performed to verify their practical separation performance. Process simulations can be carried out in tools like Aspen Plus based on the pre-selected ILs to compare the process performance of different ILs. Finally, life cycle assessments (LCAs) can be conducted to carefully examine the environmental and sustainability impacts of IL-based processes. It should be noted that LCA, a common method for measuring the environmental impact of products during all stages of their life cycle, has some inherent limitations when applied to the IL field. Although further development is still needed, LCAs can serve as the final step in designing IL-based processes. For all the property estimations involved in the multilevel screening, ML models can be used wherever models are available or could be developed based on available databases.

3.4.2 Computer-Aided IL Design

Beyond the IL screening studies, there has been a surge in publications focused on IL design. These initial research efforts primarily concentrated on designing new IL molecular structures tailored to specific target properties or characteristics. For instance, leveraging the extension of GC–COSMO–SAC, Peng et al. [40] reported two CAILDs for separation processes, specifically the extraction of benzene from cyclohexane and the capture of carbon dioxide after combustion. They utilized the GC–COSMO–SAC model for estimating thermodynamic properties and applied semi-empirical models to approximate physical properties like melting point, viscosity, and surface tension. To address the mixed-integer nonlinear programming (MINLP) challenges in CAILD, a deterministic optimization strategy was employed, utilizing the branch and bound algorithm. Subsequently, further refinements were made to the GC–COSMO–SAC method, enabling the calculation of the σ-profile for the IL groups in its reference state, along with the introduction of two correction parameters (i.e. stretching and translation parameters). A framework for CAILD was proposed based on optimization, combining separation performance with penalties for violating physicochemical property and process requirement constraints as the objective function. To facilitate the use of the improved GC–COSMO–SAC model for thermodynamic property calculations, the IL was represented as a rooted tree, where each group was considered as a set of nodes connected by branches. To solve the CAILD problem based on optimization, they developed a mixed simulated annealing-genetic algorithm (MSAGA). The proposed CAILD framework was illustrated using two extraction processes, i.e. benzene/cyclohexane and heptane/thiophene, and ILs that have better separation performance than those reported in the literature were identified. This CAILD framework was subsequently extended to study different cases of carbon dioxide capture, i.e. flue gas (CO_2/N_2), syngas (CO_2/H_2), and acid gas (CO_2/H_2S), highlighting the importance of different design objectives. Finally, a recommendation was made for the absorption–selectivity–desorption index (ASDI) based on mass, which can effectively balance different ideal thermodynamic characteristics (i.e. gas solubility, selectivity, and desorption capacity) with low energy consumption.

Zhang et al. [55] proposed a data-driven IL design strategy for carbon dioxide capture. They predicted carbon dioxide solubility using simpler and more reliable data-driven models instead of the traditional approach of combining state equations with the UNIFAC–IL model. Different batches of MINLP were based on multiple data-driven models trained from the same experimental database, and only ILs that were consistently identified across different batches were considered as reliable solutions within the effective domain of the data-driven model. The MINLP problem was encoded using GAMS27.2 and solved using the global optimization solver BARON. Through DFT calculations, the carbon dioxide capture performance of the designed ILs was ultimately confirmed, demonstrating that the proposed strategy can effectively improve design reliability.

In many IL molecular design cases, different IL characteristics exhibit conflicting trends as molecular structures change, which makes selecting the best candidate substances a dilemma. The optimal method to balance the effects of different IL characteristics is through process simulation and evaluation because the best IL for a specific process application should be based on the highest process performance [27]. Therefore, several works on CAILD have included performing process studies after IL molecular design to make final selections from shortlists of pre-screened IL candidates.

Song et al. [10] conducted a CAILD study on the extraction desulfurization of fuel oil. This study combined UNIFAC–IL extension, molecular design based on MINLP, and process evaluation based on Aspen Plus (as shown in Figure 3.5). First, they extended the UNIFAC–IL model to the extraction desulfurization system and proposed a MINLP problem based on it, with the objective function of maximizing the LLE performance of ILs. The constraints were imposed on structural feasibility and complexity, as well as thermodynamic and physical properties. The MINLP problem was solved using a generation-and-test method, gradually reducing the number of candidate ILs through different constraints, and finally, 44 candidate ILs were retained for target property evaluation. In a continuous extraction desulfurization process, the top five candidate ILs were evaluated, including an extraction tower and a distillation tower for IL regeneration. The results showed that all candidate ILs have much better process performance than the reference solvent sulfone. This CAILD framework was later extended by the research group to be used for *n*-hexane/methylcyclopentane ED, with different groups defined in the UNIFAC–IL extension to handle the proximity effects of groups in alkanes and cycloalkanes.

Wang et al. [56] combined molecular design of IL based on MINLP with TRNSYS simulation to identify candidate ILs that are promising as solar energy storage media and heat transfer fluids. The MINLP problem used a target function of thermal storage density, which integrated properties such as density, heat capacity,

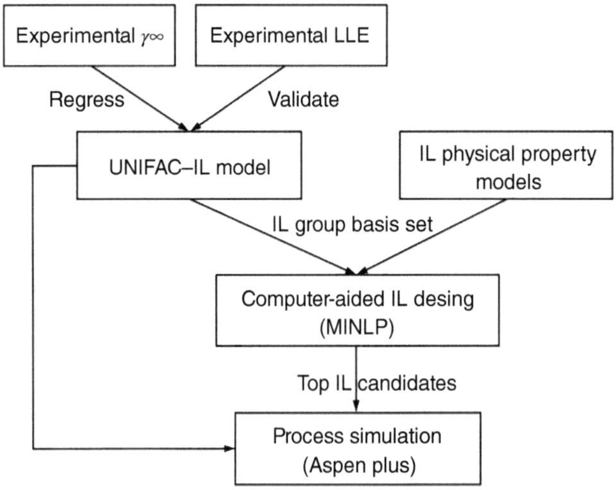

Figure 3.5 Framework for CAILD by combining UNIFAC–IL extension, MINLP optimization, and process simulation. Source: Song et al. [10]/with permission of John Wiley & Sons.

melting point, and decomposition temperature, while also considering physical property constraints such as thermal conductivity and viscosity. They solved the MINLP problem using the MSAGA mixed-integer algorithm and selected the top five IL molecules for TRNSYS simulation evaluation in long-term solar power generation systems. The results from annual system performance and long-term cost savings demonstrated the applicability, feasibility, and sustainability of the designed IL as a solar energy storage process medium.

In summary, considering the best process performance in a specific process, both IL design and process configuration design are important. Although sequential IL molecular design and process evaluation can solve this issue to some extent, the intrinsic dependence of IL characteristics on process conditions cannot be fully addressed due to strong interactions between IL and process design. In fact, in most process evaluations, only a few ILs are typically retained, and the best-performing candidate ILs may be missed during the IL design step. Therefore, simultaneously solving both IL design and process design issues is highly desirable through integrated computer-aided ionic liquid and process design (CAILPD).

Zhang et al. [57] reported a new method for integrated IL and absorption process design, where a rigorous rate-based process model is used to incorporate absorption thermodynamics and kinetics. They employed different types of models including GC models and thermodynamic models to predict the process-relevant physical, kinetic, and thermodynamic (gas solubility) properties of ILs. Combining the property models with the process models allows the integrated IL and process design problem to be formulated as an MINLP optimization problem. In the case study, the application of rigorous thermodynamic models (UNIFAC and PR) to predict the CO_2 equilibrium solubility in ILs results in a convergence failure of the MINLP problem. To address this problem, they used an ANN-based surrogate model to replace the thermodynamic models, and successfully solved the integrated design problem to the global optimality.

3.4.3 Extensions to IL Mixtures

Although ILs have opened up a vast chemical space for chemical processes, selecting ILs with practical appeal for specific industrial applications can still be challenging. This is because, in many cases, only a small subset of ILs simultaneously meet multiple desired characteristics. For example, in the screening of ILs for fuel oil extractive desulfurization, only 831 candidates out of an initial set of 36 260 cation–anion combinations satisfy the thermodynamic property requirements. After imposing physical property constraints, only 15 candidates survive. It can be imagined that the number of viable candidates will further decrease if more restrictions such as commercial availability, ease of preparation, low cost, and EHS compatibility are taken into account. In some cases, there may not even be any candidates.

Given this challenge, some researchers in the IL community have turned their attention to DSILs. Compared to classical two-ion ILs (i.e. one cation and one anion), DSILs consist of more than one cation and/or anion. Among different types of DSILs, $[C_1]_x[C_2]_{1-x}A$, $C[A_1]_x[A_2]_{1-x}$, and $[C_1A_1]_x[C_2A_2]_{1-x}$ (see Figure 3.6) are

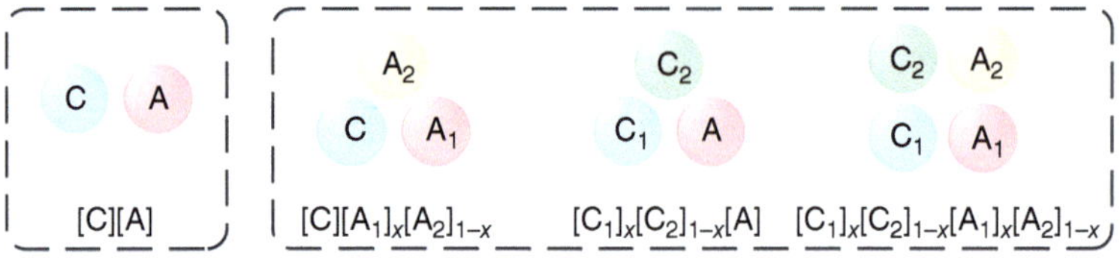

Figure 3.6 Classical two-ion IL versus three common types of DSILs (C and A represent cation and anion, and x refers to the molar ion ratio).

more practical and convenient to be prepared. Considering the simplest way to obtain such DSILs is to mix different two-ion ILs together, they are often directly treated as IL mixtures in many cases. Rogers and coworkers [58] advocate the use of DSILs instead of IL mixtures because they found that IL mixtures behave differently from molecular mixtures: the ions in a single IL lose their identity and one cannot identify which ion comes from which IL; the properties of the mixtures arise from the properties of the ions and their relative proportions rather than the individual ILs used to make the mixture. Regardless of the terminology used, it is generally accepted that DSILs are more attractive than two-ion ILs in two aspects. First, DSILs greatly expand the chemical space for screening and designing process fluids. It is estimated that there are about 10^6 possible two-ion ILs, whereas only a few hundred have been synthesized or commercialized so far. In contrast, the potential DSILs that can be generated by simply mixing two-ion ILs can be much larger; even if only using commercialized or experimental ILs, a huge DSIL space can still be accessed. Second, DSILs are a convenient method for customizing desired properties. Typically, adjusting the properties of two-ion ILs involves changing alkyl chains and/or adding functional substituents; however, the synthesis and purification of these two-ion ILs (in many cases functional ILs) are always complicated, which greatly hinders their practical applications. On the contrary, adjusting the physical and thermodynamic properties of DSILs can be easily achieved by varying the characteristics and/or proportion of the parent ILs used for their preparation.

Although DSILs have opened up a new horizon for selecting process media, there have been relatively few studies on the computer-aided screening or design of DSILs compared to the research on ILs. Song et al. [59] reported the rational design of DSILs as extractants for the separation of thiophene/n-octane as an example. They first analyzed the influence of additional ions (cation/anion) and ion ratios on C^∞ and S^∞ through the COSMO-RS model, demonstrating them as important degrees of freedom in DSIL design. Three dependencies of C^∞ on DSILs were observed (see Figure 3.7), and accordingly, DSILs were divided into three groups. In the first group, C^∞ of DSILs monotonically change between the C^∞ values of their IL parents; in the second and third groups, C^∞ of DSILs show parabolic changes and display minimum and maximum values at certain ion ratios, respectively. They then performed multilevel screening of DSILs through predictions of infinite dilution thermodynamic properties, estimations of physical properties, assessments of phase equilibrium behavior, and experimental verification. It should be noted that since there are

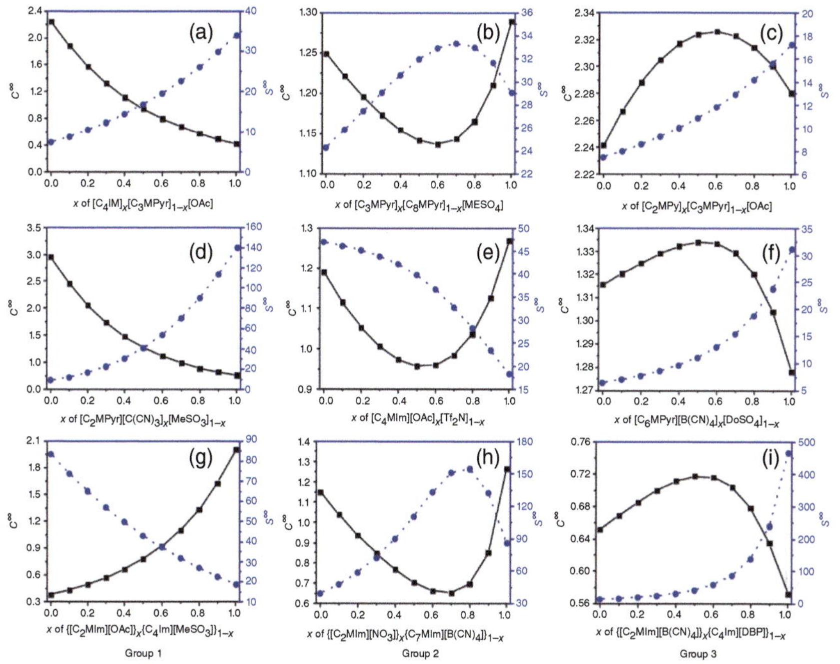

Figure 3.7 C^∞ and S^∞ of the representative DSILs as a function of the ion ratio x. Source: Song et al. [59]/with permission of John Wiley & Sons.

few QSPR models that can directly be used to evaluate T_m and η of DSILs, these two important physical properties were estimated based on the physical properties of their IL parents. Specifically, at least one of the IL parents in a DSIL should satisfy the T_m constraint condition, while η of the DSIL was estimated based on the molar average value of its IL parents. For the separation of thiophene/n-octane, $[C_2mim][OAc]_x[SCN]_{1-x}$ ($x = [0.70, 1]$).

Xie et al. [60] systematically investigated DSILs for the aromatics extraction from hydrocarbons (with o-xylene/n-octane as a model mixture) by combining a rational screening-validation of practical DSILs and a thorough microscopic mechanism exploration. They predicted key thermodynamic properties of paired DSILs from an initial pool of commercially available ILs by COSMO-RS and estimated the important physical properties of these ILs from those of corresponding parent ILs (retrieved from the experimental database or predicted by a DL model). The promising DSILs are followed up with LLE experimental tests, quantum chemistry calculation, and molecular dynamics simulation. The results of this work will be a valuable reference for guiding the design of DSILs for various applications.

3.5 Conclusion and Perspectives

As introduced in this chapter, ML-boosted rational screening of task-specific ILs has been employed by numerous researchers to explore different applications. In the stage of structure–property modeling based on ML, it is necessary to adopt an appropriate molecular representation method, and different algorithms including classic linear and nonlinear methods as well as DL algorithms can be integrated. Apart from these two fundamental pillars, an ideal structure–property model is preferable to be built on a large model development dataset (to cover a large chemical space) and rigorous modeling procedures. However, many models cannot meet both requirements simultaneously, thus leading to a compromise between the design space of IL molecules and the reliability of the model when using the CAILD approach. As the optimal process medium should be determined based on the optimal process performance, an ideal design scenario would be the integrated design of IL molecules and process systems. Nevertheless, solving the correspondingly intricate mathematical problem poses a considerable challenge. Therefore, a trade-off has to be oftentimes made between the design space of IL molecules and the complexity of the model. When handling specific applications, people should consider existing models or models that can be endeavored based on available databases to select a pragmatic molecular design strategy. Irrespective of the approach adopted, the CAILD method can often save experimental costs and time, enabling the development of better processes based on selecting appropriate ILs as advanced process media.

Despite the progress achieved thus far, there are still several future directions worth exploring. First, developing structure–property models that are more widely applicable or more accurate has always been a focus of attention, as few models satisfy both requirements simultaneously. These two aspects directly determine the chemical space that can be explored by CAILD and the reliability of the design results. Most CAILD studies must choose one or the other. To address this issue, additional data pools can be generated using theoretical tools such as quantum chemistry calculation; on the other hand, as ML methods become more prevalent and accessible, advanced data-driven methods are expected to play an increasingly important role. Second, most previous structural and property modeling efforts have mainly focused on basic physical properties and IL-related thermodynamic properties. Existing models for other important physical–chemical properties and EHS-related properties of ILs are either rare or limited to a small chemical space. This deficiency restricts the widespread application of CAILD in areas such as reaction solvents, energy storage media, and battery electrolytes among others. Moreover, as the mechanism of ILs in such applications is more complex, simulating such applications to formulate a CAILD problem still has a long way to go. Third, CAILD methods should be more closely integrated with experimental work to accelerate the practical application of ILs. A common challenge encountered by CAILD is that the best IL design may not be available commercially. In such cases, compromises often have to be made, and inferior but commercially available ILs are chosen for further testing, which are usually not as optimal as the designed candidates. This compromise not only leaves a gap in IL performance for the target process but also, in the long-term, hinders the development of IL structure–property models (as no data for new IL chemical space are obtained). Finally, as advanced ML methods for molecular property modeling (such as DNNs, CNNs, GNNs, and transformer) become more prevalent, how these models can be integrated within the CAILD framework and how the resultant problems can be effectively solved are worthwhile for future investigations.

References

1 Rogers, R.D. and Seddon, K.R. (2003). Ionic liquids – solvents of the future? *Science* 302 (5646): 792–793.

2 Dong, K., Liu, X., Dong, H. et al. (2017). Multiscale studies on ionic liquids. *Chemical Reviews* 117 (10): 6636–6695.

3 Welton, T. (1999). Room-temperature ionic liquids. Solvents for synthesis and catalysis. *Chemical Reviews* 99 (8): 2071–2083.

4 Giernoth, R. (2010). Task-specific ionic liquids. *Angewandte Chemie International Edition* 49 (16): 2834–2839.

5 Song, Z., Zhang, J., Zeng, Q. et al. (2016). Effect of cation alkyl chain length on liquid-liquid equilibria of {ionic liquids plus thiophene plus heptane}: COSMO-RS prediction and experimental verification. *Fluid Phase Equilibria* 425: 244–251.

6 Lei, Y., Yu, Z., Wei, Z. et al. (2022). Energy-efficient separation of propylene/propane by introducing a tailor-made ionic liquid solvent. *Fuel* 326: 124930.

7 Lei, Z., Dai, C., Zhu, J. et al. (2014). Extractive distillation with ionic liquids: a review. *AIChE Journal* 60 (9): 3312–3329.

8 Malik, H., Khan, H.W., Shah, M.U.H. et al. (2023). Screening of ionic liquids as green entrainers for ethanol water separation by extractive distillation: COSMO-RS prediction and aspen plus simulation. *Chemosphere* 311: 136901.

9 Lei, Y., Yu, Z., Wei, Z. et al. (2023). Structure optimization of task-specific ionic liquids targeting low-carbon-emission ethylbenzene production. *Separation and Purification Technology* 308: 122827.

10 Song, Z., Zhang, C., Qi, Z. et al. (2018). Computer-aided design of ionic liquids as solvents for extractive desulfurization. *AIChE Journal* 64 (3): 1013–1025.

11 Lei, Z., Zhang, J., Li, Q. et al. (2009). UNIFAC model for ionic liquids. *Industrial & Engineering Chemistry Research* 48 (5): 2697–2704.

12 Chen, Y., Kontogeorgis, G.M., and Woodley, J.M. (2019). Group contribution based estimation method for properties of ionic liquids. *Industrial and Engineering Chemistry Research* 58 (10): 4277–4292.

13 Song, Z., Zhou, T., Qi, Z. et al. (2020). Extending the UNIFAC model for ionic liquid–solute systems by combining experimental and computational databases. *AIChE Journal* 66 (2): e16821.

14 Müller, S. (2019). Flexible heuristic algorithm for automatic molecule fragmentation: application to the UNIFAC group contribution model. *Journal of Cheminformatics* 11 (1): 57.

15 Hukkerikar, A.S., Sarup, B., Ten Kate, A. et al. (2012). Group-contribution$^+$ (GC^+) based estimation of properties of pure components: improved property estimation and uncertainty analysis. *Fluid Phase Equilibria* 321: 25–43.

16 Hukkerikar, A.S., Kalakul, S., Sarup, B. et al. (2012). Estimation of environment-related properties of chemicals for design of sustainable processes: development of group-contribution$^+$ (GC$^+$) property models and uncertainty analysis. *Journal of Chemical Information and Modeling* 52 (11): 2823–2839.

17 Wu, K.J., Chen, Q.L., and He, C.H. (2014). Speed of sound of ionic liquids: database, estimation, and its application for thermal conductivity prediction. *AIChE Journal* 60 (3): 1120–1131.

18 Wu, K.J., Luo, H., and Yang, L. (2016). Structure-based model for prediction of electrical conductivity of pure ionic liquids. *AIChE Journal* 62 (10): 3751–3762.

19 Klamt, A. and Eckert, F. (2000). COSMO-RS: a novel and efficient method for the a priori prediction of thermophysical data of liquids. *Fluid Phase Equilibria* 172 (1): 43–72.

20 Klamt, A. (2005). *COSMO-RS: From Quantum Chemistry to Fluid Phase Thermodynamics and Drug Design*, 1–234. Elsevier.

21 Eckert, F. and Klamt, A. (2002). Fast solvent screening via quantum chemistry: COSMO-RS approach. *AIChE Journal* 48 (2): 369–385.

22 Palomar, J., Ferro, V.R., Torrecilla, J.S. et al. (2007). Density and molar volume predictions using COSMO-RS for ionic liquids. An approach to solvent design. *Industrial and Engineering Chemistry Research* 46 (18): 6041–6048.

23 Palomar, J., Torrecilla, J.S., Lemus, J. et al. (2008). Prediction of non-ideal behavior of polarity/polarizability scales of solvent mixtures by integration of a novel COSMO-RS molecular descriptor and neural networks. *Physical Chemistry Chemical Physics* 10 (39): 5967–5975.

24 Diedenhofen, M. and Klamt, A. (2010). COSMO-RS as a tool for property prediction of IL mixtures – a review. *Fluid Phase Equilibria* 294 (1–2): 31–38.

25 Lemaoui, T., Darwish, A.S., Hammoudi, N.E.H. et al. (2020). Prediction of electrical conductivity of deep eutectic solvents using COSMO-RS sigma profiles as molecular descriptors: a quantitative structure-property relationship study. *Industrial and Engineering Chemistry Research* 59 (29): 13343–13354.

26 Wang, J., Song, Z., Chen, L. et al. (2021). Prediction of CO_2 solubility in deep eutectic solvents using random forest model based on COSMO-RS-derived descriptors. *Green Chemical Engineering* 2 (4): 431–440.

27 Palomar, J., Torrecilla, J.S., Ferro, V.R. et al. (2008). Development of an a priori ionic liquid design tool. 1. Integration of a novel COSMO-RS molecular descriptor on neural networks. *Industrial and Engineering Chemistry Research* 47 (13): 4523–4532.

28 Zhao, Y., Zeng, S., Huang, Y. et al. (2015). Estimation of heat capacity of ionic liquids using $S_{\sigma\text{-profile}}$ molecular descriptors. *Industrial and Engineering Chemistry Research* 54 (51): 12987–12992.

29 Chen, G., Song, Z., and Qi, Z. (2021). Transformer-convolutional neural network for surface charge density profile prediction: enabling high-throughput solvent screening with COSMO-SAC. *Chemical Engineering Science* 246: 117002.

30 Chen, G., Song, Z., Qi, Z. et al. (2023). Generalizing property prediction of ionic liquids from limited labeled data: a one-stop framework empowered by transfer learning. *Digital Discovery* 2 (3): 591–601.

31 Mikolov, T., Sutskever, I., Chen, K. et al. (2013). Distributed representations of words and phrases and their compositionality. In: *Advances in Neural Information Processing Systems* (ed. C.J. Burges, L. Bottou, M. Welling, et al.), 26. Curran Associates, Inc.

32 Chen, G., Song, Z., Qi, Z. et al. (2021). Neural recommender system for the activity coefficient prediction and UNIFAC model extension of ionic liquid-solute systems. *AIChE Journal* 67 (4): e17171.

33 Klamt, A., Jonas, V., Bürger, T. et al. (1998). Refinement and parametrization of COSMO-RS. *Journal of Physical Chemistry A* 102 (26): 5074–5085.

34 Lin, S.T. and Sandler, S.I. (2002). A priori phase equilibrium prediction from a segment contribution solvation model. *Industrial and Engineering Chemistry Research* 41 (5): 899–913.

35 Klamt, A. and Schüürmann, G. (1993). COSMO: a new approach to dielectric screening in solvents with explicit expressions for the screening energy and its gradient. *Journal of the Chemical Society, Perkin Transactions* 2 (5): 799–805.

36 Ferreira, A.R., Freire, M.G., Ribeiro, J.C. et al. (2011). An overview of the liquid-liquid equilibria of (ionic liquid + hydrocarbon) binary systems and their modeling by the conductor-like screening model for real solvents. *Industrial and Engineering Chemistry Research* 50 (9): 5279–5294.

37 Ferreira, A.R., Freire, M.G., Ribeiro, J.C. et al. (2012). Overview of the liquid-liquid equilibria of ternary systems composed of ionic liquid and aromatic and aliphatic hydrocarbons, and their modeling by COSMO-RS. *Industrial and Engineering Chemistry Research* 51 (8): 3483–3507.

38 Qin, H., Wang, Z., Zhou, T. et al. (2021). Comprehensive evaluation of COSMO-RS for predicting ternary and binary ionic liquid-containing vapor-liquid equilibria. *Industrial and Engineering Chemistry Research* 60 (48): 17761–17777.

39 Zhou, T., Chen, L., Ye, Y. et al. (2012). An overview of mutual solubility of ionic liquids and water predicted by COSMO-RS. *Industrial and Engineering Chemistry Research* 51 (17): 6256–6264.

40 Peng, D., Zhang, J., Cheng, H. et al. (2017). Computer-aided ionic liquid design for separation processes based on group contribution method and COSMO-SAC model. *Chemical Engineering Science* 159: 58–68.

41 Zhang, J., Peng, D., Song, Z. et al. (2017). COSMO-descriptor based computer-aided ionic liquid design for separation processes. Part I: Modified group contribution methodology for predicting surface charge density profile of ionic liquids. *Chemical Engineering Science* 162: 355–363.

42 Song, Z., Li, X., Chao, H. et al. (2019). Computer-aided ionic liquid design for alkane/cycloalkane extractive distillation process. *Green Energy & Environment* 4 (2): 154–165.

43 Weidlich, U. and Gmehling, J. (1987). A modified UNIFAC model. 1. Prediction of VLE, hE, and. gamma..infin. *Industrial and Engineering Chemistry Research* 26 (7): 1372–1381.

44 Kato, R. and Gmehling, J. (2005). Systems with ionic liquids: measurement of VLE and γ^{∞} data and prediction of their thermodynamic behavior using original UNIFAC, mod. UNIFAC(Do) and COSMO-RS(Ol). *Journal of Chemical Thermodynamics* 37 (6): 603–619.
45 Nebig, S. and Gmehling, J. (2011). Prediction of phase equilibria and excess properties for systems with ionic liquids using modified UNIFAC: typical results and present status of the modified UNIFAC matrix for ionic liquids. *Fluid Phase Equilibria* 302 (1–2): 220–225.
46 Kato, R. and Gmehling, J. (2005). Measurement and correlation of vapor-liquid equilibria of binary systems containing the ionic liquids [EMIM][$(CF_3SO_2)_2N$], [BMIM][$(CF_3SO_2)_2N$], [MMIM][$(CH_3)_2PO_4$] and oxygenated organic compounds respectively water. *Fluid Phase Equilibria* 231 (1): 38–43.
47 Nebig, S., Bölts, R., and Gmehling, J. (2007). Measurement of vapor-liquid equilibria (VLE) and excess enthalpies (H^E) of binary systems with 1-alkyl-3-methylimidazolium bis(trifluoromethylsulfonyl)imide and prediction of these properties and γ^{∞} using modified UNIFAC (Dortmund). *Fluid Phase Equilibria* 258 (2): 168–178.
48 Hector, T. and Gmehling, J. (2014). Present status of the modified UNIFAC model for the prediction of phase equilibria and excess enthalpies for systems with ionic liquids. *Fluid Phase Equilibria* 371: 82–92.
49 Song, Z., Shi, H., Zhang, X. et al. (2020). Prediction of CO_2 solubility in ionic liquids using machine learning methods. *Chemical Engineering Science* 223: 115752.
50 Song, Z., Zhou, T., Zhang, J. et al. (2015). Screening of ionic liquids for solvent-sensitive extraction – with deep desulfurization as an example. *Chemical Engineering Science* 129: 69–77.
51 Qin, L., Zhang, J., Cheng, H. et al. (2016). Selection of imidazolium-based ionic liquids for vitamin E extraction from deodorizer distillate. *ACS Sustainable Chemistry and Engineering* 4 (2): 583–590.
52 Qin, H., Cheng, J., Yu, H. et al. (2022). Hierarchical ionic liquid screening integrating COSMO-RS and Aspen Plus for selective recovery of hydrofluorocarbons and hydrofluoroolefins from a refrigerant blend. *Industrial and Engineering Chemistry Research* 61 (11): 4083–4094.
53 Song, Z., Zhou, T., Qi, Z. et al. (2017). Systematic method for screening ionic liquids as extraction solvents exemplified by an extractive desulfurization process. *ACS Sustainable Chemistry and Engineering* 5 (4): 3382–3389.
54 Bechtel S, Song Z, Zhou T, et al. Integrated process and ionic liquid design by combining flowsheet simulation with quantum-chemical solvent screening. *Computer Aided Chemical Engineering* 2018, 44, 2167–72. Elsevier. Edited by Mario R. Eden, Marianthi G. Ierapetritou, Gavin P. Towler.
55 Zhang, X., Wang, J., Song, Z. et al. (2021). Data-driven ionic liquid design for CO_2 capture: molecular structure optimization and DFT verification. *Industrial and Engineering Chemistry Research* 60 (27): 9992–10000.
56 Wang, J., Tang, X., Qi, Z. et al. (2022). Ionic liquids as thermal fluids for solar energy storage: computer-aided molecular design and TRNSYS simulation. *ACS Sustainable Chemistry and Engineering* 10 (6): 2248–2261.

57 Zhang, X., Ding, X., Song, Z. et al. (2021). Integrated ionic liquid and rate-based absorption process design for gas separation: global optimization using hybrid models. *AIChE Journal* 67 (10): e17340.

58 Chatel, G., Pereira, J.F.B., Debbeti, V. et al. (2014). Mixing ionic liquids – "simple mixtures" or "double salts"? *Green Chemistry* 16 (4): 2051–2083.

59 Song, Z., Hu, X., Zhou, Y. et al. (2019). Rational design of double salt ionic liquids as extraction solvents: separation of thiophene/*n*-octane as example. *AIChE Journal* 65 (8): e16625.

60 Xie, K., Chen, J., Cheng, J. et al. (2023). Enhancing aromatics extraction by double salt ionic liquids: rational screening-validation and mechanistic insights. *AIChE Journal* 70: e18301.

4

Integration of Observed Data and Reaction Mechanisms in Deep Learning for Designing Sustainable Glycolic Acid

Xin Zhou

Ocean University of China, College of Chemistry and Chemical Engineering, 238 Songling Road, Qingdao, Shandong 266100, China

4.1 Introduction

Dwindling fossil fuel resources [1], increasingly serious waste disposal problems [2], and the accumulation of nonbiodegradable plastic waste have become significant problems for the environment and sustainable ecosystems [3, 4]. To date, the world is estimated to have produced over eight billion tons of nondegradable plastic waste [5], of which only 9% was recycled and another 12% is incinerated [6]. The remaining 79% of trash accumulated in landfills or the natural environment is thrown out and deposited on refuse landfills or aquatic ecosystems [7]. The singularly inappropriate approach to plastic solid waste directly contributes to a surge in CO_2 emissions [8, 9], which in turn contributes to global warming and brings a grievous threat to the environment and health of living things on the earth [10–12]. Worldwide, the development of degradable materials [13–15], has become the "hot spot" to deal with the ecological and environmental problems caused by nondegradable plastics [16, 17].

Poly glycolic acid (PGA) [18, 19], a synthetic polymer material with high-level biodegradability, makes it an ideal and versatile material for developing biodegradable and biomass-derived plastics to replace nondegradable plastics [20–22]. PGA also exhibits good biocompatibility and can be applied to human tissue and biomedical engineering, such as absorbable surgical sutures, biodegradable cardiovascular stents, and enhancing bone regeneration [23, 24]. The annual demand for PGA has increased rapidly [25]. The global bioresorbable polymers market size is estimated to be USD 417 million in 2022 and is projected to reach USD 688 million by 2027, at a compound annual growth rate (CAGR) of 10.5% between 2022 and 2027 [26]. Statistics show that the global demand for PGA Market is presumed to reach the market size of nearly USD 15.64 Billion by 2032 from USD 6.62 Billion in 2023 with a CAGR of 10.02% under the study period 2024–2032. [27]. However, faced with the above momentous application requirements, the production capacity of PGA is insufficient. One of the main reasons is the inadequate production capacity of PGA's high-purity monomer GA [19]. The hydroxyl and carboxyl groups in the

Applied AI Techniques in the Process Industry: From Molecular Design to Process Design and Optimization, First Edition. Edited by Chang He and Jingzheng Ren.

GA structure can be complex with other ions. This complex is difficult to remove. In addition, GA is easy to self-polymerize into low-molecular-weight glycolic acid dimer, glycolic acid tetramer, and other byproducts [28]. These characteristics of glycolic acid are difficult to purify, which will directly lead to the low purity of GA crystal [29]. If the purpose is to synthesize high-purity GA crystals, this will put forward high requirements for the synthesis and separation process of GA.

The process technologies used for GA industrial synthesis and production mainly involve the hydrolysis of hydroxyacetonitrile [30], the hydrolysis of chloroacetic acid [19], the hydrolysis of methyl glycolate [31], and the carbonylation of formaldehyde [32]. Among these GA synthetic routes, developing a green process for synthesizing glycolic acid with fewer byproducts is crucial. Our previous research indicated a few byproducts from ethylene glycol as raw material through highly efficient Pt–Mn_2O_3 nanocatalysts, including only a tiny amount of glyoxylic acid and formic acid [33, 34]. Moreover, the industrial scale-up simulation and conceptual process design also show that the selective oxidation of ethylene glycol to GA (EGtGA) has superior techno-economic performance to conventional processing routes [35]. Therefore, the direct preparation of GA from ethylene glycol with a low market price has a promising industrial application prospect. It is the premise of producing high-purity GA to identify the essential factors affecting the target product GA, improve the yield of target product GA, and reduce the content of other byproducts [36, 37]. The operation parameters of the EGtGA process, such as reaction temperature and residence time, affect the GA yields, process energy consumption, and net present value [35]. Hence, it is feasible to improve the yield and quality of GA by adjusting these key factors. In addition, other related aspects in the EGtGA process, such as temperature, catalysts [38], and separation process [39], may also substantially impact GA's yields and techno-economic performance. However, due to the high experimental costs and the time and energy cost of repeating experiments, it is a great challenge to study all the above factors and determine the optimal value of essential elements to manufacture GA with high purity.

Due to the rapid rise of artificial intelligence [40], machine learning (ML) methods have gradually evolved into an effective strategy to address complex nonlinear issues (i.e. regression and classification with high order and multiloop) [41]. The ML approach has been employed to predict the yields of complex catalytic processes in the chemical engineering field [42, 43]. Using artificial intelligence to model and predict the properties and distribution of products in the chemical engineering reaction process can significantly save time and cost. The random forest (RF) ML model with high prediction accuracy was established to predict the mass yields and properties of specific products produced by the reaction of different raw materials under other conditions [44]. Compared with the RF model and support vector machine model, the prediction accuracy of the deep neural network (DNN) model is the highest, and the average R^2 of the optimized DNN model is higher than 0.90. Thybaut and coworkers systematically compared the prediction results of ML methods in Fischer–Tropsch reaction models [45]. The results show that the deep belief network (DBN) model is superior to other verifying indicators [45, 46]. Furthermore, multiple linear regression, support vector machines, extreme gradient boosting

(XGboost), and RF models have been employed to guide the optimization of product distribution and improve the yield of target products in the catalytic conversion process [47, 48]. Based on the above research background, selecting appropriate artificial intelligence modeling methods and developing ML prediction models with good accuracy will play an essential role in improving the yield of target products in the chemical process. Hence, many studies have been carried out to make up for the gap between the prediction of input variables, including raw material properties and operating conditions, and the yield and properties of output key target products [49, 50]. To improve the accuracy of artificial intelligence modeling, a large amount of data is usually needed for training. However, in the research and development stage of the chemical process, especially in the laboratory pilot stage, it is incredibly time-consuming and costly to obtain massive experimental data (such as thousands of sets of data) required for artificial intelligence modeling. Moreover, the progress of existing artificial intelligence model research in accelerating engineering and guiding the production of polymer-grade pure glycolic acid is still limited.

In the previous work, the laboratory pilot stage, reaction kinetics solution, and conceptual process design of glycolic acid production were realized by selective oxidation with ethylene glycol as raw material and efficient Pt-based nanocatalysts [30]. Based on this, the ML prediction model of the yield and characteristics of polymerization grade glycolic acid "dual-core driven" by coupling data and reaction kinetics is successfully established in this study. The hybrid data and mechanism "dual-core driven" deep learning model, termed the HDM model, is developed to guide experimental research and optimize operating conditions. Concretely, data expansion is implemented to enrich the cube based on reaction kinetics to meet the challenge of data scarcity. In this study, more than 6000 mechanism-driven virtual data points are created and synthesized to enhance the model's prediction ability, which is the main innovation of this paper. Then, the ML model is trained to predict the yield of the process products. The relative importance of each input to the target is further analyzed. The multi-objective optimization method based on the genetic algorithm is used to optimize the expected polymer-grade glycolic acid production process (maximize the yield, purity, and economic benefits of glycolic acid as well as minimize energy consumption). Finally, experiments and process simulation verify that the optimized operating conditions can produce polymeric glycolic acid with low energy consumption, high purity, high yield, and high economic performance.

4.2 Methodology

The framework for predicting GA production performance includes five steps, as illustrated in Figure 4.1. Through the case study of the EGtGA process, the detailed implementation of our framework is further introduced.

Step 1: Database generation. The feature variables were first determined. Based on the existing experimental data, a representative reaction kinetic model, including the EGtGA process, was established and simulated to generate the yield dataset

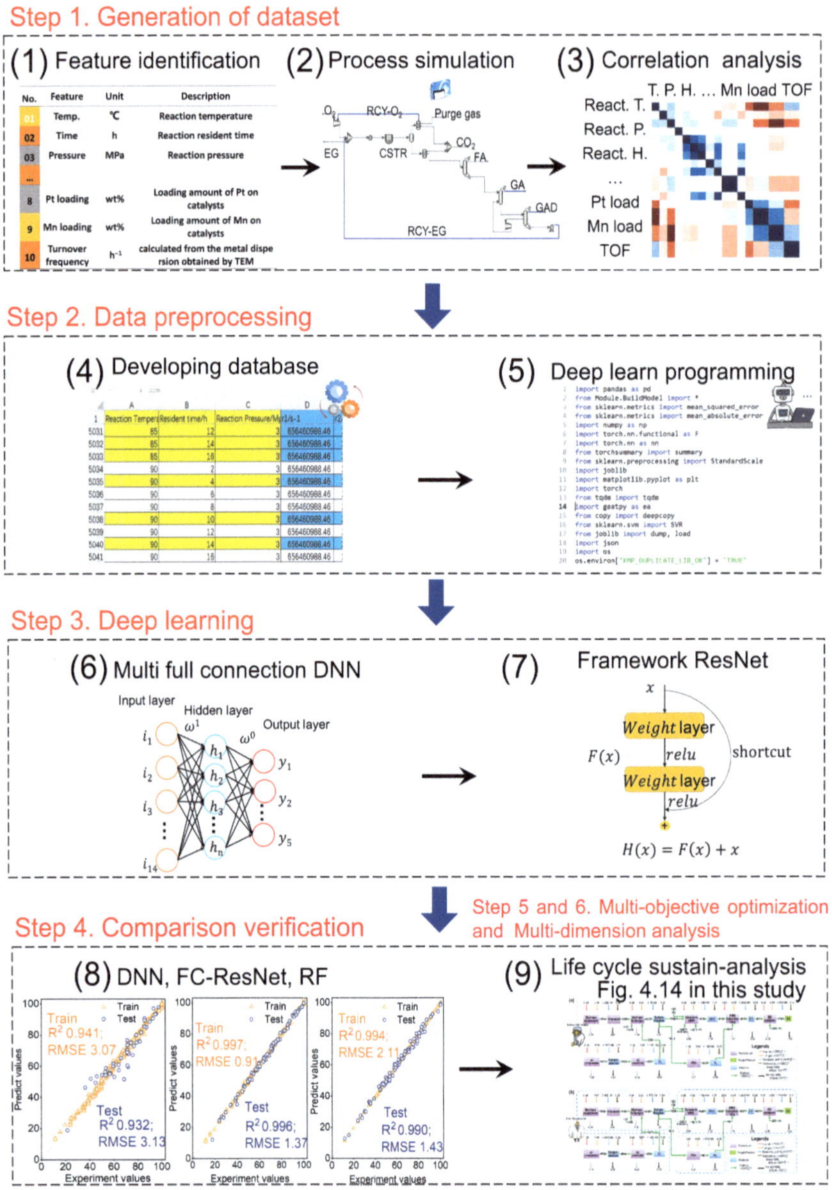

Figure 4.1 The overall research framework for predicting glycolic acid production by selective oxidation of ethylene glycol using deep learning. Source: Zhou et al. [51]/with permission from John Wiley & Sons.

of key products for steady-state simulation. The correlation between features and targets was further analyzed.

Step 2: Data preprocessing. Developing the detailed database and introductory programming (i.e. loading ML library functions and reading database files).

Step 3: Deep learning. Building ML or deep learning models and performing model training, testing, and prediction.

Step 4: Comparative analysis. The performance and robustness of various ML models, namely deep neural networks, DBNs, fully connected residual network (FC-ResNet), and RF regression, were compared to identify the most appropriate ML model for the EGtGA process.

Step 5: Optimization and prediction. Predict the yields and performance of critical products in the EGtGA process to verify the trained FC-ResNet model. The optimal operating parameters for the highest yield of GA were obtained and verified by experiments. The genetic algorithms were used to optimize the model parameters (such as iterations) of the ML model, and the optimized parameters were used to train the ML model.

Step 6: Multidimensional evaluation. The EGtGA process was deeply analyzed and multi-dimensionally evaluated using optimal operating parameters based on the life cycle, economic, and social environment framework.

4.2.1 Database Generation

The database used to train the ML model consists of many schemes with different characteristics and corresponding production performance. The first step of building the database is to select representative parameters as training features among the factors affecting GA production. We decided on 10 features that play an essential role in GA production as features (see Table 4.1). The EGtGA process model was established based on the conceptual design flow chart (see Figure 4.2a).

A total of 610 sets of experimental datasets using different reaction conditions and catalysts were collected. Because of the advantages of steady-state simulation

Table 4.1 The selected features and desired targets for deep learning.

No.	Items	Unit	Description
F1	Temperature	°C	Reaction temperature
F2	Time	hour	Resident time
F3	Pressure	MPa	Reaction pressure
F4	E_1	kJ mol^{-1}	The reaction activation energy in the first reaction
F5	E_2	kJ mol^{-1}	The reaction activation energy in the second reaction
F6	E_3	kJ mol^{-1}	The reaction activation energy in the third reaction
F7	E_4	kJ mol^{-1}	The reaction activation energy in the fourth reaction
F9	Pt loading	wt%	Pt atom loading in PtMn/MCM-41 nanocatalysts
F0	Mn loading	wt%	Mn atom loading in PtMn/MCM-41 nanocatalysts
F10	TOF	h^{-1}	Turnover frequency
F11	Conversion	%	The conversion of feedstock ethylene glycol
F12	GA yields	wt%	The yields of desired product glycolic acid
F13	GAD yields	wt%	The yields of byproduct glycolaldehyde
F14	CO_2 yields	wt%	The yields of byproduct CO_2
F15	FA yields	wt%	The yields of byproduct formic acid

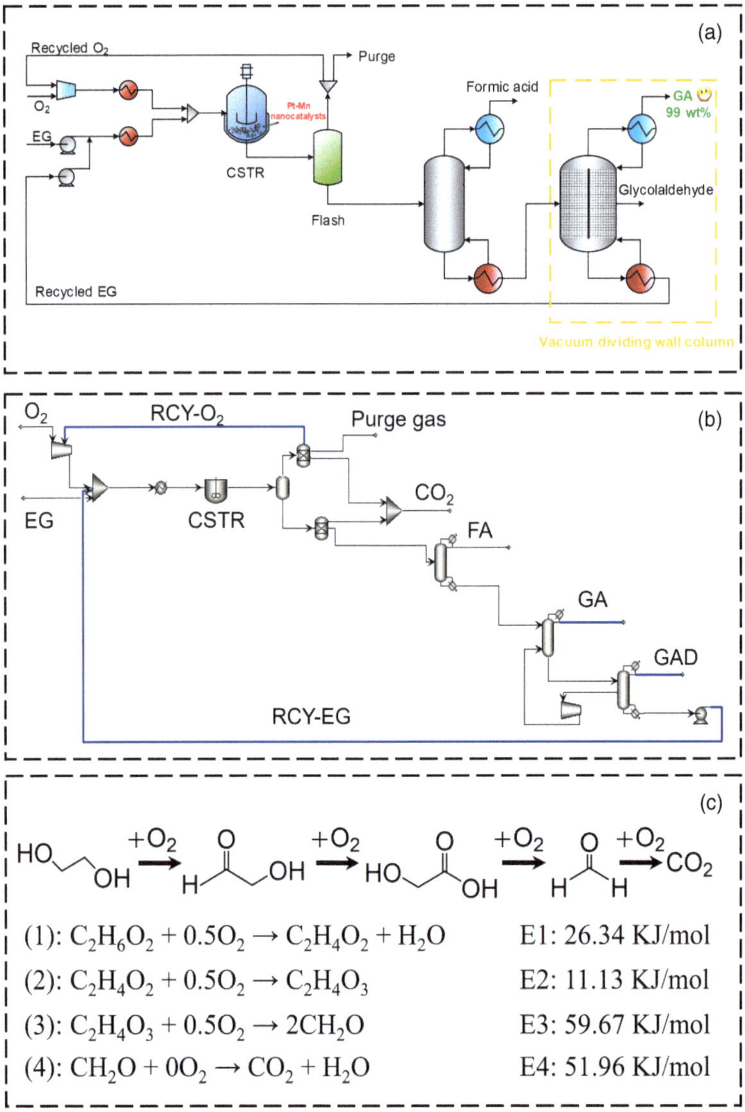

Figure 4.2 (a) conceptual design process flow chart; (b) simulation flowsheet in Aspen Plus; (c) reaction network and kinetics parameters (Pt loading 1.4 wt%; Mn loading 0.9 wt%).

and process control, Aspen Plus has been widely used in process simulation, optimization, and prediction. Hence, the general process simulation software Aspen Plus was also applied to establish the simulation model of the EGtGA process. The process simulation consists of 13 process unit modules: compressor, continuous stirred tank reactor, vacuum column, and vacuum dividing wall column, as shown in Figure 4.2b. The binary interaction parameters of stream material compositions using the non-random two liquid (NRTL) method are adopted to calculate the phase equilibrium. The reaction kinetics and process models were established based on these experimental data. By using the process model developed by Aspen Plus, abundant process simulation data can be obtained. As a result, Aspen Plus process simulation software is applied to generate the data of 6110 cases through process simulation, fill the ML models, and enhance the database. It is worth noting that the hold-out method is used to achieve the dataset partitioning [51]. A small part is taken as the test set and verification set, and the rest is taken as the training set. The dataset partitioning details are shown as follows, training set:verification set:test set = 80% : 10% : 10%. The correlation performance was identified by analyzing the relationship between essential process operating parameters (reaction temperature, reaction residence time, and reaction pressure) and conversion and GA yield. The reaction networks of the EGtGA process and reaction activation energies (Pt loading 1.4 wt%; Mn loading 0.9 wt%) are demonstrated in Figure 4.2c.

The computing environment conditions in this study are CPU: Intel (R) Core (TM) i7-10,700 CPU@2.90GHZ and GPU: NVIDIA GTX 1060. Therefore, a total of 5040 complete datasets, including experimental and simulation data, composed of process data, reaction kinetic parameters, and product yields, constitute a comprehensive database required for ML training.

4.2.2 Deep Learning

This work's ML models mainly include DNN and RF. The neural network models considered in this work include a deep neural network, a DBN, and a FC-ResNet. The integrated algorithm, RF, is illustrated in the following part.

4.2.2.1 Deep Neural Networks

The DNN is the feedforward neural network with multilayers [46]. It is one of the simplest neural network forms and has good prediction performance. These networks usually have three layers: the input layer, hidden layer, and output layer, as illustrated in Figure 4.3a. A gradient descent optimization method called backpropagation identifies or trains weights in neural networks. This procedure is represented visually in Figure 4.3b. In the feedforward neural network, the input layer has the same number of nodes as the input, and the input is directed to the subsequent layer. One or more hidden layers calculate the weighted sum of all outputs from the previous layer and regularize the weighted sum using the activation function. Activation functions include linear, sigmoid, hyperbolic tangent, exponential linear unit, rectified linear unit, and scaled exponential linear unit. The output layer repeats the same process as the hidden layer with its unique weight. Each weight in the system will be updated in each training process, and the updated amount is determined by distinguishing the prediction error from their respective weights. Using the differential chain rule, the backpropagation algorithm is computationally efficient because almost all components of each derivative are calculated during the forward transmission (i.e. prediction calculation) of the network and can be reused when updating the weight.

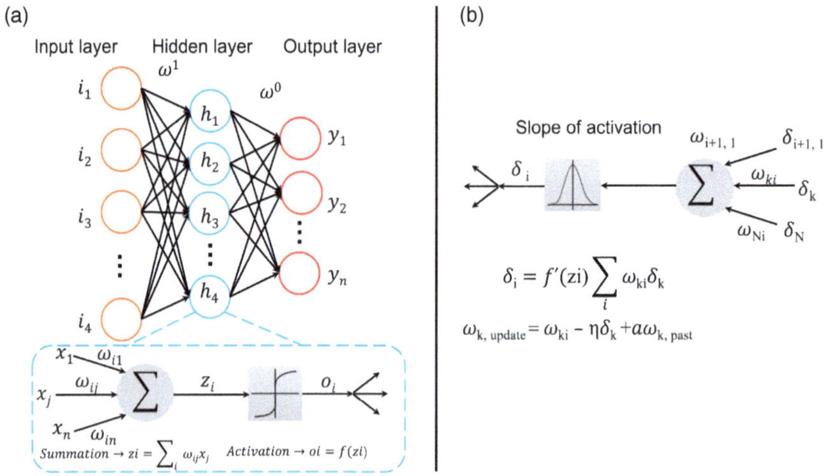

Figure 4.3 (a) Neural network structure of each layer node in DNN; (b) backpropagation calculation algorithm.

4.2.2.2 Deep Belief Networks

The DBN is a probability generation model for unsupervised and supervised learning [52]. Compared with the traditional neural network, the generation model establishes a joint distribution between observation data and labels. DBN comprises multilayer neurons divided into dominant and recessive neurons. The bottom layer represents data vectors, and each neuron represents one dimension of the data vector. The component of DBN is restricted Boltzmann machines (RBMs). The DBN model can extract features layer by layer from the original data and obtain high-level representations through the layer-by-layer stacking of RBMs. Its core is to use the layer-by-layer greedy learning algorithm to optimize the connection weight of the deep neural network, that is, first use the unsupervised layer-by-layer training method to effectively mine the fault characteristics of the equipment to be diagnosed, and then, based on adding the corresponding classifier, optimize the fault diagnosis capability of the DBN through the reverse supervised tuning. The unsupervised layer-by-layer training can learn complex nonlinear functions by directly mapping data from input to output, which is also the key to its robust feature extraction ability. The training goal of RBM is to make the Gibbs distribution fit the input data as much as possible, that is, to make the Gibbs distribution represented by the RBM network as close as possible to the distribution of input samples. The Kullback–Leibler (KL) distance between the sample distribution and the edge distribution of the RBM network can be used to express the difference between them. Because the deep network quickly falls into local optimization, and the selection of initial parameters significantly impacts where the network finally converges, DBN training is divided into pretraining and fine-tuning. RBM performs layer-by-layer unsupervised training on the deep network and takes the parameters obtained from each layer of training as the initial parameters of neurons in each layer of the deep network. This parameter is in a good position in the deep network parameter space, pretraining. After RBM trains the initial values of the deep network parameters layer by layer, it trains the deep network with the traditional backpropagation (BP) algorithm. In this way, the parameters of the deep network eventually converge to a good position, that is, fine-tuning. In the pretraining, the unsupervised layer-by-layer learning method is used to learn the parameters. First, the data vector x and the first hidden layer are used as an RBM to train the RBM parameter. Then, the RBM parameter is fixed, h1 is used as the visible layer vector, h2 is used as the hidden layer vector, and the second RBM is trained.

The whole neural network can generate training data according to the maximum probability by training the weights between its neurons. DBN can be used not only to identify features and classify data but also to generate data. DBN comprises multilayer neurons divided into dominant and recessive neurons. The graphic element accepts input and the implicit part extracts features. Therefore, hidden elements also have an alias called feature detectors. The connection between the top two layers is undirected and forms associative memory. There are connections between other lower layers and directed connections up and down. The bottom layer represents data vectors, and each neuron represents one dimension of the data vector. RBM and its training process are shown in Figure 4.4a. The method of training DBN is carried out layer-by-layer. The hidden layer of the previous RBM provides input to the next visible layer. Figure 4.4b illustrates a schematic diagram of a DBN composed of three RBM layers. In each layer, the data vector is used to infer the hidden layer, and then the hidden layer is regarded as the data vector of the next layer. Each RBM can be used as a separate cluster. RBM has only two layers of neurons. One layer is called the visible layer, composed of visible units used to input training data. The other layer is called the hidden layer. Accordingly, it is composed of hidden units used as feature detectors. The visible layer of the lowest RBM receives the input characteristic data, and the hidden layer of the highest RBM is connected to the backpropagation layer to obtain the model output.

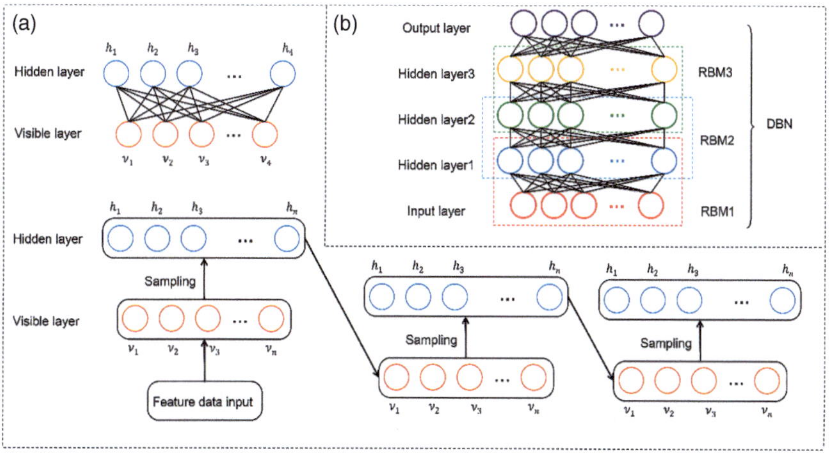

Figure 4.4 (a) Network structure and training process of restricted Boltzmann machine; (b) the structure of DBN composed of three-layer RBM.

4.2.2.3 Fully Connected Residual Networks

The model accuracy is continuously improved with the continuous increase in network layers. However, the network reaches saturation when the network layers increase to a certain number. Training and testing accuracy will decline rapidly, indicating that the deep network with more layers is challenging to train. The deeper the network layers are, the worse the prediction effect is because the neural network will continuously propagate the gradient in the backpropagation process. When the network layers are more profound, the gradient will gradually disappear in the propagation process (if the sigmoid function is used, for the signal with an amplitude of 1, the gradient will decay to the original 0.25 every time it is transmitted back, and the more layers, the more severe the attenuation). As a result, the weight of the previous network layer cannot be effectively adjusted. He et al. proposed the residual network (ResNet) [53]. The ResNet model can continue improving the network's training and test accuracy when the original network is close to saturation. Adding ResNet structure to the network can solve the problematic network convergence caused by deepening the network layers and helping the whole network converge faster. Figure 4.5a shows a 4-layer fully connected network with 14 nodes in the input layer and 5 nodes in the output layer. The general network structure is shown in Figure 4.5b, and the ResNet structure is shown in Figure 4.5c. When the ResNet is constructed in the form of a full connection, it constitutes a FC-ResNet. It is difficult to directly let some layers fit a potential identity mapping function (see Eq. (4.1)), which may be why it is challenging to train the deep neural network. However, if the network is designed as Eq. (4.2), we can transform it into learning a residual function, as defined in Eq. (4.3). When $F(x)$ equals zero, it constitutes an identity map. If the residual is not zero, the accumulation layer learns new features based on the input features to perform better.

$$H(x) = x \tag{4.1}$$

$$H(x) = F(x) + x \tag{4.2}$$

$$F(x) = H(x) - x \tag{4.3}$$

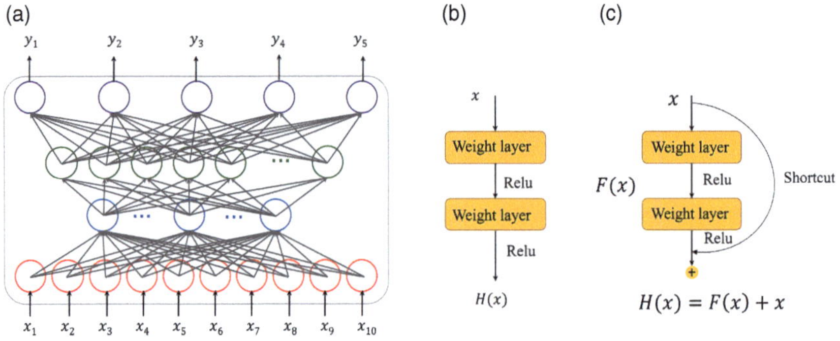

Figure 4.5 (a) 4-Layer fully connected network with 14 nodes in the input layer and 5 nodes in the output layer; (b) structure unit of general network structure; (c) structure unit of ResNet unit.

Suppose an existing shallow network (Shallow Net) has reached saturation accuracy. Then several identity mapping layers ($y = x$, output equals input) are added to it so that the depth of the network is increased. At least the error will not increase. That is, the deeper network should not increase the error on the training set. The inspiration of the deep ResNet is to use identity mapping to directly transmit the previous layer's output to the rear mentioned here.

4.2.2.4 Random Forest

RF is composed of multiple decision trees [54]. Each decision tree is different. When building the decision tree, we randomly select some samples from the training data, and we will not use all the features of the data but randomly select some features for training. Each tree uses different samples and features, and the training results are also different. RF applies the general technique of bootstrap aggregating to tree learners. As depicted in Figure 4.6, assuming that the training data includes $\boldsymbol{B}$ inputs ($X = X_1, \ldots, X_B$) and responses ($y = y_1, \ldots, y_B$), bagging could select a random sample back and forth. It replaces the training data $\boldsymbol{B}$ times to train the tree f_B on X_B, and Y_b ($b = 1, \ldots, b$). When constructing the regression tree, m features or input variables are randomly selected from P features or input variables ($m < P$), called feature bagging. Each regression tree grows freely without restriction until it cannot continue to split. That is, it reaches the minimum node size. Finally, $\boldsymbol{B}$ regression trees form the RF algorithm. The advantage of RF is quite outstanding. It has the following

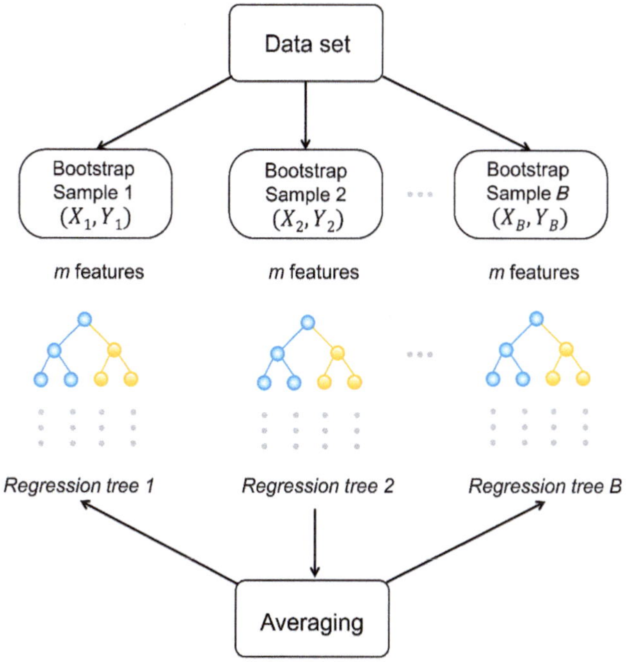

Figure 4.6 Flowsheet of random forest method.

benefits: (i) using an integrated algorithm with high accuracy, (ii) not easy to have overfitting (random samples and random characteristics), (iii) strong anti-noise ability (noise refers to abnormal data), and (iv) processing high-dimensional data with many features.

4.2.3 Optimization and Prediction

A genetic algorithm is based on natural selection and genetics [55, 56]. Since the 1990s, it has been widely used because of its high efficiency and strong robustness. In the genetic algorithm, the candidate solution group of an optimization problem evolves toward a better solution. Evolution starts with a group of randomly generated individuals, an iterative process. The population in each iteration is called a generation. In each generation, the adaptability of each individual in the population is evaluated. More suitable individuals are randomly selected from the current population, and each individual's genome is modified (recombined, possibly random mutation), forming a new generation. The next-generation candidate solutions are then used in the next iteration. Usually, when the generated algebra reaches the maximum population's satisfactory fitness level, the algorithm terminates.

In this study, the genetic algorithm and HDM models are combined to optimize the model parameters of the four deep learning models. It effectively searches the minimum objective function and avoids the calculation and analysis of each response. This paper uses the high-performance and practical Python genetic algorithm toolbox to optimize.

4.2.4 Life Cycle Multidimensional Evaluation

Although recent advances in life cycle methodology have expanded the range of environmental impacts considered, including biodiversity and ecosystem services [57], when considering social and livelihood factors, the application trend of life cycle assessment in these critical aspects is still limited. Life cycle sustainability assessment (LCSA) can assess the impact of target products, processes, and supply chain decisions on the ecological environment for projects from multiple dimensions [58]. In this study, LCSA can quantitatively provide a feasible framework for the proposed dual-core-driven process model.

To analyze the impact of the proposed HDM model on society, people's livelihood, and the environment, the multidimensional sustainability assessment method of the life cycle proposed in our previous work is adopted. In this method, greenhouse gas (GHG) emissions and fossil energy demand (FED) are integrated into the comparison range of gross domestic product (GDP, US$) generated based on the HDM model [59, 60].

4.3 Results and Discussion

4.3.1 Data Analysis and Statistics Before Modeling

4.3.1.1 Analysis of Experimental Data

Figure 4.7 depicts the detailed data (including feature categories and ranges) with a residence time of eight hours in the experimental dataset. Other experimental data with residence times of 4–12 hours are depicted in Appendix. The X-axis and Y-axis of the three-dimensional image are reaction temperature and pressure, respectively. According to the composition analysis of chromatographic data, the reaction temperature is the main characteristic factor affecting the conversion and other key indicators. The conversion rate fluctuates in a wide range, ranging from 0% to 100%. The two basic characteristic categories are residence time and reaction pressure, the other two key indicators to determine the product distribution in the EGtGA process. GA yields, glyoxalic acid (GAD) yields, formic acid (FA) yields, and CO_2 yields range from zero to the upper limit of 9–89%.

The above results show that this study's preliminary experimental dataset is 610 groups. The experimental data is extensive enough to cover a broader range of product distributions and support constructing a mechanism model. The database of the "dual-core-driven" deep learning HDM model also includes parameters such as reaction rate constants and activation energy.

The reaction temperature of the HDM model was 20–90 °C, the reaction pressure was 1–3 MPa, and the residence time was 4–12 hours. By comprehensively analyzing Figure 4.7 and the database, we can intuitively conclude that the reaction temperature is the most crucial factor affecting the product yield of the EGtGA process. In terms of the yield of the target product, the GA yield is between 0.73% and 75.85%. The yield of CO_2 and formic acid from the EGtGA process is between 0.11% and 54.11% and 0.01–9.43%. Due to the different reaction conditions of the HDM model, the product yield distribution has a large span. This also strengthens our determination to perform the HDM model to identify further and tune essential factors affecting the EGtGA process to produce the highest yield of glycolic acid product.

4.3.1.2 Data Dependence Analysis

Figure 4.8 illustrates the heat map of the linear regression correlation coefficient matrix between the selected features and the desired targets. The deep blue and light navy blue grids (absolute value $\geq$0.80) show strongly positive and negative correlations, respectively. It can be seen that some areas on both sides of the diagonal of the grid are relatively deep, indicating that there is a linear relationship between the conversion and immediate product yield of the EGtGA process and these features. For example, the conversion rates of the HDM model represent strongly positive correlations with reaction temperature (see F1 in Table 4.1), in which the correlation coefficient value is 0.913.

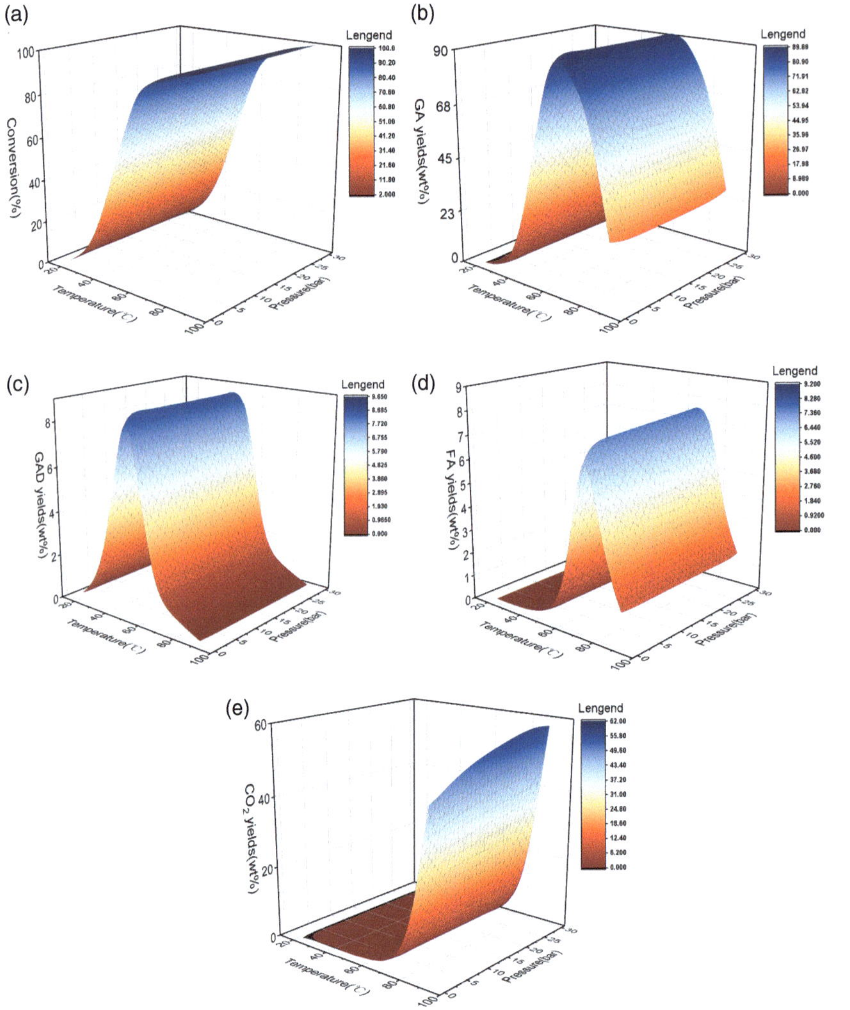

Figure 4.7 Data analysis and statistics with the reaction condition resident time eight hours: (a) the changing trend of conversion rates with reaction temperature and pressure; (b) the changing trend of GA yields with reaction temperature and pressure; (c) the changing trend of GAD yields with reaction temperature and pressure; (d) the changing trend of FA yields with reaction temperature and pressure; (e) the changing trend of CO_2 yields with reaction temperature and pressure.

4.3.2 Model Comparison and Feature Important Analysis

4.3.2.1 Model Comparison

Further analysis is carried out based on the above four HDM models to provide more details about each target's prediction performance and input characteristics' role. Figures 4.9–4.13 show each target's predicted and actual values. The setting of model parameters greatly influences the prediction results of R^2 and root-mean-square error (RMSE) in the ML model. Unreasonable learning rates

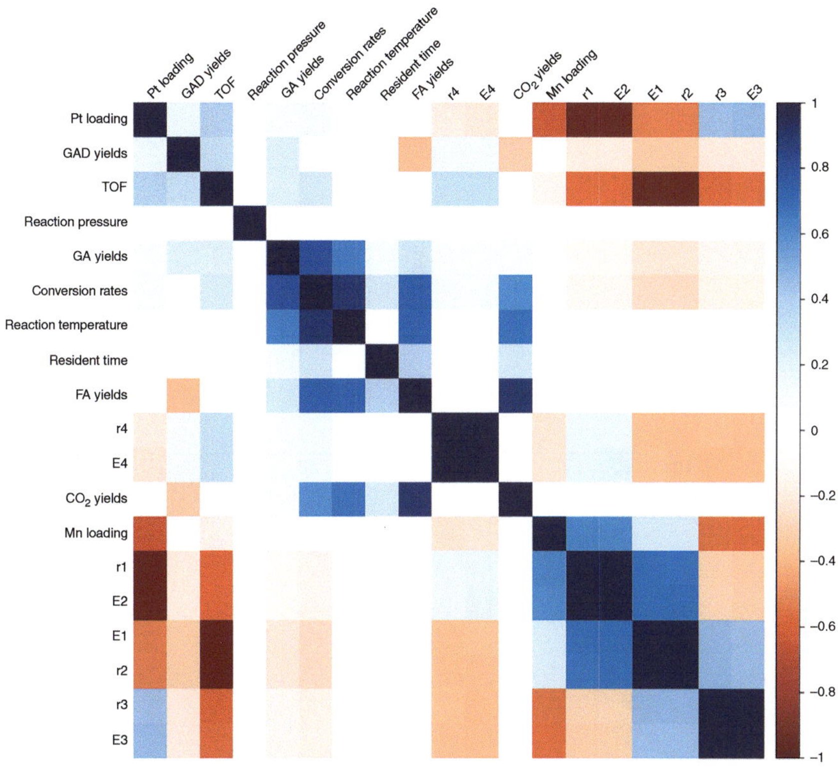

Figure 4.8 Heat map of the linear regression correlation coefficient matrix between the selected features and the desired targets.

and iteration times are straightforward to produce overfitting. Therefore, DNN, DBN, and FC-ResNet have the same number of neural network iterations, 500, and the neural network learning rate is 0.0001 in this study. In the RF method, the max_depth and n_estimators are 10 and 100, respectively. The experiment and prediction values of convention rates (training and testing) in the HDM model by DNN, DBN, FC-ResNet, and RF are shown in Figure 4.9a–d. The conversion prediction result of the EGtGA process is in the range of 20–97%. It is worth noting that the R^2 of the "dual-core driven" HDM model for the conversion prediction is 0.932–0.996. Moreover, a comprehensive comparison of Figure 4.9 shows that the R^2 of the FC-ResNet method is the highest at 0.996, which is also the lowest of its RMSE of 1.37.

The experiment and prediction values of GA yields in the EGtGA process (training and testing) in the HDM model by DNN, DBN, FC-ResNet, and RF are shown in Figure 4.10a–d. The GA yield prediction results of the EGtGA process are in the range of 0.02–79.5%. It is worth noting that the R^2 of the "dual-core driven" HDM model for the conversion prediction is 0.902–0.988. Similarly, a comprehensive comparison of Figure 4.10 shows that the R^2 of the FC-ResNet method is the highest at 0.988, which is also the lowest of its RMSE of 1.95.

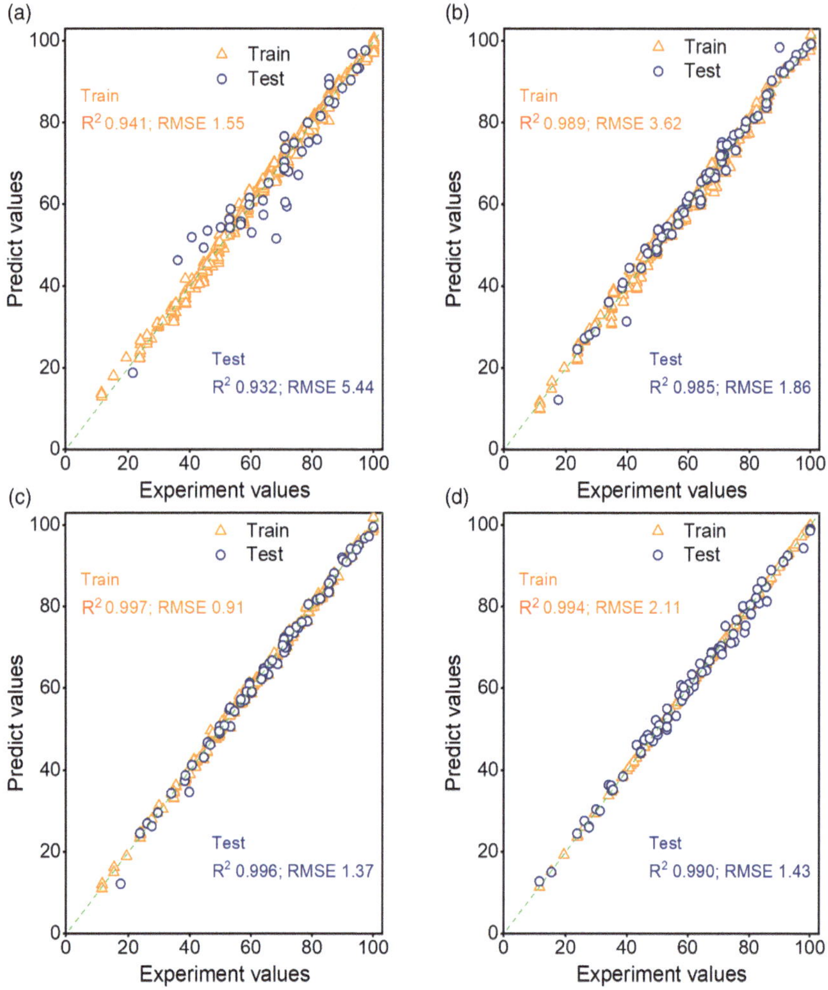

Figure 4.9 Experiment and prediction values of EGtGA convention rates (training and testing) in the HDM model based on (a) DNN, (b) DBN, and (c) FC-ResNet, (d) RF.

Figure 4.11 depicts the experiment and prediction values of GAD yields in the EGtGA process (training and testing) in the HDM model by DNN, DBN, FC-ResNet, and RF. The absolute value of GAD yield prediction results of the EGtGA process is in the range of 0.02–9.21%. It is worth noting that the R^2 of the "dual-core driven" HDM model for the prediction is 0.913–0.996. Similarly, a comprehensive comparison shows that the R^2 of the FC-ResNet method is the highest at 0.996, which is also the lowest of its RMSE of 0.17.

Figure 4.12 illustrates the experiment and prediction values of FA yields in the EGtGA process (training and testing) in the HDM model by DNN, DBN, FC-ResNet, and RF. A comprehensive analysis shows that the prediction results of DNN and DBN are worse than those of FC-ResNet and RF. The absolute value of FA yield

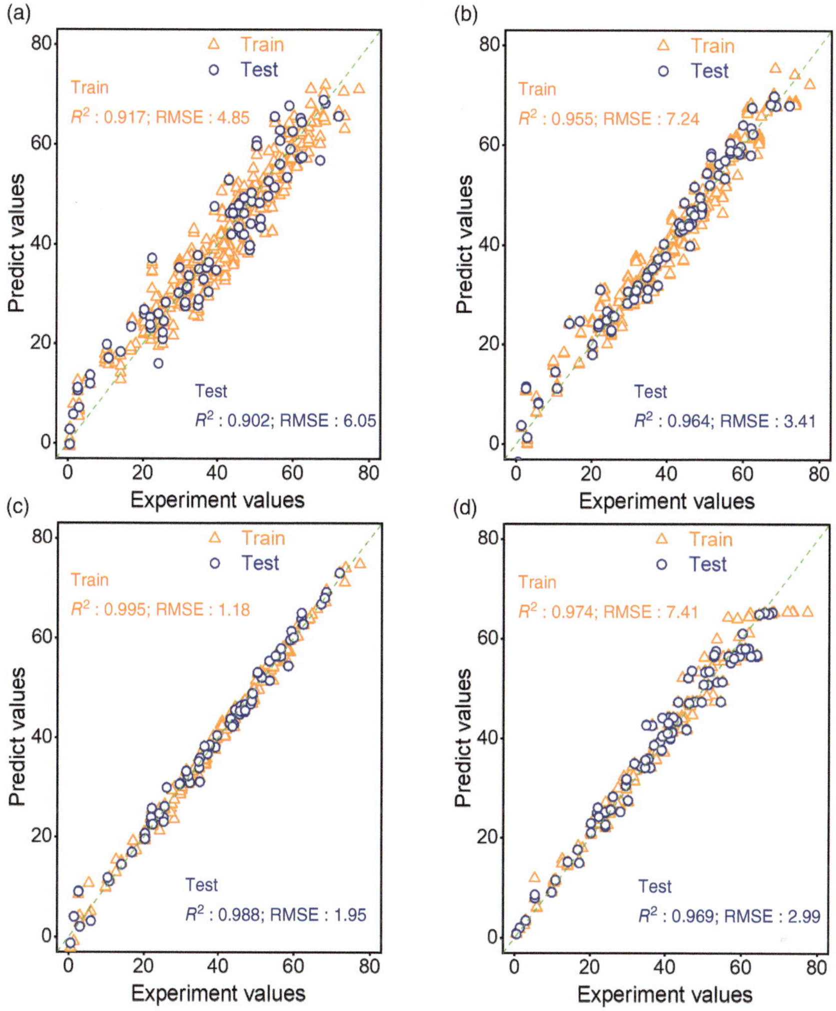

Figure 4.10 Experiment and prediction values of GA yields in the EGtGA process (training and testing) using the HDM model based on (a) DNN, (b) DBN, and (c) FC-ResNet, (d) RF.

prediction results of the EGtGA process is in the range of 0.24–19.13%. It is worth noting that the R^2 of the "dual-core driven" HDM model for the conversion prediction is 0.899–0.995. Similarly, a comprehensive comparison depicts that the R^2 of the FC-ResNet method is the highest at 0.995, which is also the lowest of its RMSE of 0.35.

The experiment and prediction values of FA yields in the EGtGA process (training and testing) in the HDM model by DNN, DBN, FC-ResNet, and RF are demonstrated in Figure 4.13. Similar to Figure 4.12, it can be seen visually that the DNN and DBN prediction results are also worse than those of FC-ResNet and RF. The absolute value of CO_2 yield prediction results of the EGtGA process is 0.12–77.29%. It is worth noting that the R^2 of the "dual-core driven" HDM model for the conversion prediction

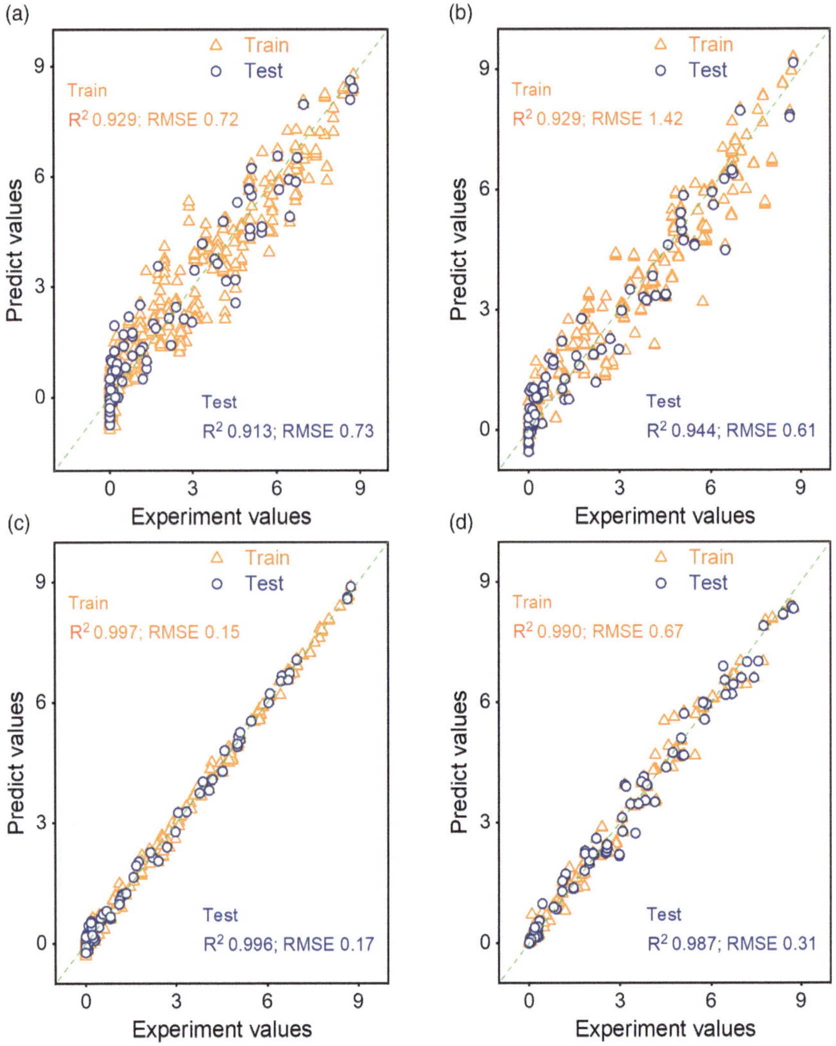

Figure 4.11 Experiment and prediction values of GAD yields in the EGtGA process (training and testing) using the HDM model based on (a) DNN, (b) DBN, and (c) FC-ResNet, (d) RF.

is 0.960–0.998. A comprehensive comparison shows that the R^2 of the FC-ResNet method is the highest at 0.998, which is also the lowest of its RMSE of 0.84.

This study further adopts the "dual-core driven" models to perform the prediction and provides the calculation results of feature importance. There are apparent differences between the method of calculating feature weight by the deep neural networks and the method of calculating RF. The idea of the neural network to estimate the importance of features is based on a permutation algorithm. The core idea of this algorithm is to train a neural network first. When calculating the importance of the *i*th feature, disrupt the *i*th feature, and then calculate the index change after the disruption. The importance calculation method of the corresponding features of RF

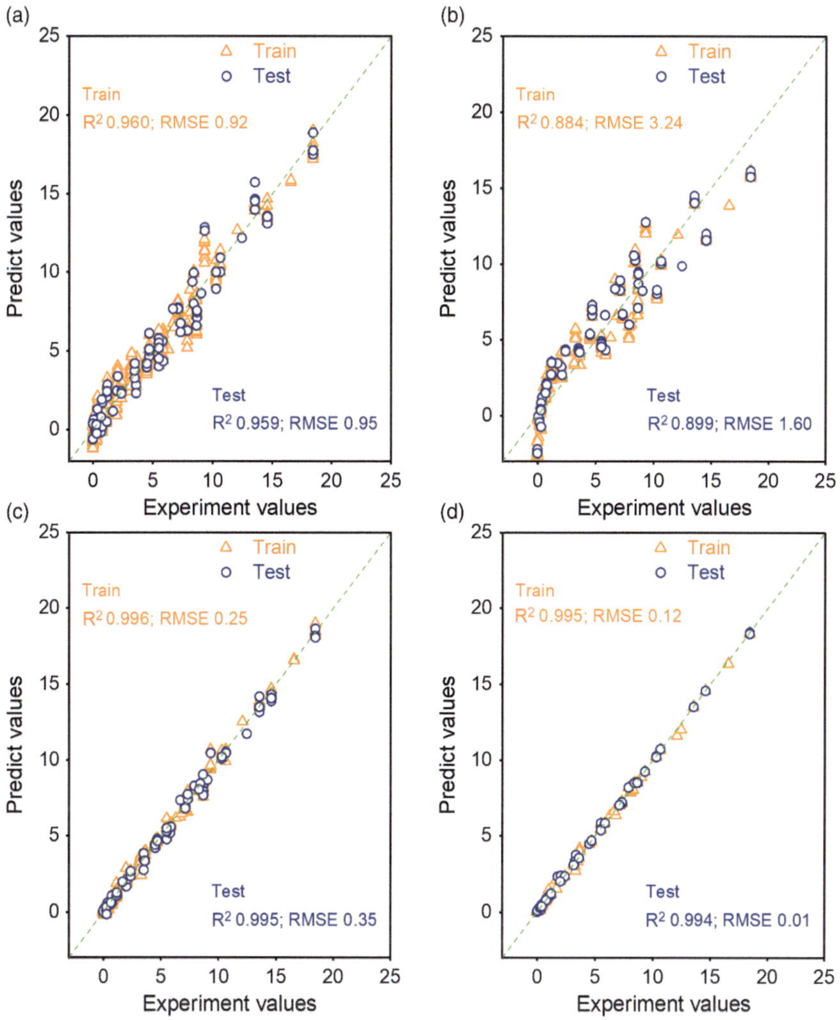

Figure 4.12 Experiment and prediction values of FA yields in the EGtGA process (training and testing) using the HDM model based on (a) DNN, (b) DBN, and (c) FC-ResNet, (d) RF.

is: if the accuracy outside the bag decreases significantly after adding noise to a feature randomly, it shows that this feature has a significant impact on the classification results of samples, that is, its importance is relatively high.

4.3.2.2 Feature Important Analysis

Figure 4.14 depicts each input feature's importance to the HDM model's five targets. It can be found that there is a certain difference between the results of calculating the importance of characteristics in the neural network like dual-core driven model and RF. Figure 4.14a–c intuitively shows that reaction activation energy and catalyst features (including Pt loading, Mn loading, and TOF) are more critical than reaction

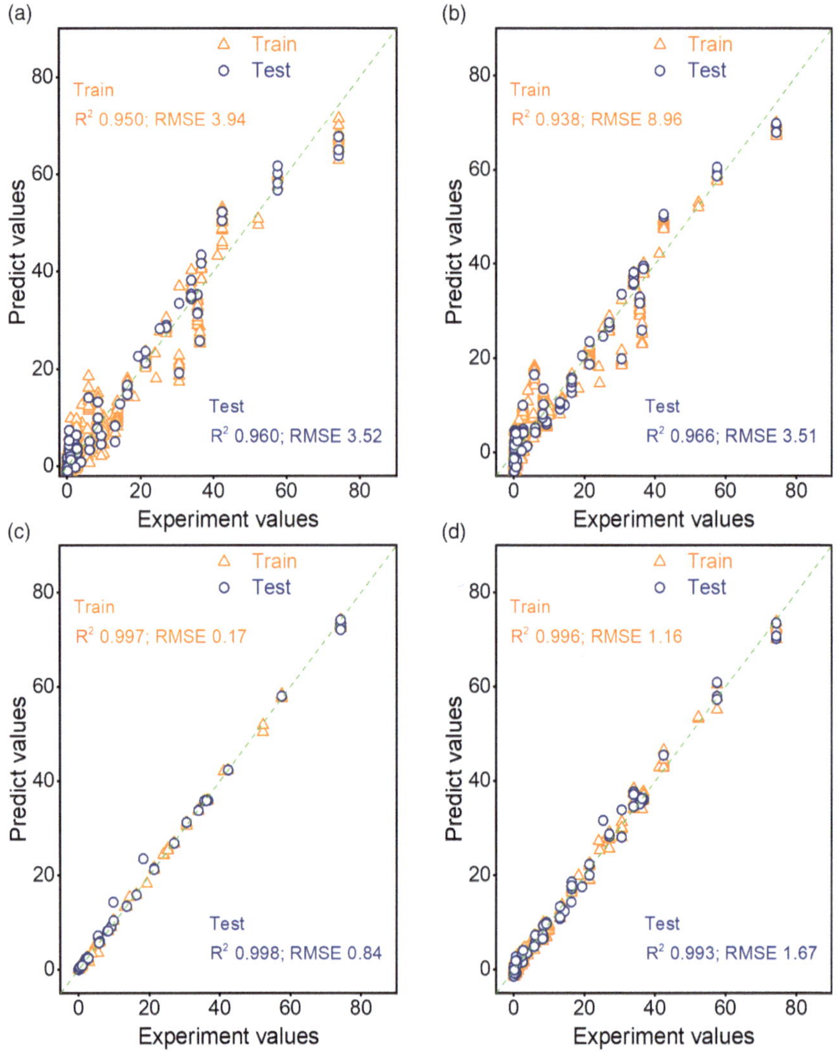

Figure 4.13 Experiment and prediction values of CO_2 yields in the EGtGA process (training and testing) using the HDM model based on (a) DNN, (b) DBN, and (c) FC-ResNet, (d) RF.

conditions. The sum of specific feature importances (reaction activation energy, catalyst composition, and TOF) is 70–71%, which are the dominant and decisive factors. Furthermore, the results of all feature importances show that the activation energy of the second step reaction (see Figure 4.2c) is the most important feature. This also confirms our previous research results. The second step reaction is the rate control step in the process of ethylene glycol oxidation to glycolic acid.

From Figure 4.14, it can be found that the reaction temperature is the dominant factor. The differences between the two kinds of algorithms cause the above differences. Similarly, the sum of specific feature importance (reaction activation energy,

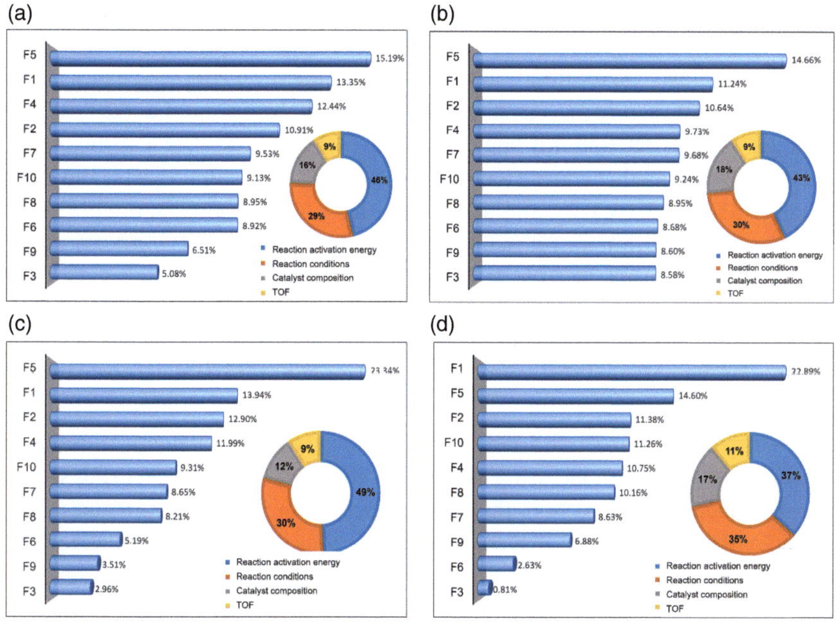

Figure 4.14 Results of feature importances for: (a) DNN, (b) DBN, (c) FC-ResNet, and (d) RF.

catalyst composition, and TOF) is about 65% in Figure 4.14d, which are also the dominant factors. They all clearly showed that the reaction activation energy, catalyst composition, and TOF were the main features affecting the conversion and yield of the EGtGA reaction. Since our study is committed to providing a "dual-core drive" model that can predict and optimize the industrial application of the EGtGA process in the future, we have chosen features that are biased toward the macro in the subsequent optimization process. Hence, catalyst compositions (Pt loading and Mn loading) and reaction conditions (reaction temperature, reaction time, and reaction pressure) were selected as decision variables of the multi-objective optimization.

4.3.3 Performance and Feature Analysis of the Optimized FC-ResNet–GA Model

By comprehensively analyzing four deep learning algorithms (DNN, DBN, FC-ResNet, and RF), it can be intuitively concluded that compared with the other three methods, FC-ResNet performs very well in both the training set and the test set. For the training set, the RMSE of the FC-ResNet algorithm for predicting conversion rate, GA yield, GAD yield, FA yield, and CO_2 yield are 1.55, 1.18, 0.15, 0.25, and 0.17, respectively. For the test set, the RMSE of the FC-ResNet algorithm for prediction conversion, GA yield, GAD yield, FA yield, and CO_2 yield are 1.37, 1.95, 0.17, 0.25, and 0.17, respectively. For the experimental data, the RMSE of the FC-ResNet algorithm for both the training set and test set is the lowest, showing excellent prediction performance. Therefore, the model parameters of FC-ResNet, by integrating the GA, were optimized in this section. Moreover, the optimized

FC-ResNet algorithm (termed FC-ResNet–GA) will also be applied to train and test 6720 datasets (including 610 mechanism datasets and 6110 experimental datasets).

Figure 4.15 shows the training and test results for the simulation and experiment data in the EGtGA process using the HDM model based on the FC-ResNet–GA model. As illustrated in this figure, the RMSE of the FC-ResNet–GA model for the training set to predict conversion rate, GA yield, GAD yield, FA yield, and CO_2 yield are 1.53, 1.93, 0.03, 0.05, and 0.86, respectively. For the test set, the RMSE of the FC-ResNet algorithm for prediction conversion, GA yield, GAD yield, FA yield, and CO_2 yield are 1.12, 1.29, 0.06, 0.07, and 0.56, respectively.

4.3.4 Process Multi-objective Optimization and Experimental Verification

Further investigation is carried out, and multi-objective optimization of process parameters is implemented using the HDM model based on the FC-ResNet–NSGA-III model. In this section, this study uses Python 3.8 to program and apply the NSGA-III algorithm in Pymoo to perform multi-objective optimization of process parameters and then solve the problem. Finally, the Pareto solution set that meets the requirements is obtained. The objective functions and constraints are shown in Eq. (4.4).

$$\begin{aligned}
\text{Obj: } \max\ &: F_{\text{conversion}} = F_{\lambda}(T, P, H, \text{Pt}_{\text{loading}}, \text{Mn}_{\text{loading}}) \\
\max\ &: F_{\text{GA}_{\text{yields}}} = F_{\mu}(T, P, H, \text{Pt}_{\text{loading}}, \text{Mn}_{\text{loading}}) \\
\min\ &: F_{\text{CO}_2\text{yields}} = F_{\mu}(T, P, H, \text{Pt}_{\text{loading}}, \text{Mn}_{\text{loading}}) \\
\text{s.t. } & 20 \leq T \leq 100 \\
& 1.0 \leq P \leq 3.0 \\
& 4 \leq H \leq 16 \\
& 1.3 \leq \text{Pt}_{\text{loading}} \leq 1.4 \\
& 0.2 \leq \text{Mn}_{\text{loading}} \leq 0.9
\end{aligned} \tag{4.4}$$

The results of Pareto sets (mainly including reaction conditions and supported catalyst composition for experimental verification, such as reaction temperature, reaction pressure, reaction time, Pt loading, and Mn loading) for the process multi-objective optimization using the NSGA-III algorithm and experiment verification are illustrated in Figure 4.16. It should be pointed out that the results of Pt and Mn loading in the Pareto sets are 1.40 and 0.90, respectively. The experimental verification of repeatability is performed by adopting the feasible solution in the Pareto solution set. The relative errors between experiment and simulation repeatability verification of EG conversion and GA yield are about 4%, while the relative errors between experiment and simulation repeatability verification of GAD yield, CO_2 yield, and FA yield are about 10%. The main reason is that GAD yields, CO_2 yields, and FA yields are all in the range of low values (0–2%), and a small absolute error could bring about relatively significant errors. More detailed results of Pareto sets and experiment verification are shown in Appendix.

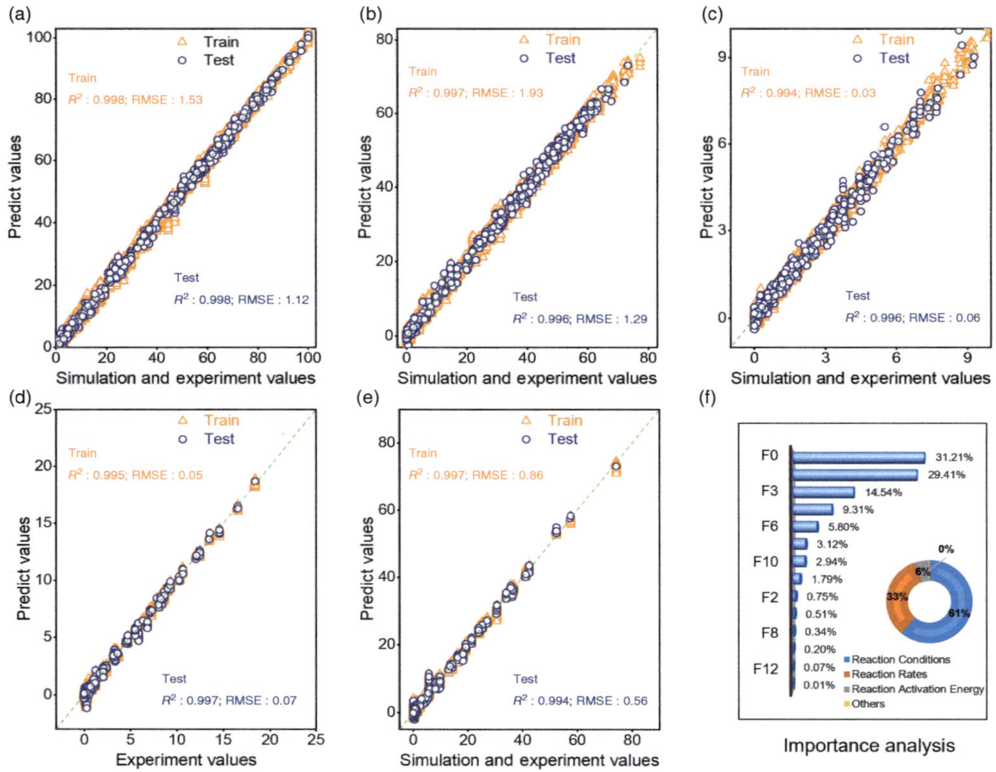

Figure 4.15 The train and test results for the simulation + experiment data in the EGtGA process using the HDM model based on the FC-ResNet-GA model: (a) conversion, (b) GA yields, and (c) GAD yields, (d) FA yields, (e) CO_2 yields, and (f) importance analysis.

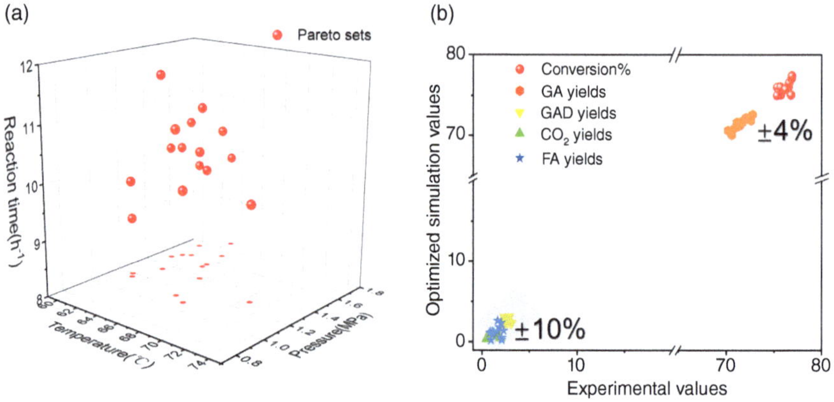

Figure 4.16 (a) The results of Pareto sets for the process multi-objective optimization using the NSGA-III algorithm and (b) experiment verification.

4.3.5 LCSA Based on the Optimized Parameters

4.3.5.1 Original Life Cycle Framework

Further LCSA investigation is implemented based on the optimized parameters. The original framework to execute the life cycle assessment is shown in Figure 4.17 [61, 62]. It should be mentioned that the black dashed box indicates the existing foundation, while the red implementation box indicates that the execution starts from the current step. As demonstrated, one of the feasible solutions applied in this study is reaction temperature, reaction pressure, reaction time, Pt loading, and Mn loading from the Pareto-optimal set, which are 63.53 °C, 1.54 MPa, 9.81 hours, 1.40 wt%, and 0.90 wt%, respectively. The selected feasible solution from the Pareto-optimal set is used to carry out the process simulation.

4.3.5.2 Life Cycle Inventory Analysis

Detailed and comprehensive life cycle inventory analysis is a crucial procedure for implementing LCSA, which can intuitively give critical data such as nonrenewable resources and energy consumption in different life cycle stages. These inventory data will lay a solid foundation for implementing sustainable life cycle interpretation and evaluation. Figure 4.18 demonstrates the life cycle assessment results of inventory values, which contain FED and GHG emissions in the form of direct and indirect for creating the one million GDP. Figure 4.18a shows the life cycle inventory analysis before optimization for the ethylene glycol (EG) to GA process using biomass as feedstock. The simulation data and inventory data are from our previous work [33, 34]. The simulation and inventory data of life cycle inventory analysis after optimization are depicted in Figure 4.18b. By comprehensively comparing the data before and after optimization, it can be intuitively concluded that the use of biomass raw materials has been reduced by about 3% by using the optimization strategy proposed in this study.

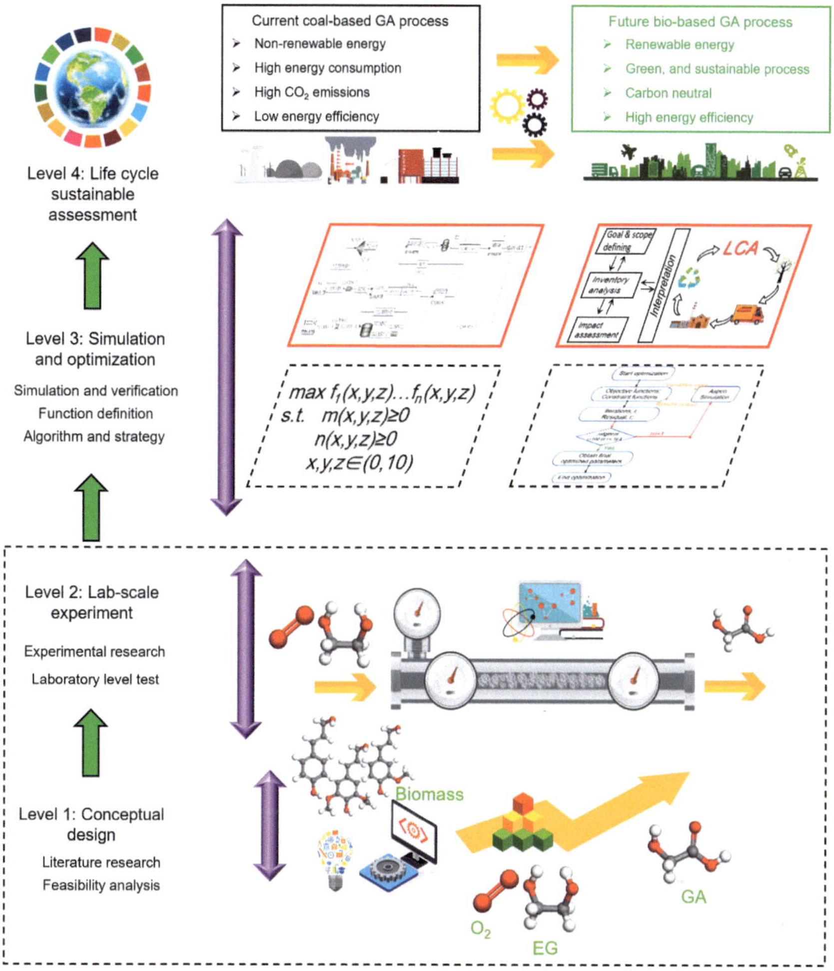

Figure 4.17 The framework of life cycle multi-objective sustainable optimization. The black dashed box indicates the existing foundation, while the red implementation box indicates that the execution starts from the current step.

4.3.5.3 Life Cycle Sustainable Interpretation and Assessment

Figure 4.19 illustrates the life cycle sustainable interpretation and evaluation results based on the inventory analysis from Figure 4.18. After optimization, the EG to GA process identifies obvious techno-economic and environmental advantages. As shown in Figure 4.19a, the FED after optimization is 6201.91 MJ (one million GDP)$^{-1}$, which has decreased by 2.96%. The GHG emissions before optimization are 39.26 tCO_2eq (one million GDP)$^{-1}$, 1.03 times larger than the emissions after optimization (see Figure 4.19b). The breakdown analysis of FED and GHG is illustrated in Figure 4.19c,d. The decomposition results of each unit's life cycle sustainability analysis are carried out to recognize further the bottleneck units hidden in the EGtGA process. The syngas production process contributes the most

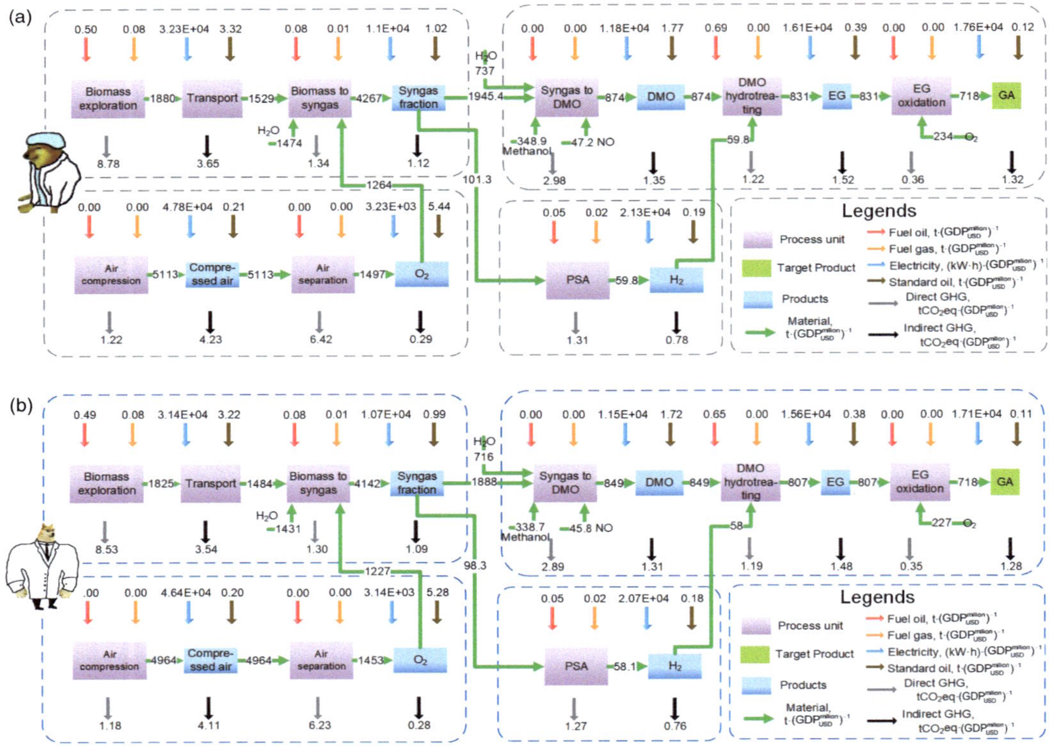

Figure 4.18 The life cycle inventory analysis for the EG to GA process using biomass as feedstock: (a) before optimization; (b) after optimization.

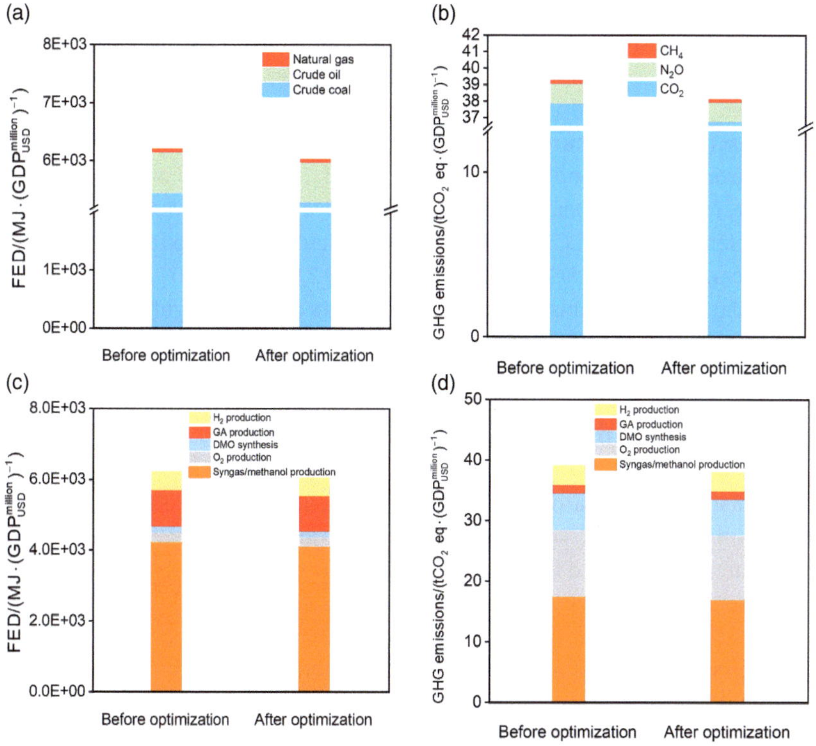

Figure 4.19 Overall resource consumption, greenhouse gas emission results, and the breakdown distribution analysis of FED and GHG for the EG to GA process: (a) Overall results of FED; (b) Overall results of GHG; (c) breakdown results of FED; (d) breakdown results of GHG.

to FED and GHG emissions, accounting for about 68% of the whole FED and 45% of the whole GHG, respectively. The sustainability analysis results intuitively indicate that syngas production is the most extensive bottleneck process. Therefore, optimizing parameters in the syngas production process and developing energy recovery technology is virtual to enhance the process performance.

4.4 Conclusion

The kinetic reaction mechanism, catalyst properties, and reaction conditions in the cases of selective oxidation at low temperatures without alkali for bio-GA production were collected to construct a hybrid deep learning framework driven by data and reaction mechanisms for predicting sustainable glycolic acid production performance. The FC-ResNet model exhibited superior performance for the prediction of conversion rate, GA, and byproduct yields. The feature importance was further analyzed. Reaction activation energy and other features (Pt loading, Mn loading, and TOF) are far less critical than reaction conditions and rates. The reaction conditions are the dominant and decisive factor, the influence of the reaction rate is low, and other characteristics are far less critical than reaction conditions. The HDM model based on the FC-ResNet–NSGA-III model is implemented with multi-objective optimization of process parameters. The results of Pareto sets were obtained, and the optimized models were experimentally validated. The selected parameters, including reaction temperature, reaction pressure, reaction time, Pt loading, and Mn loading from the Pareto-optimal set are 63.53 °C, 1.54 MPa, 9.81 hours, 1.40 wt%, and 0.90 wt%, respectively. Detailed and comprehensive life cycle inventory analysis was also implemented. The LCSA further identifies that using the optimized operating parameters, the FED, and GHG have decreased by 2.96% and 3.00%, respectively. The hybrid data and mechanism (HDM) model offers a novel insight and strategy to accelerate the engineered selective oxidation for desired GA production.

4.A Pareto Optimization Set

Pareto-optimal solution is a multi-objective intelligent algorithm. Pareto-optimal solution is a multi-objective solution that will filter out a relatively optimal set of solutions. Pareto will be used to find a reasonably optimal solution or optimal solution. In this study, the Pareto-optimal solution set is shown in Figure 4.A.1. The Pareto-optimal solution set only provides noninferior solutions of the problem to multiple objectives. In this study, the reaction conditions in the red box are selected as the optimal solution.

The Pareto optimal solution set

No.	T/℃	P/MPa	Time/h	Conversion /%	GA yields/wt%	GAD yields/wt%	CO2 yields/wt%	FA yields/wt%
1	65.78777	1.1239	10.92227	76.03089	72.27078	2.40588	0.43295	1.82813
2	61.57536	1.34112	11.60197	75.08177	70.90087	3.07605	0.56579	0.26328
3	60.68384	1.56996	10.10116	76.05948	71.1715	2.47649	0.58336	1.22428
4	63.53179	1.54423	9.81039	75.65345	71.92389	2.96367	0.50261	1.19956
5	60.35063	1.74099	9.61755	76.56252	71.84082	2.96217	0.53524	2.65163
6	63.23443	1.697	10.42723	76.05497	71.23965	3.04535	0.57041	0.528
7	66.99601	1.08077	9.96749	76.01943	70.07562	2.74071	0.55147	1.26351
8	70.54725	1.29416	9.72585	75.06751	71.67536	2.32328	0.54088	2.38578
9	60.28999	1.51214	10.11617	75.85429	71.54163	2.51111	0.53804	1.30112
10	60.48161	1.19401	9.04879	77.07577	71.57523	2.52865	0.5861	1.42394
11	61.27586	1.59901	10.58374	75.475	71.1364	2.50975	0.52772	1.13526
12	64.99013	1.3686	10.33422	75.00947	70.63095	2.42292	0.53167	0.91636
13	61.9774	1.88286	9.73417	77.41898	72.63333	3.10547	0.54491	0.28695
14	63.56154	1.50479	10.99166	76.03309	71.50273	3.01703	0.59697	0.92128
15	60.87355	1.16102	9.77736	75.76438	72.12208	2.81071	0.54463	0.53906

Figure 4.A.1 The Pareto optimization Set.

4.B Experimental Data

Data for reaction residence times of 4–12 hours was established and shown in Figure 4.B.1 and 4.B.2.

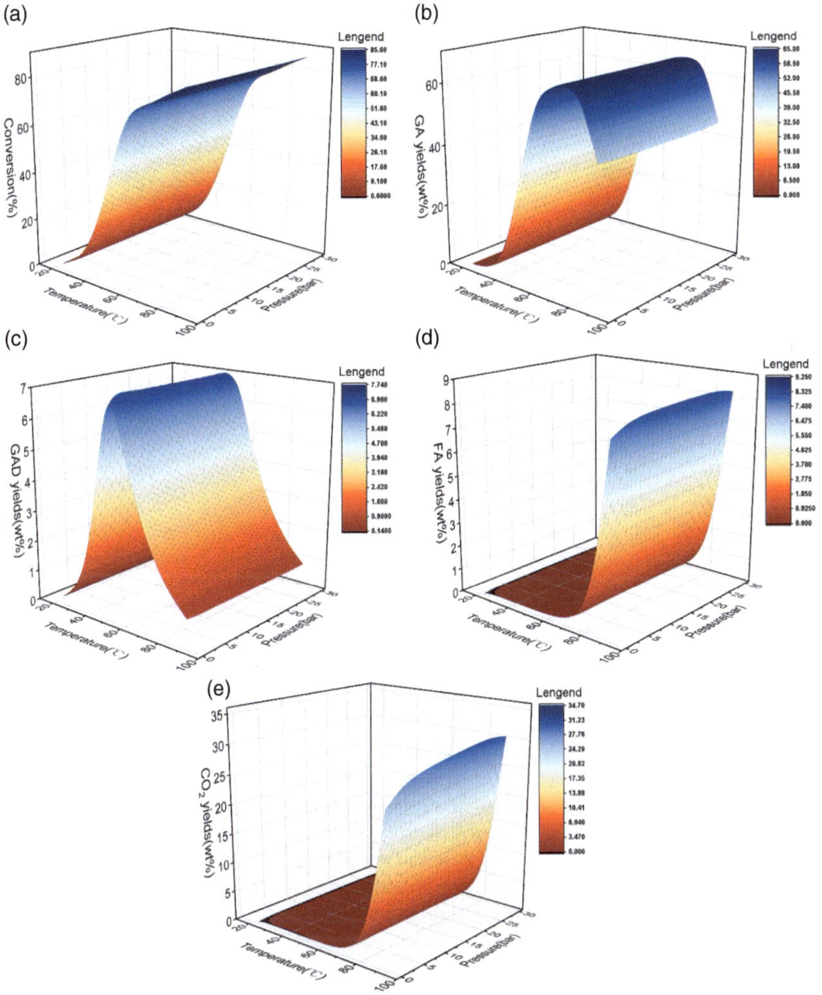

Figure 4.B.1 Data analysis and statistics with the reaction condition resident time four hours: (a) the changing trend of conversion rates with reaction temperature and pressure; (b) the changing trend of GA yields with reaction temperature and pressure; (c) the changing trend of GAD yields with reaction temperature and pressure; (d) the changing trend of FA yields with reaction temperature and pressure; (e) the changing trend of CO_2 yields with reaction temperature and pressure.

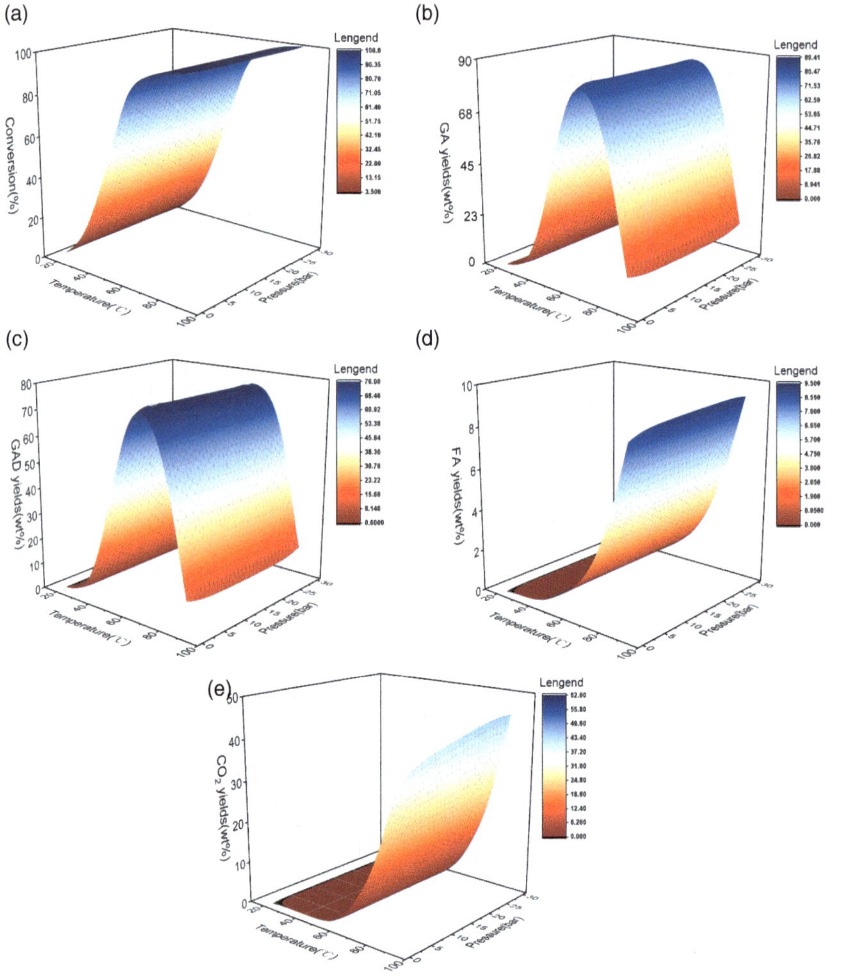

Figure 4.B.2 Data analysis and statistics with the reaction condition resident time 12 hours: (a) the changing trend of conversion rates with reaction temperature and pressure; (b) the changing trend of GA yields with reaction temperature and pressure; (c) the changing trend of GAD yields with reaction temperature and pressure; (d) the changing trend of FA yields with reaction temperature and pressure; (e) the changing trend of CO_2 yields with reaction temperature and pressure.

4.C Construction Method of Process Simulation Database Using Reaction Mechanism

On the basis of the above reaction mechanism identification, we analyzed the possible reaction path of the Pt catalyst used in the selective oxidation of ethylene glycol to glycolic acid. The reaction path diagram is shown in Figure 4.C.1.

Based on the reaction path shown in Figure 4.C.1, the reaction kinetics parameters in this study are obtained by regression based on the reaction mechanism and

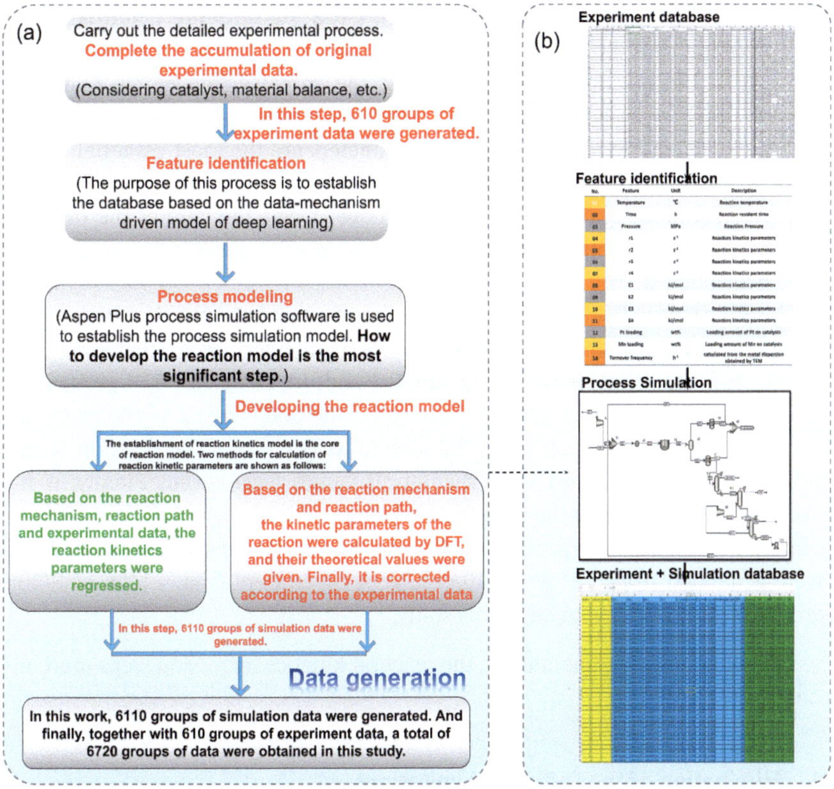

Figure 4.C.1 The reaction networks in the EGtGA process.

a large number of experimental data. The detailed steps of the calculation process of the reaction kinetics parameters we performed are shown below. The detailed description is shown as follows (the description and Figure 4.C.2):

(a) Carry out the detailed experimental process. Complete the accumulation of original experimental data. It should be noted that in this step catalyst, material balance, etc. are considered. In this step, 610 groups of experiment data were generated.
(b) The next step is the feature identification. In this step, the purpose is to establish the database based on the data-mechanism-driven model of deep learning.
(c) Following is the process simulation (process modeling). In this procedure, Aspen Plus process simulation software is used to establish the simulation model. For more details on the process simulation of the EGtGA process, we

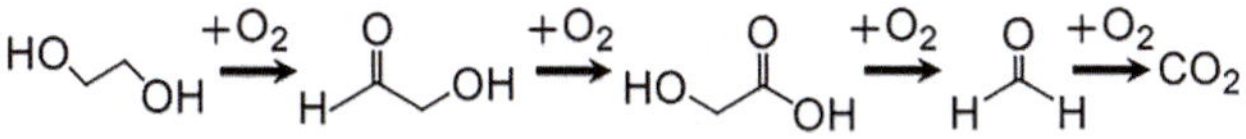

Figure 4.C.2 The detailed flowsheet of the database generation procedure.

recommend referring to our previous work (**ACS Sustainable Chemistry & Engineering 2021 9 (32), 10948–10962.**).

(d) Developing the reaction model to achieve the enhancement of mechanism-assisted database is significant in this step. Furthermore, determining reaction path and calculating reaction kinetics parameters are the most essential steps in the step of developing the reaction model.

(e) The establishment of a reaction kinetics model forms the core of the overall reaction model. Two methods for calculation of reaction kinetic parameters are shown as follows:

The first method is based on the reaction mechanism, reaction path, and experimental data; the reaction kinetics parameters were regressed. In this study, the first method was adopted and, in this step, 6110 groups of simulation data were generated. The second method is based on the reaction mechanism and reaction path; the kinetic parameters of the reaction were calculated by DFT, and their theoretical values were given. Finally, it is corrected according to the experimental data.

(f) Finally, the database was generated. In this work, 6110 groups of simulation data were generated. Together with 610 groups of experiment data, a total of 6720 groups of data were obtained in this study.

The following further elaborates on the reaction kinetics regression steps used in this chapter.

4.C.1 Elimination of the Diffusion Limitations

In this part, we will use the PtMn/MCM-41 (*Pt loading 1.4 wt%; Mn loading 0.9 wt%; termed as In-70*) catalyst as the example. More detailed information is also shown in our previous work (***Applied Catalysis B: Environmental 2021, 284, 119803***). The PtMn/MCM-41 (*Pt loading 1.4 wt%; Mn loading 0.9 wt%*) was selected to study the internal and external mass transfer. It is ensured that all the catalysts were evaluated with negligible mass transfer limitations.

(1) Effect of external diffusion on EG oxidation
(Reaction conditions: 60 °C, 1 MPa O_2, 0.15 M EG, 0.1 g Cat.)
   (a) At 700 rpm, (20 minutes) $R_{\text{oxidation}} = 0.47\ \text{kmol m}^{-3}\ \text{h}^{-1}$;
   (b) At 1000 rpm, (20 minutes) $R_{\text{oxidation}} = 0.51\ \text{kmol m}^{-3}\ \text{h}^{-1}$;
   (c) At 1200 rpm, (20 minutes) $R_{\text{oxidation}} = 0.49\ \text{kmol m}^{-3}\ \text{h}^{-1}$;

(2) Interphase mass transfer limitation for oxygen
   (a) Gas–liquid mass transfer limitation ***(J. Catal. 2016, 337, 272–283; Chem. Eng. Process, 2004, 43, 823–830; J. Catal. 2008, 257, 1–4; Three phase catalytic reactors, Ramachandran & Chaudhari, 1983; J. Chem. Eng. Data 1984, 29, 286–287)***:

$$\frac{R_{\text{oxidation}} d_{\text{bubble}}}{6 \cdot \varepsilon \cdot k_{g-l} \cdot C_{O_{2,b}}} = \frac{0.51\ (\text{kmol m}^{-3}\ \text{h}^{-1}) \cdot 0.000\,002\ (\text{m})}{6 \cdot 0.09 \cdot 1.44\ (\text{m h}^{-1}) \cdot 0.0078\ (\text{kmol m}^{-3})} = 1.7 \times 10^{-4} < 0.1$$

(b) Liquid–solid mass transfer limitation ***(J. Catal. 2016, 337, 272–283; Three phase catalytic reactors, Ramachandran and Chaudhari, 1983; MCM-41, assume 1 μm, see Figure 4.C.3)***:

$$\frac{R_{\text{oxidation}}\rho_{\text{p}} \cdot d_{\text{p}}}{6 \cdot \omega_{\text{cat}} \cdot k_{1-s} \cdot C^{*}_{\text{O}_{2,\text{b}}}} = \frac{0.51\,(\text{kmol m}^{-3} \cdot \text{h}^{-1}) \cdot 143\,(\text{kg m}^{-3}) \cdot 10^{-6}\,(\text{m})}{6 \cdot 8\,(\text{kg m}^{-3}) \cdot 72.1\,(\text{m h}^{-1}) \cdot 0.0078\,(\text{kmol m}^{-3})} = 2.7 \times 10^{-6} < 0.1$$

(c) Internal diffusion ***(J. Catal. 2016, 337, 272–283; Three phase catalytic reactors, Ramachandran and Chaudhari, 1983; Perry's Handbook: see tables 5–16)***:

$$\frac{R_{\text{oxidation}}\rho_{\text{p}} \cdot d_{\text{p}}^2}{4 \cdot \omega_{\text{cat}} \cdot D_{\text{e}} \cdot C^{*}_{\text{O}_2}} = \frac{0.51(\text{kmol m}^{-3} \cdot \text{h}^{-1}) \cdot 143\,(\text{kg m}^{-3}) \cdot (10^{-6}\,\text{m})^2}{4 \cdot 8\,(\text{kg m}^{-3}) \cdot 0.094\,(\text{m}^2\,\text{h}^{-1}) \cdot 0.0078\,(\text{kmol m}^{-3})} = 3.1 \times 10^{-9} < 1$$

(3) Interphase mass transfer limitation for glycerol

(a) Liquid–solid transfer limitation ***(J. Catal. 2016, 337, 272–283; AIChE J. 1980, 26, 177–201; Three phase catalytic reactors, Ramachandran and Chaudhari, 1983)***:

$$\frac{R_{\text{oxidation}}\rho_{\text{p}} \cdot d_{\text{p}}}{6 \cdot \omega_{\text{cat}} \cdot k_{1-s} \cdot C_{\text{gly}}} = \frac{0.51\,(\text{kmol m}^{-3}\,\text{h}^{-1}) \cdot 143\,(\text{kg m}^{-3}) \cdot 10^{-6}(\text{m})}{6 \cdot 8\,(\text{kg m}^{-3}) \cdot 0.6\,(\text{m h}^{-1}) \cdot 0.15\,(\text{kmol m}^{-3})} = 1.7 \times 10^{-5} < 0.1$$

(b) Intraparticle transfer limitation ***(J. Catal. 2016, 337, 272–283; AIChE J. 1980, 26, 177–201; Three phase catalytic reactors, Ramachandran &***

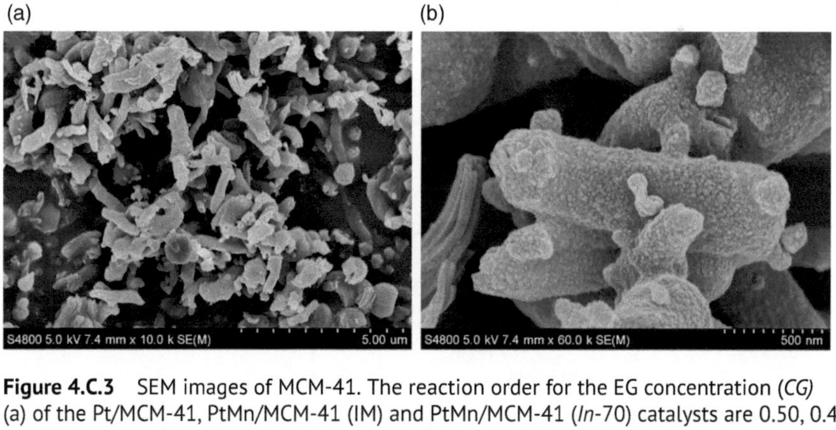

Figure 4.C.3 SEM images of MCM-41. The reaction order for the EG concentration (*CG)* (a) of the Pt/MCM-41, PtMn/MCM-41 (IM) and PtMn/MCM-41 (*In*-70) catalysts are 0.50, 0.44 and 0.38, respectively. This suggests that the PtMn catalysts have stronger adsorption ability for EG than monometallic Pt/MCM-41 catalyst, and the Pt-Mn2O3 interfacial active site in the PtMn/MCM-41 (In-70) catalyst further promotes the adsorption of EG, which is well consistent with the DFT calculation results. Meanwhile, the PtMn/MCM-41 (In-70) catalyst displays the minimum reaction order for O_2 pressure (PO) (b) (0.26) compared with the other two catalysts, meaning that the PtMn/MCM-41 (In-70) catalyst strongly adsorbs oxygen, probably due to the relatively faster replenishment rate of the consumed surface molecular oxygen [33]/with permission of Elsevier.

Chaudhari, 1983):

$$\frac{d_p}{6}\left[\frac{(m+1)\cdot R_{\text{oxidation}}\rho_p}{2\cdot\omega_{\text{cat}}\cdot D_e\cdot C_{\text{gly}}}\right]^{0.5}$$
$$=\frac{10^{-6}(\text{m})}{6}\left[\frac{0.51\,(\text{kmol m}^{-3}\,\text{h}^{-1})\cdot 143\,(\text{kg m}^{-3})}{8\,(\text{kg m}^{-3})\cdot 0.036\,(\text{m}^2\,\text{h}^{-1})\cdot 0.15\,(\text{kmol m}^{-3})}\right]^{0.5}$$
$$=6.9\times 10^{-6}<0.2$$

4.C.2 Reaction Kinetics

As shown in Figure 4.C.4, the power function type reaction kinetic equation was employed to investigate the reaction orders for EG and O_2 over the PtMn/MCM-41 (*Pt loading 1.4 wt%; Mn loading 0.9 wt%, termed as In-70*) and Pt/MCM-41 catalysts

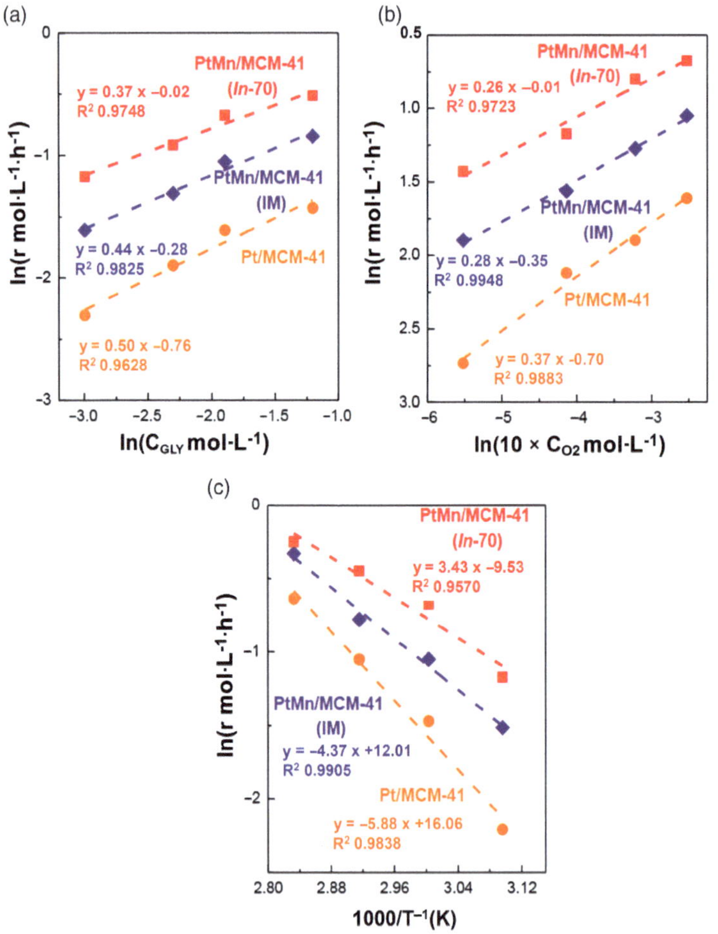

Figure 4.C.4 The effects of (a) EG concentration and (b) oxygen pressure on the oxidation of EG. Reaction conditions: 25 ml of EG aqueous solution (0.15 M), 0.1 g Cat., $T = 60\,°C$, stirring speed = 1000 rpm. (c) Apparent activation energy for EG conversion.

(*Pt loading 1.3 wt%*). The equation can be expressed as:

$$r = -\frac{dC_0}{dt} = A\ \exp\left(\frac{-Ea}{RT}\right) C_G^a P_O^b$$

C_{EG} and P_O are the initial concentration of ethylene glycol and O_2 pressure (mol l^{-1}) respectively; a and b are the corresponding reaction order. r, T, A, R, and Ea are the initial reaction rate of glycerol (in mol l^{-1} h^{-1}), reaction temperature (K), the pre-exponential factor, ideal gas constant (8.314×10^{-3} kJ mol^{-1} K^{-1}), and activation energy (kJ mol^{-1}). It should be noted that the concentration of oxygen at the surface equals its bulk concentration (calculated by Henry's law) under the reaction conditions (***J. Phys. Chem. C 2010, 114, 1164–1172; J. Chem. Thermodyn. 2000, 32, 1145; J. Phys. Chem. 1996, 100, 5597***).

References

1 Birner, B., Severinghaus, J., Paplawsky, B., and Keeling, R. (2022). Increasing atmospheric helium due to fossil fuel exploitation. *Nature Geoscience* 15: 346–348.

2 Lee, K., Jing, Y., Wang, Y., and Yan, N. (2022). A unified view on catalytic conversion of biomass and waste plastics. *Nature Reviews Chemistry* 6: 635–652.

3 Gao, Z., Ma, B., Chen, S. et al. (2022). Converting waste PET plastics into automobile fuels and antifreeze components. *Nature Communications* 13: 3343–3351.

4 Cabernard, L., Pfister, S., Oberschelp, C., and Hellweg, S. (2022). Growing environmental footprint of plastics driven by coal combustion. *Nature Sustainability* 5: 139–148.

5 Tatrari, G., Karakoti, M., Tewari, C. et al. (2021). Solid waste-derived carbon nanomaterials for supercapacitor applications: a recent overview. *Materials Advances* 2: 1454–1484.

6 Wen, Z., Xie, Y., Chen, M., and Dinga, C. (2021). China's plastic import ban increases prospects of environmental impact mitigation of plastic waste trade flow worldwide. *Nature Communications* 12: 425.

7 Zhou, H., Ren, Y., Li, Z. et al. (2021). Electrocatalytic upcycling of polyethylene terephthalate to commodity chemicals and H_2 fuel. *Nature Communications* 12: 4679.

8 Maity, A., Chaudhari, S., Titman, J., and Polshettiwar, V. (2020). Catalytic nanosponges of acidic aluminosilicates for plastic degradation and CO_2 to fuel conversion. *Nature Communications* 11: 3828.

9 Kwon, D., Jung, S., Moon, D. et al. (2022). Strategic management of harmful chemicals produced from pyrolysis of plastic cup waste using CO_2 as a reaction medium. *Chemical Engineering Journal* 437: 135524.

10 Zhuang, Z., Mohamed, B., Li, L., and Swei, O. (2022). An economic and global warming impact assessment of common sewage sludge treatment processes in North America. *Journal of Cleaner Production* 370: 133539.

11 Ribeiro, F., Nascimento, F., and Silva, M. (2022). Environmental performance analysis of railway infrastructure using life cycle assessment: selecting pavement projects based on global warming potential impacts. *Journal of Cleaner Production* 365: 132558.

12 Lee, J., Bhagwat, S., Kuanyshev, N. et al. (2023). Rewiring yeast metabolism for producing 2,3-butanediol and two downstream applications: techno-economic analysis and life cycle assessment of methyl ethyl ketone (MEK) and agricultural biostimulant production. *Chemical Engineering Journal* 451: 138886.

13 Feist, J., Lee, D., and Xia, Y. (2022). A versatile approach for the synthesis of degradable polymers via controlled ring-opening metathesis copolymerization. *Nature Chemistry* 14: 53–58.

14 Yang, L., Wang, Y., Yuan, J. et al. (2022). Construction of covalent-integrated MOFs@COFs composite material for efficient synergistic adsorption and degradation of pollutants. *Chemical Engineering Journal* 446: 137095.

15 Tang, X., Ma, S., Xu, S. et al. (2023). Effects of different pretreatment strategies during porous carbonaceous materials fabrication on their peroxydisulfate activation for organic pollutant degradation: focus on mechanism. *Chemical Engineering Journal* 451: 138576.

16 Zhou, N., Dai, L., Lyu, Y. et al. (2022). A structured catalyst of ZSM-5/SiC foam for chemical recycling of waste plastics via catalytic pyrolysis. *Chemical Engineering Journal* 440: 135836.

17 Yue, X., Zhang, F., Wu, L. et al. (2022). Upcycling of blending waste plastics as flexible growing substrate with superabsorbing property. *Chemical Engineering Journal* 435: 134622.

18 Li, T., Tan, Z., Tang, Z. et al. (2022). One-pot chemoenzymatic synthesis of glycolic acid from formaldehyde. *Green Chemistry* 24: 5064–5069.

19 Samantaray, P., Little, A., Haddleton, D. et al. (2020). Poly(glycolic acid) (PGA): a versatile building block expanding high performance and sustainable bioplastic applications. *Green Chemistry* 22: 4055–4081.

20 Jem, K. and Tan, B. (2020). The development and challenges of poly(lactic acid) and poly(glycolic acid). *Advanced Industrial and Engineering Polymer Research* 3: 60–70.

21 Niu, D., Xu, P., Sun, Z. et al. (2021). Superior toughened bio-compostable poly(glycolic acid)-based blends with enhanced melt strength via selective interfacial localization of in-situ grafted copolymers. *Polymer* 235: 124269.

22 Yeo, T., Ko, Y., Kim, E. et al. (2021). Promoting bone regeneration by 3D-printed poly(glycolic acid)/hydroxyapatite composite scaffolds. *Journal of Industrial and Engineering Chemistry* 94: 343–351.

23 Yan, H., Wang, Z., Li, L. et al. (2021). DOPA-derived electroactive copolymer and IGF-1 immobilized poly(lactic-*co*-glycolic acid)/hydroxyapatite biodegradable microspheres for synergistic bone repair. *Chemical Engineering Journal* 416: 129129.

24 Kim, B., Ko, Y., Yeo, T. et al. (2019). Guided regeneration of rabbit calvarial defects using silk fibroin nanofiber-poly(glycolic acid) hybrid scaffolds. *ACS Biomaterials Science & Engineering* 5: 5266–5272.

25 Zhu, T., Yao, D., Li, D. et al. (2021). Multiple strategies for metabolic engineering of *Escherichia coli* for efficient production of glycolate. *Biotechnology and Bioengineering* 118: 4699–4707.

26 Bioresorbable Polymers Market by Type (Polylactic acid (PLA), Polyglycolic acid (PGA), Polylactic-co-glycolic acid (PLGA), Polycaprolactone(PCL)), application (orthopedic devices, drug delivery), and Region - Global Forecast to 2027. https://www.marketsandmarkets.com/Market-Reports/bioresorbable-polymer-market-235258717.html (accessed Jan 2023).

27 Global Polyglycolic Acid (PGA) Market Report By Form (Fibers, Films, Others (Plate, Rod, and Composites)), By End-use Industry (Medical, Oil & Gas, Packaging, Others (Civil Engineering, Agriculture, and Filter)) And By Regions - Industry Trends, Size, Share, Growth, Estimation and Forecast, 2023-2032. https://www.valuemarketresearch.com/report/polyglycolic-acid-pga-market (accessed July 2024).

28 Yu, Q., Hua, X., Zhou, X. et al. (2022). A roundabout strategy for high-purity glycolic acid biopreparation via a resting cell bio-oxidation catalysis of ethylene glycol. *Green Chemistry* 24: 5142–5150.

29 Huang, Q., Xie, C., Li, Y. et al. (2017). Thermodynamic equilibrium of hydroxyacetic acid in pure and binary solvent systems. *The Journal of Chemical Thermodynamics* 108: 76–83.

30 Shi, Y., Sun, H., Cao, H. et al. (2011). Synergistic extraction of glycolic acid from glycolonitrile hydrolysate. *Industrial and Engineering Chemistry Research* 50: 8216–8224.

31 Zhao, D., Zhua, T., Lia, J. et al. (2021). Poly(lactic-*co*-glycolic acid)-based composite bone-substitute materials. *Bioactive Materials* 6: 346–360.

32 Lee, S., Kim, J., Lee, J., and Kim, Y. (1993). Carbonylation of formaldehyde over ion exchange resin catalysts. 1. Batch reactor studies. *Industrial and Engineering Chemistry Research* 32: 253–259.

33 Yan, H., Yao, S., Wang, J. et al. (2021). Engineering Pt–Mn_2O_3 interface to boost selective oxidation of ethylene glycol to glycolic acid. *Applied Catalysis B: Environmental* 284: 119803.

34 Yan, H., Zhao, M., Feng, X. et al. (2022). PO_4^{3-} coordinated robust single-atom platinum catalyst for selective polyol oxidation. *Angewandte Chemie, International Edition* 61: e202116059.

35 Zhou, X., Zha, M., Cao, J. et al. (2021). Glycolic acid production from ethylene glycol via sustainable biomass energy: integrated conceptual process design and comparative techno-economic–society–environment analysis. *ACS Sustainable Chemistry & Engineering* 9: 10948–10962.

36 Xu, S., Xiao, Y., Zhang, W. et al. (2022). Relay catalysis of copper-magnesium catalyst on efficient valorization of glycerol to glycolic acid. *Chemical Engineering Journal* 428: 132555.

37 Kang, N., Kim, M., Baek, K. et al. (2022). Photoautotrophic organic acid production: glycolic acid production by microalgal cultivation. *Chemical Engineering Journal* 433: 133636.

38 Kim, D., Oh, L., Tan, Y. et al. (2021). Enhancing glycerol conversion and selectivity toward glycolic acid via precise nanostructuring of electrocatalysts. *ACS Catalysis* 11: 14926–14931.

39 Kim, H., Kim, Y., Lee, D. et al. (2017). Coproducing value-added chemicals and hydrogen with electrocatalytic glycerol oxidation technology: experimental and techno-economic investigations. *ACS Sustainable Chemistry & Engineering* 5: 6626–6634.

40 Silver, D., Huang, A., Maddison, C. et al. (2016). Mastering the game of Go with deep neural networks and tree search. *Nature* 529: 484–489.

41 Yang, H., Li, J., Lim, K. et al. (2022). Automatic strain sensor design via active learning and data augmentation for soft machines. *Nature Machine Intelligence* 4: 84–94.

42 Li, J., Zhang, L., Li, C. et al. (2022). Data-driven based in-depth interpretation and inverse design of anaerobic digestion for CH_4-rich biogas production. *ACS ES&T Engineering* 2: 642–652.

43 Leonard, K., Hasan, F., Sneddon, H., and You, F. (2021). Can artificial intelligence and machine learning be used to accelerate sustainable chemistry and engineering? *ACS Sustainable Chemistry & Engineering* 9: 6126–6129.

44 Li, J., Zhang, W., Liu, T. et al. (2021). Machine learning aided bio-oil production with high energy recovery and low nitrogen content from hydrothermal liquefaction of biomass with experiment verification. *Chemical Engineering Journal* 425: 130649.

45 Chakkingal, A., Janssens, P., Poissonnier, J. et al. (2022). Multi-output machine learning models for kinetic data evaluation: a Fischer–Tropsch synthesis case study. *Chemical Engineering Journal* 446: 137186.

46 Shi, S. and Xu, G. (2018). Novel performance prediction model of a biofilm system treating domestic wastewater based on stacked denoising auto-encoders deep learning network. *Chemical Engineering Journal* 347: 280–290.

47 Wu, Z., Rincon, D., Luo, J., and Christofides, P. (2021). Machine learning modeling and predictive control of nonlinear processes using noisy data. *AIChE Journal* 67: e17164.

48 Zhan, N. and Kitchin, J. (2022). Uncertainty quantification in machine learning and nonlinear least squares regression models. *AIChE Journal* 68: e17516.

49 Wang, Z., Su, Y., Shen, W. et al. (2019). Predictive deep learning models for environmental properties: the direct calculation of octanol–water partition coefficients from molecular graphs. *Green Chemistry* 21: 4555–4565.

50 Yang, A., Su, Y., Wang, Z. et al. (2021). A multi-task deep learning neural network for predicting flammability-related properties from molecular structures. *Green Chemistry* 23: 4451–4465.

51 Ouyang, Y., Vandewalle, L., Chen, L. et al. (2022). Speeding up turbulent reactive flow simulation via a deep artificial neural network: a methodology study. *Chemical Engineering Journal* 429: 132442.

52 Hinton, G. and Osindero, S. (2006). A fast learning algorithm for deep belief nets. *Neural Computation* 18: 527–1554.

53 He, K., Zhang, X., Ren, S., and Sun, J. (2016). Deep residual learning for image recognition. *Proceedings of the IEEE Conference on Computer Vision and Pattern Recognition*, Las Vegas, NV, USA. IEEE. pp. 770–778.

54 Cheng, F., Belden, E., Li, W. et al. (2022). Accuracy of predictions made by machine learned models for biocrude yields obtained from hydrothermal liquefaction of organic wastes. *Chemical Engineering Journal* 442: 136013.

55 Lira, J., Riella, H., Padoin, N., and Soares, C. (2022). Computational fluid dynamics (CFD), artificial neural network (ANN) and genetic algorithm (GA) as a hybrid method for the analysis and optimization of micro-photocatalytic reactors: NOx abatement as a case study. *Chemical Engineering Journal* 431: 133771.

56 Kabak, E., Cagcag, Y., Aydın, T., and Turan, N. (2022). Prediction and optimization of nitrogen losses in co-composting process by using a hybrid cascaded prediction model and genetic algorithm. *Chemical Engineering Journal* 437: 135499.

57 Tao, Y., Rahn, C.D., Archer, L.A., and You, F. (2021). Second life and recycling: energy and environmental sustainability perspectives for high-performance lithium-ion batteries. *Science Advances* 7 (45): eabi7633.

58 Tian, X., Stranks, S., and You, F. (2021). Life cycle assessment of recycling strategies for perovskite photovoltaic modules. *Nature Sustainability* 4: 821–829.

59 Tao, Y., Steckel, D., Klemeš, J., and You, F. (2021). Trend towards virtual and hybrid conferences may be an effective climate change mitigation strategy. *Nature Communications* 12: 7324.

60 Niaz, H., Shams, M., Liu, J., and You, F. (2022). Mining bitcoins with carbon capture and renewable energy for carbon neutrality across states in the USA. *Energy & Environmental Science* 15: 3551–3570.

61 Zhou, X., Yan, H., Feng, X. et al. (2021). Producing glyceric acid from glycerol via integrating vacuum dividing wall columns: conceptual process design and techno-economic-environmental analysis. *Green Chemistry* 23: 3664–3676.

62 Zhou, X., Yang, Q., Yang, S. et al. (2022). One-step leap in achieving oil-to-chemicals by using a two-stage riser reactor: molecular-level process model and multi-objective optimization strategy. *Chemical Engineering Journal* 444: 136684.

63 Zhou, X., Li, Z., Feng, X. et al. (2023). A hybrid deep learning framework driven by data and reaction mechanism for predicting sustainable glycolic acid production performance. *AIChE Journal* 69 (7): e18083.

5

Innovation of Gas Separation Processes: Integrating Computational MOF Design and Adsorption Process Optimization

Xiang Zhang[1] and Teng Zhou[2]

[1] *East China University of Science and Technology, School of Chemical Engineering, State Key laboratory of Chemical Engineering, 130 Meilong Road, Shanghai, 200237, China*
[2] *The Hong Kong University of Science and Technology (Guangzhou), Sustainable Energy and Environment Thrust, Nansha, Guangzhou, 511400, China*

5.1 Introduction

Pressure swing adsorption (PSA) is widely used for gas separation. It exploits different affinities of gases on solid adsorbents at different pressures to achieve separations. Currently, various porous materials are used as adsorbents, such as metal–organic framework (MOF) and zeolite. MOFs are formed via the self-assembly of molecular building blocks (i.e. metal nodes and organic linkers) in different topologies. The large variety of building blocks offers a huge design space [1]. In addition, MOFs own many superior properties such as ultrahigh porosity and controllable pore geometry. Thus, MOFs have great potential to achieve remarkable progress in gas separation [2].

MOFs can be adopted for separation via four alternative mechanisms: equilibrium separation, kinetic separation, molecular sieving, and gate-opening separation. Equilibrium separation, based on different species' equilibrium gas loadings, is the most common method [3]. To date, numerous MOFs with distinct isotherm characteristics have been synthesized in the laboratory by tailoring MOF chemistry and structure. This trial-and-error experimental approach is time-consuming and inefficient. Importantly, many MOFs cannot lead to good process performance [4]. As depicted in Figure 5.1, the design of a MOF for use in PSA processes is a multiscale design problem that incorporates the interlinked material, phase, and process levels. Variations of both material and process operating parameters affect the adsorption behavior of the gas–solid system and thus jointly determine the process performance. With this picture in mind, a computational approach is highly desirable to expedite the identification of promising MOFs that can best serve the PSA process [6, 7].

In the literature, computational screening methods have been employed to find suitable adsorbents. The most popular one is high-throughput screening [8–13]. A large set of synthesized or hypothetical adsorbents are first collected and their

Applied AI Techniques in the Process Industry: From Molecular Design to Process Design and Optimization, First Edition. Edited by Chang He and Jingzheng Ren.

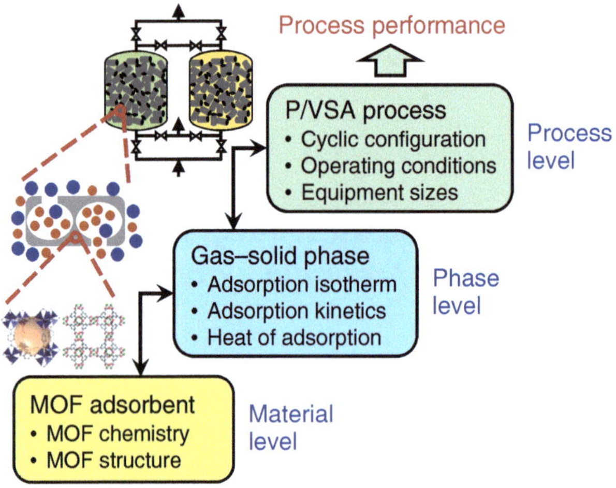

Figure 5.1 Multiscale structure of the pressure/vacuum swing adsorption process. Source: [5]/with permission of John Wiley & Sons.

adsorption and diffusion properties are computed using molecular simulation. Then, performance metrics (e.g. working capacity and selectivity) are calculated to identify top-ranked candidates for process evaluation. Although this approach works for a large adsorbent database, those widely used metrics can neither tell whether separation requirements are satisfied nor rank the adsorbents correctly in terms of real process performance [14–16]. Alternatively, hypothetical adsorption isotherms can be sampled for process optimization. With the results, surrogate models that correlate isotherm characteristics with process performance can be constructed for adsorbent screening [17–19]. This method generates an adsorbent ranking that is more consistent with process performance. However, like other screening methods, it applies to adsorbents with known chemistry and structure. The insights from the phase and process levels cannot be used to guide new adsorbent design.

Unlike screening, the multiscale design problem can be formulated as an optimization problem. Each level is represented by different models. In this way, design variables at material and process levels can be simultaneously optimized. Such a strategy has been widely applied for integrated solvent and process design [20–24]. Its success lies in the foundation of the computer-aided molecular design method, molecular structure–property relationship models, and structural feasibility rules, etc. For adsorption processes, Khurana and Farooq [25, 26] integrated the phase and process levels. Isotherm characteristics and process operating conditions were simultaneously optimized for carbon capture. Optimal isotherm shapes are obtained, which can serve as targets for adsorbent design. However, without considering the valid design space of adsorbent materials, the isotherm targets may not correspond to any feasible adsorbents. Importantly, it is unknown which adsorbent chemistry and structure are favored for the process. To tackle these issues, material design should be incorporated into process optimization. Unfortunately, none has

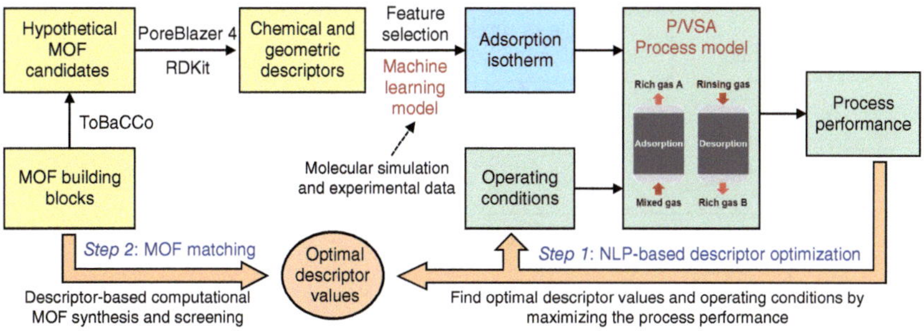

Figure 5.2 General methodology for integrated metal–organic framework and pressure/vacuum swing adsorption process design. Source: [29]/with permission of John Wiley & Sons.

been done on this topic yet. The major challenge is the lack of mathematical models to guarantee the structural feasibility of the adsorbents and to predict the adsorbent properties. For instance, hypothetical MOFs are usually assembled using computational tools (e.g. ToBaCCo), and their structural feasibility is verified via *ab initio* calculations [27]. To determine the MOF adsorbent properties, a computationally expensive molecular simulation is needed [28]. Clearly, these computations are impractical to be embedded into a mathematical optimization problem.

In this chapter, focusing on equilibrium separation, an integrated MOF and PSA process design framework is proposed in Figure 5.2 to tackle the above challenges. It consists of two steps: descriptor optimization and MOF matching. In the first step, a MOF is represented as a set of descriptors. Several data science techniques are adopted to select proper descriptors, define valid design space, and build data-driven models for predicting the adsorption isotherms. This produces an explicit formulation of the integrated design problem to enable the simultaneous optimization of MOF descriptors and process operating conditions for maximizing the process performance. In the second step, the aim is to computationally synthesize and screen hypothetical MOFs for enhancing equilibrium-based gas separation. The design guidelines and data-driven models obtained from the first step are well-adopted for designing new MOFs that can eventually result in better separation performance using optimized PSA processes. The framework is demonstrated on the separation of propene (PE) and propane (PA), which is currently achieved with energy-intensive cryogenic distillation. This chapter is organized as follows. For the first step, detailed material–property relationships are constructed, and the integrated design problem is formulated and solved. For the second step, detailed property–performance relationships are developed for fast MOF screening, and MOF building blocks are specifically extracted to perform *in silico* MOF design and subsequent screening.

5.2 Step One: Descriptor Optimization

The workflow for the first step is illustrated in Figure 5.3. First, representative MOF descriptors are selected to serve as design variables. Meanwhile, the selected

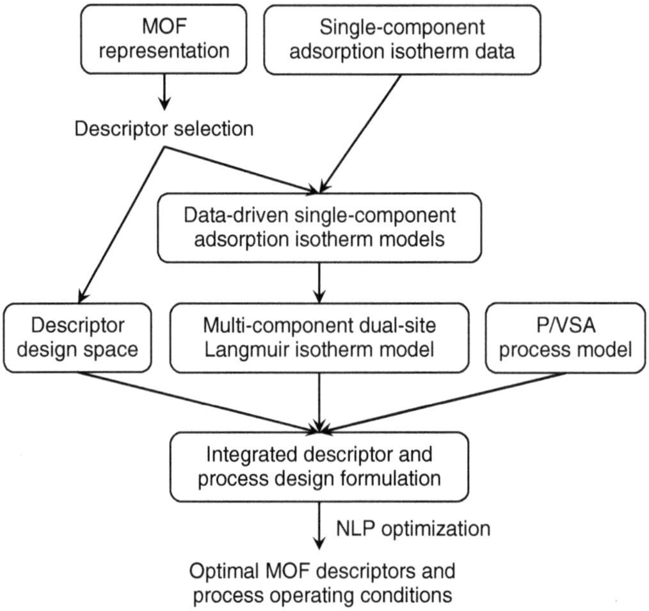

Figure 5.3 Workflow of the descriptor optimization step. Source: [5]/with permission of John Wiley & Sons.

descriptors are used to develop data-driven models for predicting single-component adsorption loadings. Then, the data-driven models are employed to derive multi-component dual-site Langmuir (DSL) isotherm models. Based on the generated descriptor dataset, specific models are defined to denote the valid design space of descriptors. Afterward, by integrating the design space model, the DSL isotherm model, and the PSA process model, the integrated MOF descriptor and PSA process design task is formulated as a nonlinear programming (NLP) problem. This problem is solved to identify the optimal descriptor values and process operating conditions. The detailed procedures and models are elaborated below.

5.2.1 Material–Property Relationship of MOFs

5.2.1.1 MOF Representation

Every MOF has a unique chemical composition (metal node and organic linker) and/or structure (pore geometry and topology). For adsorption, MOF chemistry influences the interaction with the adsorbate. For instance, some metals (e.g. copper and zinc) can form open metal sites (OMSs), leading to a stronger interaction with olefins than with paraffins [30, 31]. In addition, since adsorbates tend to occupy the void spaces of an MOF during adsorption, pore geometry plays a crucial role in determining the adsorption capacity. Thus, chemical and geometric descriptors are required to cover the effects of MOF chemistry and pore geometry on the adsorption properties [32]. For differentiating MOF topologies, topological descriptors can be computed. However, those descriptors are hard to interpret and trace back to specific topology [33]. Additionally, only a small number of well-known topologies can be found in synthesized MOFs [34]. Considering the low interpretability and limited topology variations, topological descriptors are not included.

Fourteen geometric descriptors are considered, including void fraction, volumetric surface area, and largest cavity diameter. These descriptors can capture the important geometric features of porous materials like MOFs [35]. Moreover, Gharagheizi et al. [36] recently defined 5009 chemical descriptors to describe MOF chemistry. Their definitions are adjusted from molecular descriptors that have been widely used in cheminformatics. The current study employs the 5009 descriptors directly.

Four hundred and seventy-one real MOFs are selected from the CoRE MOF database where stable and synthesizable MOFs with available atomic coordinates are collected [37]. These 471 MOFs include considerable diversity of chemistry, geometry, and topology. For each MOF, the 14 geometric descriptors have already been computed using the open-source package Zeo++ with a probe radius of 1.86 Å [38]. Additionally, the molecular graph of MOFs containing atom positions and connections is read using the cheminformatics package RDkit to derive the 5009 chemical descriptors accordingly [39].

5.2.1.2 Data-Driven Model for Single-Component Adsorption Isotherm

Data-driven models are developed to predict single-component adsorption loadings using MOF descriptors. A series of adsorption isotherm data is needed for model regression. The pure species isotherms of PE and PA on 471 MOFs have been computed by Tang et al. [40] using grand canonical Monte Carlo (GCMC) simulations at 300 K over a pressure range from 0.1 to 156 bar.

Although it is useful to include a large number of descriptors, this leads to great difficulties in developing data-driven models. Since the effect of each descriptor on deciding the adsorption selectivity and capacity varies, the most influential descriptors can be selected to build simple and reliable data-driven models [41]. Figure 5.4 shows a systematic procedure for descriptor selection.

- First, the descriptors whose values are equal to zero for more than 95% of the 471 MOFs are removed. After this, 83 descriptors remained.
- Second, the remaining descriptors are normalized to fall within the range of [0, 1]. Typically, descriptors with limited variation have limited predictive capability. Therefore, descriptors with variances less than 0.02 are excluded, resulting in a remaining set of 52 descriptors.
- Third, when two descriptors are closely correlated, only one is needed. Here, if the correlation coefficient between two descriptors exceeds 0.9, the descriptor with a larger total correlation coefficient with others is eliminated. After this, 43 descriptors remained.
- Fourth, the recursive feature elimination is applied to reduce the number of descriptors [42]. Considering the 43 descriptors as input and adsorption loadings as output, the importance of each descriptor is computed using the random forest algorithm. Clearly, for different gas adsorbates, the sensitivities are different. After determining the importance of descriptors, the least important descriptors

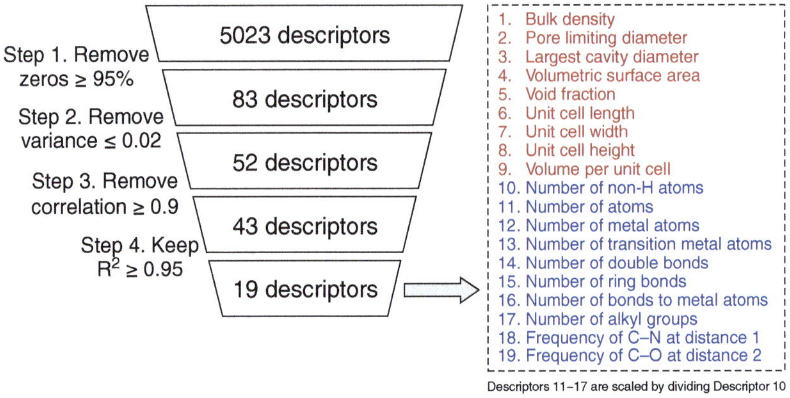

Figure 5.4 Detailed procedure for MOF descriptor selection. Source: [5]/with permission of John Wiley & Sons.

are recursively removed, and the others are sent to train ANN models to predict the isotherm data. For each trial, fivefold cross-validation is performed to tune the hyperparameters (i.e. number of layers and neurons). As expected, the model accuracy gradually decreases when reducing the number of descriptors. To ensure R^2 is larger than 0.95, in total 19 different descriptors are finally identified and shown in Figure 5.4. They consist of 9 geometric and 10 chemical descriptors.

Based on the 19 descriptors, two new ANN models are trained to predict the single-component loadings of PE (qs_E) and PA (qs_A) at 300 K and different pressures, respectively. Fivefold cross-validation is performed to find the optimal hyperparameters. Each model has one input, one hidden, and one output layer. The tansig and purelin activation functions are applied in the hidden and output layers, respectively. The number of neurons in the hidden layer are 25 for the qs_E model and 30 for the qs_A model. The two models are summarized below, and the involved parameters are given at https://github.com/zx2012flying/Integrated-MOF-and-PVSA-Process-Design.

$$qs_E = \text{ANN}_1(y_1, \ldots, y_{19}, P) \text{ at } T = 300\,\text{K} \tag{5.1}$$

$$qs_A = \text{ANN}_2(y_1, \ldots, y_{19}, P) \text{ at } T = 300\,\text{K} \tag{5.2}$$

where y_1–y_{19} are the 19 descriptor values and P is the pressure. The R^2 and MAE of both models are larger than 0.95 and less than 0.6, respectively. These indicate that the two models can provide reliable predictions on the single-component adsorption isotherms of PA and PE.

5.2.1.3 Multicomponent Dual-Site Langmuir Isotherm Model

Adsorption-based gas separation occurs in a multicomponent environment. To predict multicomponent adsorption equilibria, the DSL model is applied using single-component model parameters that are fitted from single-component adsorption loadings. This type of DSL model can provide good predictions for many systems including PE/PA [43].

The DSL model parameters are derived as follows. Given four pressures $P_i = \{0.01, 0.1, 1, 10\}$ bar, the corresponding single-component PE and PA adsorption loadings are predicted by

$$qs_{E,i} = \mathrm{ANN}_1(y_1, \ldots, y_{19}, P_i) \text{ at } T = 300\,\mathrm{K} \tag{5.3}$$

$$qs_{A,i} = \mathrm{ANN}_2(y_1, \ldots, y_{19}, P_i) \text{ at } T = 300\,\mathrm{K} \tag{5.4}$$

Meanwhile, the equilibrium adsorption loadings of PE and PA ($q_{E,i}^{eq}$ and $q_{A,i}^{eq}$) at pressure P_i can be calculated from the multicomponent DSL model by assuming that only one component exists.

$$q_{E,i}^{eq} = \frac{Q_E^1 \cdot b_E^1 \cdot P_i \cdot y_E}{1 + b_E^1 \cdot P_i \cdot y_E + b_A^1 \cdot P_i \cdot y_A} + \frac{Q_E^2 \cdot b_E^2 \cdot P_i \cdot y_E}{1 + b_E^2 \cdot P_i \cdot y_E + b_A^2 \cdot P_i \cdot y_A} \times y_E = 1,\ y_A = 0 \tag{5.5}$$

$$q_{A,i}^{eq} = \frac{Q_A^1 \cdot b_A^1 \cdot P_i \cdot y_A}{1 + b_E^1 \cdot P_i \cdot y_E + b_A^1 \cdot P_i \cdot y_A} + \frac{Q_A^2 \cdot b_A^2 \cdot P_i \cdot y_A}{1 + b_E^2 \cdot P_i \cdot y_E + b_A^2 \cdot P_i \cdot y_A} \times y_E = 0,\ y_A = 1 \tag{5.6}$$

where y_E and y_A are the molar fractions of PE and PA in the gas phase, respectively. Eqs. (5.7) and (5.8) ensure that the adsorption loadings calculated from the two models match with each other. With these, the eight DSL model parameters (Q_E^1, b_E^1, Q_E^2, b_E^2, Q_A^1, b_A^1, Q_A^2, and b_A^2) can be obtained.

$$\left(q_{E,i}^{eq} - qs_{E,i}\right)^2 \leq 0.005 \tag{5.7}$$

$$\left(q_{A,i}^{eq} - qs_{A,i}\right)^2 \leq 0.005 \tag{5.8}$$

5.2.2 Integrated Optimization of MOF Descriptors and PSA Operating Conditions

5.2.2.1 Descriptor Design Space

Nineteen descriptors and two ANN models are identified from pure data without any thermodynamic basis. Additional constraints are added to confine the solutions inside the valid design space of the models. Data-driven models are valid in the domain that is densely covered by training data [44]. Any predictions outside the domain should be prevented. Here, the approach used is to build a one-class support vector machine (SVM) classifier with the radial basis function [45]. After importing the descriptor dataset, the kernel coefficient β is tuned to 2×10^{-9} to ensure that over 99.5% of the training data are inside the validity domain. The resulting classifier or boundary is written as

$$f(\mathbf{y}) = \sum_{j=1}^{N} w_j \cdot \varphi_j(\mathbf{y}) + \rho > 0 \quad \mathbf{y} = [y_1, \ldots, y_{19}] \tag{5.9}$$

$$\varphi_j(\mathbf{y}) = e^{-\beta \cdot \|\mathbf{y} - \mathbf{SV}_j\|^2} \quad \mathbf{SV}_j = [\mathrm{SV}_{1,j}, \ldots, \mathrm{SV}_{19,j}] \tag{5.10}$$

where N is the total number of fitted support vectors, ρ is the intercept, and w_j is the coefficient of the jth support vector $\mathbf{SV}_j$. These parameters were given in the GitHub shown below.

Moreover, certain logic constraints have to be satisfied. Equation (5.11) ensures that each descriptor must be within its lower and upper bounds, namely minimal and maximal values in the training dataset.

$$\boldsymbol{y}^{\mathrm{L}} \le \boldsymbol{y} \le \boldsymbol{y}^{\mathrm{U}} \quad \boldsymbol{y} = [y_1, \dots, y_{19}] \tag{5.11}$$

The largest cavity diameter (y_3) should exceed the pore-limiting diameter (y_2). The volume of a single unit cell (y_9) should not exceed the cubic cell volume (i.e. $y_6 \cdot y_7 \cdot y_8$). The number of atoms (y_{10}) should be larger than the number of non-hydrogen atoms (y_{11}). The number of metal atoms (y_{12}) must be larger than the number of transition metal atoms (y_{13}). Finally, the total number of bonds linked to metal atoms (y_{16}) cannot exceed six times the number of metal atoms.

$$y_3 \ge y_2 \tag{5.12}$$

$$y_9 \le y_6 \cdot y_7 \cdot y_8 \tag{5.13}$$

$$y_{10} \ge y_{11} \tag{5.14}$$

$$y_{12} \ge y_{13} \tag{5.15}$$

$$6y_{12} \ge y_{16} \tag{5.16}$$

5.2.2.2 P/VSA Process Model

A one-bed and four-step PSA process is considered. A complete process cycle consists of pressurization, adsorption, rinsing, and desorption [46]. For pressurization, the feed gas is compressed to a high pressure and fed into the bottom of the column. The top is closed and the internal pressure increases. At the adsorption stage, the supply of feed gas is continued, where PE is selectively adsorbed and a PA-rich stream leaves from the top. Then, a small fraction of the purified PE stream is re-fed to the column to rinse the PA-rich gas. At the rinsing step, the input gas compositions are presumably constant and equal to the average compositions of the purified PE stream. Finally, a low pressure is given at the bottom and the bed starts a desorption step where the pressure is reduced to a lower level and the released PE is collected at the bottom. In the end, the PE output is stopped and a new cycle starts. For simplification, the following assumptions are made [47, 48]:

- The process is constantly operated at 300 K.
- Adsorbent properties (e.g. particle size and porosity) are uniform throughout the column.
- Both PE and PA follow the ideal gas law and the gas viscosity is assumed to be constant.
- Axial dispersion in the bed is neglected. No variations of pressure and gas concentrations in both gas and solid phases in the radial direction.

- Adsorption kinetics are approximated by linear driving force models. As only the equilibrium separation of PE/PA is studied, lumped mass transfer coefficients are considered and pre-fixed.
- The axial pressure drop is calculated using the Ergun equation.

Based on these assumptions, the mathematical model for the four-step PSA process includes the component mass balances, the total mass balance, and the mass transfer kinetics are given in the GitHub. Six operating conditions are design variables at the process level. They are the durations of the four steps (t_{PR}, t_{AD}, t_{RI}, and t_{DE}) and two operating pressures (P_{low} and P_{high}). Since the energy consumption is a critical factor for assessing the process performance, the objective function is defined to minimize the total energy consumption per ton of the purified PE product:

$$\min\ E = \frac{E_{PR} + E_{\mathrm{AD}} + E_{\mathrm{RI}} + E_{\mathrm{DE}}}{F_{\mathrm{E}}^{out}} \tag{5.17}$$

where E_{PR}, E_{AD}, E_{RI}, and E_{DE} are the energy consumption for pressurization, adsorption, rinsing, and desorption, respectively. $F_{\mathrm{E}}^{\mathrm{out}}$ is the weight of purified PE produced in one process cycle. The equations for calculating these variables are given in the GitHub [49].

Moreover, Eq. (5.18) shows that PE purity is equal to the average PE molar fraction ($y_{\mathrm{E}}^{\mathrm{out}}$) in the outlet stream of desorption. PE recovery is expressed in Eq. (5.19). The bottom and top of the column are defined as $z = 0$ and $z = L$, respectively. L denotes the column length. u is the interstitial velocity, P represents the total pressure, and $y_{\mathrm{E}}^{\mathrm{in}}$ is the molar fraction of PE in the feed gas.

$$\xi_{\mathrm{E}} = y_{\mathrm{E}}^{\mathrm{out}} = \frac{\int_0^{t_{\mathrm{DE}}} (u \cdot P \cdot y_{\mathrm{E}})|_{z=0} \mathrm{d}t}{\int_0^{t_{\mathrm{DE}}} (u \cdot P)|_{z=0} \mathrm{d}t} \tag{5.18}$$

$$\theta_{\mathrm{E}} = \frac{\int_0^{t_{\mathrm{DE}}} (u \cdot P \cdot y_{\mathrm{E}})|_{z=0} \mathrm{d}t - P_{\mathrm{high}} y_{\mathrm{E}}^{\mathrm{out}} \int_0^{t_{\mathrm{RI}}} u|_{z=0} \mathrm{d}t}{P_{\mathrm{high}} y_{\mathrm{E}}^{\mathrm{in}} \left(\int_0^{t_{PR}} u|_{z=0} \mathrm{d}t + \int_0^{t_{\mathrm{AD}}} u|_{z=0} \mathrm{d}t \right)} \tag{5.19}$$

The purity and recovery of PA can be calculated based on the overall input–output mass balances.

5.2.2.3 Integrated Design Formulation

The integrated MOF and PSA process design problem is formulated and solved as an NLP problem. Nineteen MOF descriptors denoting the material variations and six process operating conditions are simultaneously optimized to minimize the energy consumption. Given pre-specified equipment sizes, the energy consumption as well as product purity and recovery are calculated from the PSA process models. PE recovery and purity should be constrained according to industrial requirements.

$$\min_{\boldsymbol{x},\boldsymbol{y}} E = \frac{E_{PR} + E_{\mathrm{AD}} + E_{\mathrm{RI}} + E_{\mathrm{DE}}}{F_{\mathrm{E}}^{\mathrm{out}}} \quad \text{Objective function} \tag{5.20}$$

s.t. Equation (5.9)–(5.16) Descriptor design space

Equation (5.3)–(5.8) Multicomponent DSL isotherm model

Equation (5.18)–(5.19) P/VSA process model

$\theta_E \geq \theta_E^L \quad \xi_E \geq \xi_E^L$ Separation specifications

$\boldsymbol{x} = [t_{PR}, t_{AD}, t_{RI}, t_{DE}, P_{high}, P_{low}]$ Process operating conditions

$\boldsymbol{y} = [y_1, \dots, y_{19}]$ Representative MOF descriptors

This optimization problem cannot be solved directly. A spatial and temporal discretization of PDEs is carried out. The entire column is divided into 20 identical sections and the finite volume method with the upwind difference scheme is applied for spatial discretization. In addition, the backward Euler's scheme is used for temporal discretization. The durations of the pressurization and rinsing steps are discretized into 10 identical time intervals while those of the other two steps are divided into 20 identical time intervals. Clearly, optimizing the durations of the four steps (i.e. t_{PR}, t_{AD}, t_{RI}, and t_{DE}) is equivalent to the optimization of the corresponding time-interval duration (i.e. Δt_{PR}, Δt_{AD}, Δt_{RI}, and Δt_{DE}). After the discretization, the PSA process model is described by a set of algebraic equations that are directly coded in GAMS for NLP optimization.

5.2.3 Results

The separation of an 85 : 15 PE/PA mixture is studied. The feed gas is at 2 atm. The feed gas has been cooled to 300 K, since the isotherm models are fitted at 300 K. Moreover, the purity of PE must be higher than 99%. Additionally, PE recovery is usually less than 30% when 99% PE is produced using PSA processes with no more than five steps [50]. Thus, the minimal PE recovery is set to 30%. The PSA process is modeled on a pilot–plant basis. Table 5.1 lists the input parameters [51].

5.2.3.1 Benchmark Process Using Cu-BTC

The copper-based Cu-BTC has been recognized as one of the suitable MOFs for PE/PA separation due to its large PE adsorption capacity [52]. Here, a benchmark PSA process using Cu-BTC is optimized for comparison. The multicomponent DSL model of Cu-BTC is fitted using the experimental data [53]. The single-component isotherms are plotted in Figure 5.5a. The six process operating conditions are optimized to minimize energy consumption. The optimization problem is solved in GAMS 27.2 using CONOPT4. Different initializations are offered, and the best local optimum found is given in the second column of Table 5.2. The high pressure for adsorption is 1 atm when the initial selectivity $S_{PE/PA}$ is equal to 6.6. The low pressure for desorption is optimized to 0.011 atm and the minimal energy consumption is 505 kWh electricity per ton PE.

5.2.3.2 Optimal MOF and Process from Integrated Design

Under the pre-specified requirements on PE purity and recovery, the integrated design problem in Eq. (5.20) is solved. The solver CONOPT4 and different initial

Table 5.1 Input parameters for the P/VSA-based PE/PA separation.

Molecular weight of PE ($g\,mol^{-1}$)	42.1	Molecular weight of PA ($g\,mol^{-1}$)	44.1
Gas viscosity ($kg\,m^{-1}\,s^{-1}$)	8×10^{-6}	Isentropic coefficient of gas	1.15
Adsorbent particle density ($kg\,m^{-3}$)	703	Adsorbent particle radius (m)	5×10^{-3}
Mass transfer coefficient for PA (1/s)	0.83	Mass transfer coefficient for PE (1/s)	0.61
Bed porosity	0.43	Vacuum pump efficiency	0.75
Bed length (m)	2	Compressor efficiency	0.75
Bed diameter (m)	0.4		
Pressurization duration (s)	1–50	Adsorption duration (s)	1–200
Rinsing duration (s)	1–100	Desorption duration (s)	1–600
High pressure (bar)	1–10	Low pressure (bar)	0.01–1

Source: [5]/with permission of John Wiley & Sons.

values are used. It takes about 150 seconds to find the optimal solution. The optimal MOF refers to a hypothetical MOF (h-MOF). As listed in Table 5.2, the process operating conditions are different from those of the benchmark. The high pressure is 2 atm, which is equal to the initial pressure of the feed gas. The desorption occurs at 1 atm. Under such pressures, no gas compression is needed at pressurization and adsorption steps and no vacuum evacuation is required for desorption. Only a small amount of electricity is used at the rinsing step to re-compress PE to the high pressure. The overall energy consumption is 12.8 kWh of electricity per ton PE. This is much lower than that of the benchmark process.

The large process variations are attributed to the significant differences in adsorption isotherms. As shown in Figure 5.5a, Cu-BTC has a poor PE desorption capability. A tough vacuum condition (i.e. 0.011 atm) is needed to meet the purity and recovery requirements. This leads to a large energy consumption at the rinsing and desorption stages. For the h-MOF, the isotherms can provide a better condition for adsorption and desorption. The PE loading difference between 1 and 2 atm reaches $0.7\,mol\,kg^{-1}$. Such a gap at desirable pressures facilitates PE desorption and recovery.

With the isotherm parameters of the h-MOF, the initial selectivity at the adsorption condition (2 atm and 300 K) is calculated to be 33.9. This is much larger than that of Cu-BTC, due to its much less PA adsorption capacity. Such a large selectivity helps achieve a high PE purity. On the other hand, due to the large PE adsorption capacity of Cu-BTC, the benchmark process has a higher PE production rate than our optimal process, although it needs a longer duration for one process cycle. Based on the above analysis, several adsorbent design guidelines are extracted. First, a large selectivity and small PA adsorption capacity are beneficial for better PE purity and recovery. Second, the PE adsorption capacity has a substantial effect on its production rate. This agrees with the previous literature findings [25].

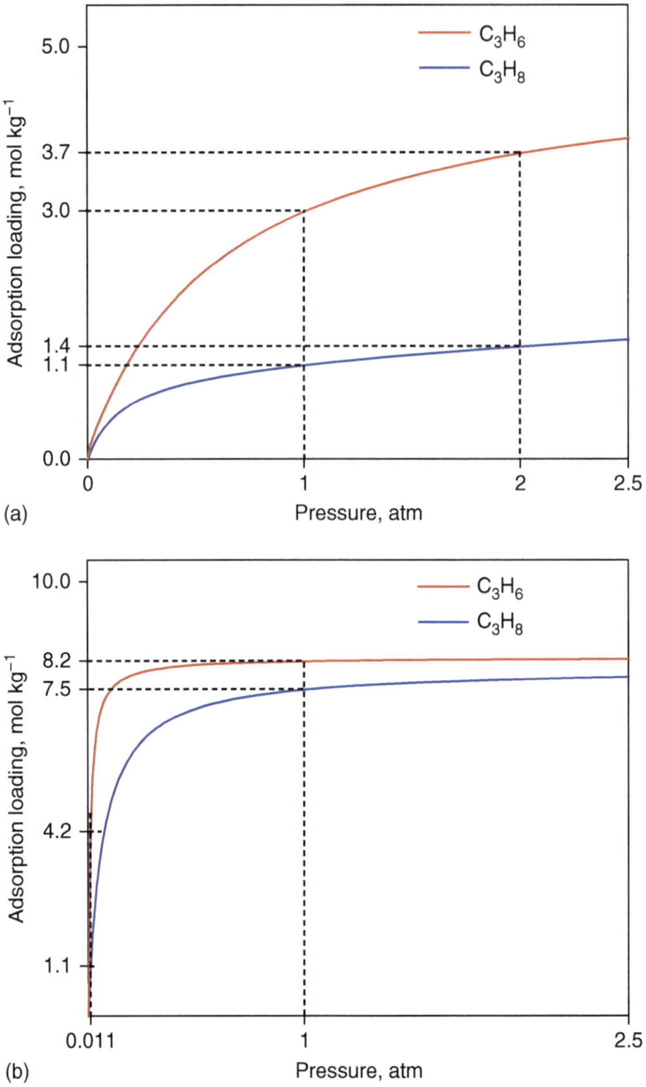

Figure 5.5 Adsorption isotherms at 300 K for (a) Cu-BTC and (b) Optimal h-MOF. Source: [5]/with permission of John Wiley & Sons.

Finally, to enable an energy-efficient separation, the adsorbent should offer a good adsorption–desorption balance that can better match the feed gas conditions.

Table 5.3 lists the optimal MOF descriptor values. Important information can be drawn for the subsequent MOF matching. For the pore geometry, a relatively small cavity diameter and volumetric surface area are preferred. The former improves the selectivity of PE over PA, and the latter may result in a small adsorption capacity of PA, helping increase the selectivity as well. For MOF chemistry, using transition metals is desirable. Transition metals (e.g. copper and zinc) can introduce OMS and

Table 5.2 Optimal results of the Cu-BTC based process and the h-MOF based process.

	Benchmark	Optimal process	Process using *SMOF-1*	Process using *SEYDUW*
PR duration (s)	22.5	13.0	12.2	30
AD duration (s)	71.5	22.1	30.3	10.1
RI duration (s)	77.0	5.6	6.4	5.0
DE duration (s)	600	600	600	600
High pressure (atm)	1	2	1.44	2
Low pressure (atm)	0.011	1	0.3	0.41
PE purity	0.99	0.99	0.99	0.99
PE recovery	0.34	0.30	0.30	0.30
PE production rate (mol s^{-1})	0.16	0.06	203.6	117.8
Initial selectivity at high pressure	6.6	33.9	10.1	7.2
Energy consumption at pressurization (kWh per ton PE)	0	0	0	0
Energy consumption at adsorption (kWh per ton PE)	1.2	0	0	0
Energy consumption at rinsing (kWh per ton PE)	184.3	12.8	24.68	46.15
Energy consumption at desorption (kWh per ton PE)	319.3	0	47.75	45.84
Total energy consumption (kWh per ton PE)	504.8	12.8	72.43	91.99

Source: [5, 29]/with permission of John Wiley & Sons.

π-complexation, which lead to stronger interactions with PE than with PA. Regarding the organic linker, it is better to use the groups consisting of rings and double bonds as well as C–N bonds. Finally, it is not recommended to use organic linkers forming C–O pairs at topological distance 2 (i.e. with C–X–O chains), such as benzoic acid and furan.

5.3 Step Two: MOF Matching

5.3.1 Material–Property Relationship of MOFs

5.3.1.1 Property–Performance Relationship for PE/PA Separation

Referring to the method from Iyer and Hasan [28], property–performance relationships of adsorbents are constructed by sampling various isotherms followed by process optimizations. By doing so, the obtained results can reveal the inherent correlations between adsorption equilibrium and process feasibility. To be consistent with

Table 5.3 Optimal MOF descriptors.

Descriptor	Value	Descriptor	Value
Bulk density ($g\,cm^{-3}$)	1.9	Number of atoms	230
Pore-limiting diameter (Å)	5.9	Number of metal atoms	46
Largest cavity diameter (Å)	7.0	Number of transition metal atoms	46
Volumetric surface area ($m^2\,cm^{-3}$)	279.4	Number of double bonds	14
Void fraction	0.82	Number of ring bonds	332
Unit cell length a (Å)	7.2	Number of bonds to metal atoms	18
Unit cell length b (Å)	11.0	Number of alkyl groups	5
Unit cell length c (Å)	7.5	Frequency of C–N at distance 1	0.45
Volume per unit cell (Å^3)	589.1	Frequency of C–O at distance 2	0
Number of non-H atoms	230		

Source: [5]/with permission of John Wiley & Sons.

step one, the binary-component DSL isotherm model in Eqs. (5.21) and (5.22) is used to represent adsorption equilibria, respectively.

$$q_E^{eq} = \frac{Q_E^1 \cdot b_E^1 \cdot P \cdot y_E}{1 + b_E^1 \cdot P \cdot y_E + b_A^1 \cdot P \cdot y_A} + \frac{Q_E^2 \cdot b_E^2 \cdot P \cdot y_E}{1 + b_E^2 \cdot P \cdot y_E + b_A^2 \cdot P \cdot y_A} \tag{5.21}$$

$$q_A^{eq} = \frac{Q_A^1 \cdot b_A^1 \cdot P \cdot y_A}{1 + b_E^1 \cdot P \cdot y_E + b_A^1 \cdot P \cdot y_A} + \frac{Q_A^2 \cdot b_A^2 \cdot P \cdot y_A}{1 + b_E^2 \cdot P \cdot y_E + b_A^2 \cdot P \cdot y_A} \tag{5.22}$$

where the subscripts E and A denote PE and PA, respectively. q^{eq} is the adsorption loading at pressure P. y is the molar fraction in the gas phase. Q and b are DSL model parameters. Clearly, the thermodynamic properties of any MOF can be characterized by the eight model parameters. From the process perspective, the eight parameters determine the adsorption process performance. Different combinations of the eight parameters were sampled for PSA optimization. All the model parameters refer to 300 K. The Latin hypercube sampling approach was used to generate samples randomly and uniformly. Additional constraints were added to ensure that the samples were distributed around the optimal isotherms obtained in the first step. The detailed constraints are given below.

- In the optimal isotherms, the PE and PA adsorption loadings at 2 atm are 3.7 and 1.4 $mol\,kg^{-1}$, respectively. Thus, the saturated adsorption capacities Q_E^1 and Q_E^2 were sampled within 0.01–6 $mol\,kg^{-1}$, while Q_A^1 and Q_A^2 vary within 0.01–4 $mol\,kg^{-1}$. The affinity parameters b_E^1, b_E^2, b_A^1, and b_A^2 were allowed to vary between 0.01 and 20 (1/bar). In addition, the samples whose PE and PA adsorption loadings at 2 atm exceed 5 and 2.5 $mol\,kg^{-1}$, respectively, were discarded.
- Since only PE-selective MOFs are studied, the adsorption capacity on PE should be larger than that on PA when pressure exceeds 0.1 atm.

- Considering the separation specifications (i.e. 99% PE purity with no less than 30% recovery), the isotherm samples were discarded when the difference of the associated PE and PA adsorption loadings at 2 atm is less than 0.5 mol kg^{-1}.

Considering the computational burdens of complex PSA process optimization, 400 samples were generated. The PSA process configuration and models used are the same as those above.

$$
\begin{aligned}
\min\ & E\left(P_{\mathrm{AD}}, P_{\mathrm{DE}}, t_{\mathrm{PR}}, t_{\mathrm{AD}}, t_{\mathrm{RI}}, t_{\mathrm{DE}}\right) \\
\text{s.t.}\ & \frac{\partial y_i}{\partial t} = f_1\left(y_i, u, P, q_{\mathrm{E}}, q_{\mathrm{A}}\right) \quad \text{gas phase component mass balances} \\
& \frac{\partial P}{\partial t} = f_2\left(P, u, q_{\mathrm{E}}, q_{\mathrm{A}}\right) \quad \text{total mass balance} \\
& \frac{\partial q_i}{\partial t} = f_3\left(q_i^{\mathrm{eq}}, q_i\right) \quad \text{solid-phase balances and mass transfer kinetics} \\
& f_4\left(q_i^{\mathrm{eq}}, P, y_i\right) = 0 \quad \text{DSL isotherm models} \\
& f_5\left(P, u, y_{\mathrm{A}}, y_{\mathrm{E}}\right) = 0 \quad \text{pressure drop} \\
& f_6\left(y_i, q_i, P, u\right) = 0 \quad \text{cyclic steady state and boundary conditions} \\
& f_7\left(\xi_{\mathrm{E}}, t_{\mathrm{DE}}, u, P, y_{\mathrm{E}}\right) = 0 \quad \text{purity specification} \\
& f_8\left(\theta_{\mathrm{E}}, t_{\mathrm{PR}}, t_{\mathrm{AD}}, t_{\mathrm{RI}}, t_{\mathrm{DE}}, u, P, y_{\mathrm{E}}\right) = 0 \quad \text{recovery specification}
\end{aligned} \tag{5.23}
$$

400 NLP optimization problems were solved. Not all the isotherms can satisfy the separation specifications. After analyzing optimization results, two feasibility maps were obtained and shown in Figure 5.6. Figure 5.6a shows the feasible regions with respect to the single-component adsorption loadings at 2 atm. Within the sampled space, when the PA adsorption loading at 2 atm (q_{A}^2) exceeds 2 mol kg^{-1}, the isotherm samples cannot achieve a feasible separation. Therefore, as written in Eq. (5.4), q_{A}^2 is expected to be less than 2 mol kg^{-1} for feasible PE/PA separation.

$$q_{\mathrm{A}}^2 \leq 2 \tag{5.24}$$

Figure 5.6b shows the feasible map with respect to adsorption uptakes at 0.01 atm ($q^{0.01}$) that are Henry's constants. The samples can be roughly divided into feasible and infeasible regions. After a linear classification, a linear classifier model (see black dot line) was obtained to separate the whole region into two. With the extracted slope and intercept, the constraint in Eq. (5.25) must be satisfied to ensure that a sample is located in the feasible region.

$$q_{\mathrm{E}}^{0.01} \geq 3.185 \times q_{\mathrm{A}}^{0.01} - 0.027 \tag{5.25}$$

Clearly, the two constraints represent the quantitative correlations between adsorption properties and process feasibility. The single-component PE and PA adsorption loadings at 300 K can be directly predicted with the 19 MOF descriptors and pressure using the established ANN models.

5.3.1.2 Validation with 471 CoRE MOFs

To validate the reliability of the two constraints, the process feasibility of the 471 CoRE MOFs was investigated. The single-component isotherm models of the 471

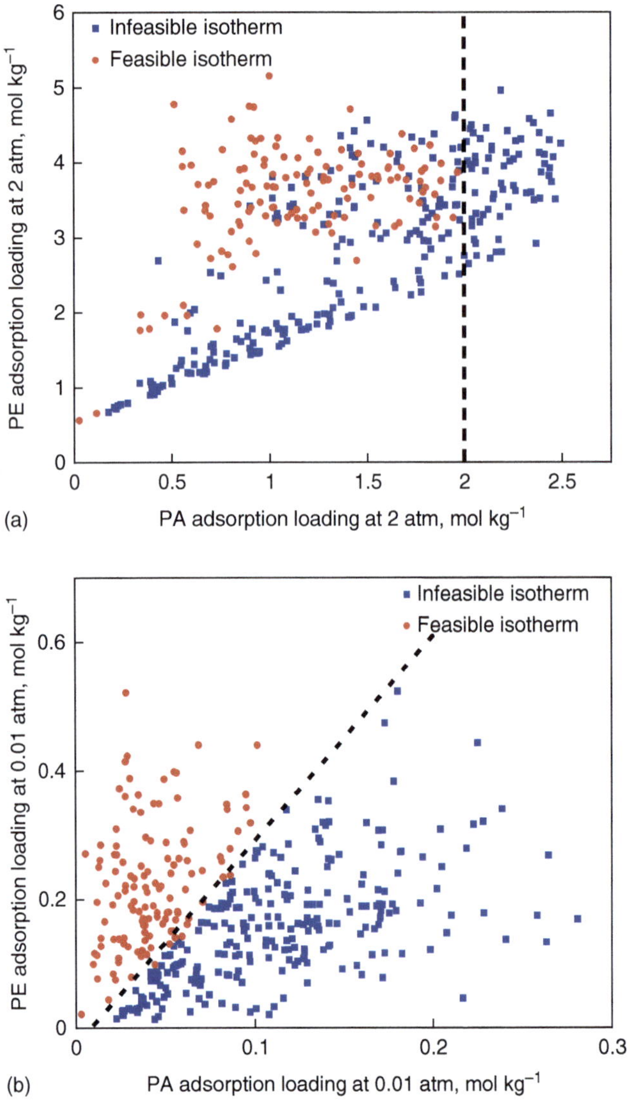

Figure 5.6 Feasibility maps based on (a) adsorption loadings at 2 atm and (b) adsorption loadings at 0.01 atm for deriving property-performance relationships. Source: [29]/with permission of John Wiley & Sons.

CoRE MOFs were fitted via collected adsorption loadings. The fitting was performed using the *lmfit* package with the Levenberg–Marquardt algorithm. The resulting models were used to predict the PE and PA adsorption uptakes at 0.01 and 2 atm. Based on the predicted loadings, the 471 CoRE MOFs were verified against the two feasibility constraints. 15 CoRE MOFs meet the 2 constraints.

Moreover, PSA process optimization was performed for the 471 CoRE MOFs, which gives the true process performance. The multicomponent DSL isotherm models in Eq. (5.2) were derived and imported into the PSA process model for optimization. Finally, 9 CoRE MOFs can meet the required separation specifications. Eight out of nine CoRE MOFs meet the two feasibility constraints except for *FIQCEN* (i.e. Cu-BTC), since its PA adsorption uptake at 2 atm is 7.76 mol kg^{-1}, which breaks the first feasibility constraint. In summary, the results demonstrate that the two constraints can filter out MOFs that are infeasible to achieve the pre-specified separation requirements. The best of the 471 CoRE MOFs is *SEYDUW*, which refers to (μ3-*N*-(phosphonomethyl)iminodiacetato)-diaqua-yttrium-monohydrate [54]. The corresponding minimum energy consumption is 91.99 kWh per ton PE. This can act as a new benchmark for the subsequent computational MOF design.

5.3.2 From Computational MOF Design to Model-Based MOF Screening

As the core of the second step, new MOFs were computationally designed for PE/PA separation. Proper MOF building blocks were first identified. Then, hypothetical MOFs were synthesized and *in silico* screened to identify top candidates leading to better PE/PA separation performance.

5.3.2.1 Identification of MOF Building Blocks

For MOFs, at least one type of metal node must exist, and some topologies need to be constructed with more than one type of organic node. Different types of metal/organic nodes have different coordination numbers that determine the number of organic linkers connected to them. Due to the complex MOFs structures, there are multiple ways to dissect MOFs into different building blocks [55]. Figure 5.7 shows a general procedure for extracting MOF building blocks for *in silico* MOF synthesis. For an MOF in the form of *cif* file, the open-source program *ToposPro* can be utilized to analyze its topology, and its metal/organic nodes can be identified as well. Afterward, MOFs can be viewed in material studio to identify organic linkers. Meanwhile, the nodes and linkers can be manually extracted after deleting all other atoms in the MOF using material studio. The extracted nodes and linkers can be used as building blocks for computational MOF synthesis. Following the above procedures, the topology, nodes, and organic linkers of the 471 CoRE MOFs were analyzed, identified, and extracted. The detailed structural decompositions are summarized in the GitHub folder. Notably, the following heuristics were considered to discard unusable building blocks.

- The Reticular Chemistry Structure Resource database collected over 2300 MOF topologies. Each topology is named as a three-letter symbol with or without an extension (i.e. *pqr*-a and *pqr*). Here, only the extracted topologies included in the RCSR database were considered, since only those topologies are available in the state-of-the-art MOF synthesis tool [55].
- Two-dimensional topologies were discarded due to their scarcity.
- The organic linkers and nodes that establish more than two connections or involve double/triple bonds with a single metal/organic node were eliminated. This is

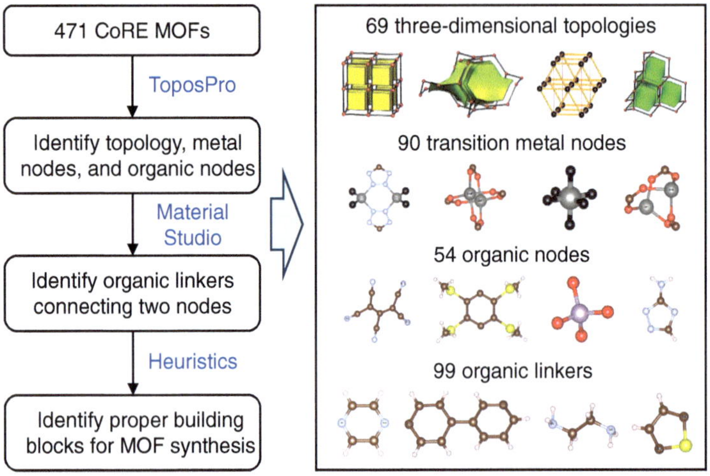

Figure 5.7 Generation of MOF building blocks from the 471 CoRE MOFs. Source: [29]/with permission of John Wiley & Sons.

because the current MOF synthesis tool only permits single-bond connections between linkers and nodes, as well as between nodes themselves [30].

- The organic nodes with one atom connecting to two different metal nodes were discarded since such nodes usually lead to infeasible MOFs.
- Asymmetric metal nodes with over 25 atoms were discarded since those metal nodes rarely exist in MOFs.

Regarding the PE/PA separation, the design guidelines from the first step were taken into account.

- Kept the metal nodes with transition metals. Transition metals are expected to introduce OMS and π-complexation, which can result in stronger interactions with PE over PA.
- Remove organic nodes and linkers with C–O pairs at topological distance 2 (C–X–O chains).
- The organic nodes and linkers consisting of rings, double bonds, or C—N bonds were retained.

In total, 69 topologies, 90 metal nodes, 54 organic nodes, and 99 organic linkers were obtained from the 471 CoRE MOFs. The corresponding "*cif*" files have been uploaded to the GitHub folder.

5.3.2.2 *In Silico* Synthesis of Hypothetical MOFs

The abundance of building blocks obtained above offers an extensive design space. Due to the limited computational resources, only the 27 topologies requiring one type of node (i.e. metal node) were adopted. In other words, none of the organic nodes were considered for MOF synthesis. Table 5.4 lists the three-letter symbols of the 27 topologies and the coordination of the required metal node (i.e. the number of free connections). Given this simplification, Figure 5.8 shows the workflow

Table 5.4 27 topologies for computational MOF design in the PE/PA separation example.

Topology	Coordination of metal node	Topology	Coordination of metal node
srs	3	uni	4
ths	3	uog	4
atn	4	usf	4
cag	4	bnn	5
cds	4	sqp	5
dft	4	acs	6
dia	4	lcy	6
dmp	4	pcu	6
lon	4	rob	6
lvt	4	sol	6
qtz	4	sxb	6
sod	4	bcu	8
sra	4	hex	8
unc	4		

Source: [29]/with permission of John Wiley & Sons.

for the computational MOF synthesis and screening. First, using the finalized building blocks, hypothetical MOFs were generated and their descriptors were calculated. Then, model-based screening was performed to identify promising MOFs. Their feasibilities were further assessed using rigorous GCMC simulations. Finally, PSA process optimization was performed to identify the real optimal MOFs. The detailed procedures and models are elaborated below.

The *ToBaCCo 3* was employed for *in silico* MOF synthesis. Hypothetical MOFs were constructed on the basis of topological templates. Each template consists of information on the number, coordination, and symmetry of the metal nodes. Node symmetry represents the spatial arrangement of connected organic linkers. During construction, the coordination compatibility between topology and metal node is tested. A topology is incompatible with a metal node when its node coordination is different from that of the metal node. Based on MOF chemistry, some other factors (node symmetry, bond connectivity, etc.) are checked for sound feasibility. The detailed algorithms in *ToBaCCo* can be found in previous publications [55, 56]. Here, the 90 metal nodes and 99 organic linkers in the form of *cif* files were imported into the *nodes* and *edges* folders in *ToBaCCo*, respectively. The 27 topological templates are found in the *ToBaCCo* template database. With these, the differential evolution algorithm in *ToBaCCo* was used for structural scaling. As a result, 45 472 hypothetical MOFs were generated and stored in the form of *cif* files.

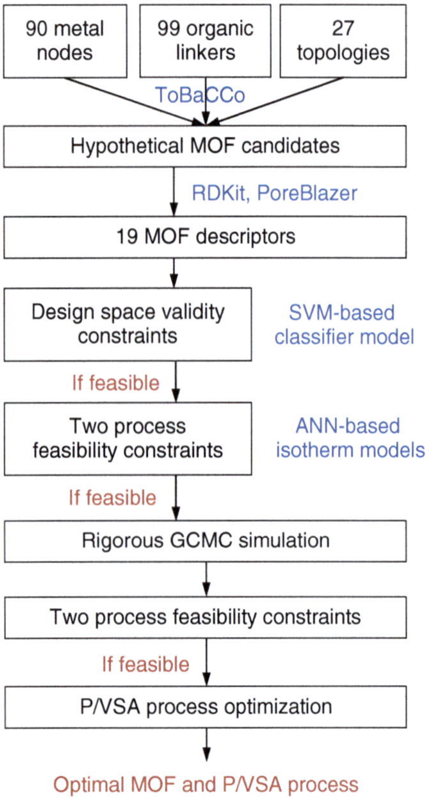

Figure 5.8 Workflow of the computational MOF synthesis and screening. Source: [29]/with permission of John Wiley & Sons.

Given the 45 472 hypothetical MOFs, the candidates leading to the best process performance should be identified. For this, calculating adsorption isotherms is a prerequisite. It is computationally demanding to perform rigorous GCMC simulations for all the candidates. Alternatively, the ANN-based isotherm models obtained in the first step can be applied for a model-based screening where only the 19 descriptors need to be calculated in advance. The open-source toolbox *OpenBabel* was used to convert the original MOF *cif* files into molfiles (*.mol*) and XYZfiles (*.xyz*) [57]. The molfiles were analyzed using the cheminformatics package *RDKit* to get detailed information on atoms, bonds, and groups. Accordingly, the 10 chemical descriptors can be calculated based on their definitions given in the first step. Additionally, the XYZfiles and the *PoreBlazer 4* program were applied for computing geometric descriptors [58]. Depending on the topology, the computed geometric descriptors were scaled to a primitive cell level. On average, it takes around five minutes for a single CPU node to compute the 19 descriptors for a single MOF. The descriptors of the 45 472 hypothetical MOFs have been uploaded to the GitHub folder.

5.3.2.3 MOF Screening via Validity and Feasibility Constraints

Model-based MOF screening was performed using the material–property–performance relationships. First, the values of the 19 descriptors were sent to verify whether they were in the validity domain of the ANN-based models. Similar to the first step, the constraints on the lower and upper bounds (Eq. (5.11)) and the one-class SVM classifier model (Eqs. (5.9) and (5.10)) are considered here. After applying the validity domain constraints, 4269 MOF candidates were retained. For these MOFs, their PE and PA adsorption uptakes at 0.01 and 2 atm were predicted using the ANN models in Eqs. (5.1) and (5.2). With the predicted uptakes, the process feasibility constraints in Eqs. (5.24) and (5.25) were tested for all the 4269 MOFs. Only 364 MOFs are left and considered feasible for further studies. These 364 MOFs and their corresponding descriptors are tabulated in GitHub.

5.3.2.4 MOF Screening via GCMC Simulations

The single-component PE and PA adsorption loadings at two pressure levels (0.01 and 2 atm) and 300 K were predicted with rigorous GCMC simulations for the remaining 364 MOFs. The open-source software *RASPA* was used for computation [59]. All MOFs were rigid, while PA and PE were flexible including intramolecular interactions for bonds, angles, and torsions. 10^5 Monte Carlo cycles were simulated including 5×10^4 initialization and 5×10^4 equilibrium cycles. Each cycle includes four possible motions (translation, rotation, reinsertion, and swap) with equal probability. The fugacity coefficient was 1. The Ewald summation method was applied to calculate electrostatic interactions with a relative precision of 10^{-6}. The interaction energies were computed through the Lennard–Jones (LJ) potentials. The LJ parameters for adsorbent atoms were taken from the universal force field. The adsorbates were combinations of CH_3, CH_2, and CH. Their LJ parameters were from the TraPPE-UA force field. The Lorentz–Berthelot mixing rule was chosen and the cutoff distance was defined as 12 Å.

The GCMC results are given in the GitHub folder. Only six MOFs meet the two feasibility constraints in Eqs. (5.24) and (5.25). These six MOFs were sent to PSA process optimization for quantifying their process performance. To do this, GCMC simulations were carried out to calculate the adsorption loadings of PE and PA at five different pressure levels (i.e. 0.01, 0.1, 1, 2, and 5 atm) and 300 K. The results were used to fit the DSL isotherm models in Eq. (5.2) for process optimization. Again, for the six MOFs, their complete GCMC results are provided in the GitHub folder.

5.3.2.5 MOF Screening via PSA Process Optimization

After fitting the DSL parameters for the six MOFs, six isothermal PSA optimizations were performed. Three of the six candidates meet the 99% purity and 30% recovery requirements. The three computationally synthesized MOFs are named as *SMOF-1*, *SMOF-2*, and *SMOF-3*. Figure 5.9 shows *SMOF-1* and the corresponding projection to the *X*–*Y* plane. It consists of a 6-coordinated iron (Fe) metal node and pyrazine ($C_4H_4N_2$) organic linker assembled in the *sxb* topology. The metal node and organic linker are extracted from CoRE MOF *LEMNOH* and *FUDQIF*, respectively.

5.3.2.6 Optimal Results of SMOF-1

For *SMOF-1*, the optimal PSA process operating conditions are listed in the fourth column of Table 5.2. The adsorption is performed at 1.44 atm, which is lower than the

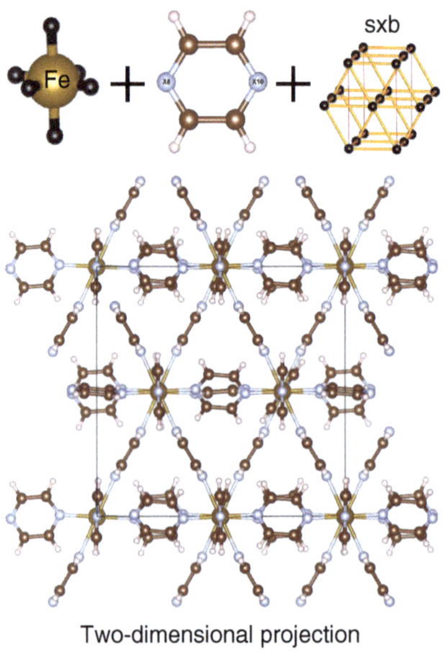

Figure 5.9 Computationally synthesized SMOF-1 (visualized by VESTA software). Source: [29]/with permission of John Wiley & Sons.

initial pressure 2 atm of the feed gas but higher than the atmospheric pressure. Thus, no gas compression is needed at the pressurization and adsorption steps. Desorption occurs in a vacuum condition of 0.3 atm. Energy is consumed for gas evacuation during desorption and for gas compression at the rinsing step. Comparing the results of *SMOF-1* with *SEYDUW*, it is found that their operating pressures are very different. For *SEYDUW*, the high and low pressures are 2 and 0.41 atm, respectively. The energy consumed at the rinsing step is 46.15 kWh per ton PE, which is much larger than that of *SMOF-1* (24.68 kWh per ton PE). This is because *SMOF-1* has a smaller gap between the high and low pressures and hence less energy is needed to re-compress the gases during rinsing.

In fact, the root reasons for different results can be explained by their different adsorption isotherms. As shown in Figure 5.10, *SMOF-1* exhibits a larger gap of PE and PA isotherms than *SEYDUW*. The initial selectivity of *SMOF-1* at the adsorption condition is calculated to be 10.1, larger than that of *SEYDUW*. This contributes to the production of high-purity PE products. Furthermore, SMOF-1 offers improved adsorption and desorption conditions. The PE loading difference between 1.44 and 0.3 atm already reaches 0.58 mol kg^{-1}, facilitating the attainment of the desired 30% PE recovery. By contrast, for *SEYDUW*, a large pressure drop between adsorption and desorption (i.e. from 2 to 0.41 atm) is needed to ensure that enough PE is adsorbed and recovered. This directly leads to a higher energy consumption at the rinsing step. The third column of Table 5.2 lists the optimal process target obtained from the first step. Clearly, the process performance led by *SMOF-1* is worse than that of the optimal process target. In other words, there are still substantial spaces for finding new MOFs whose isotherms can be closer to the optimal isotherm target.

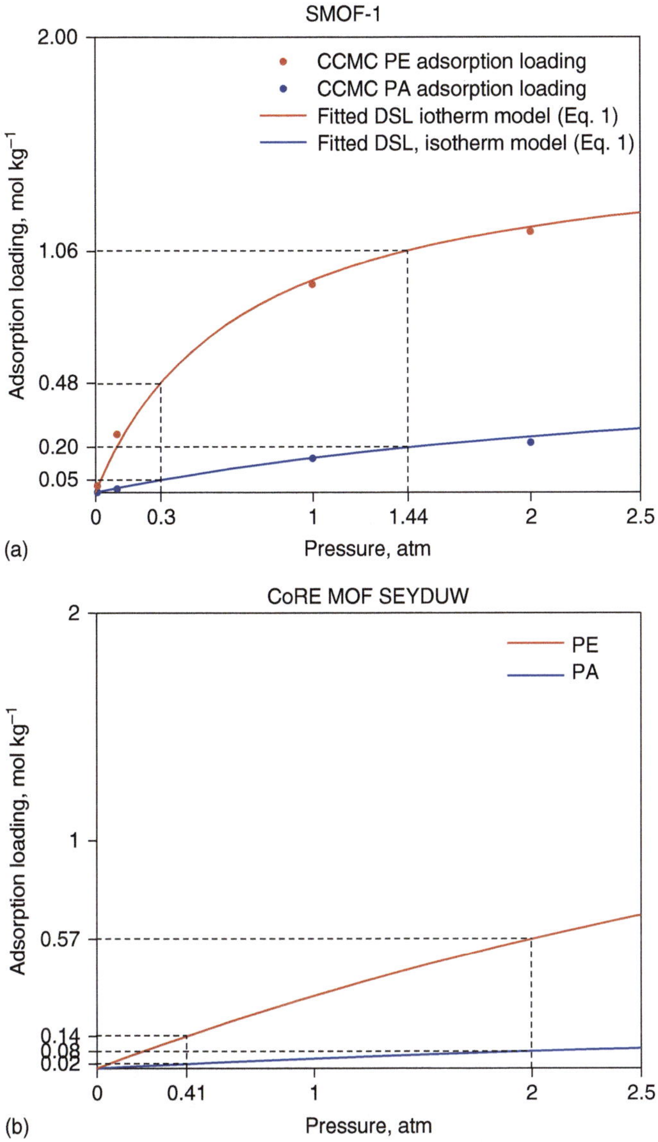

Figure 5.10 Single-component adsorption isotherms for (a) *SMOF-1*, (b) *SEYDUW*. Source: [29]/with permission of John Wiley & Sons.

5.4 Conclusion

This chapter presents a new two-step framework for integrated MOF and PSA process design. In the first step, MOFs are denoted as a set of selected descriptors serving as design variables. Data-driven models are developed to describe the valid descriptor design space and to predict the adsorption isotherms. Combining the validity model, isotherm model, and PSA process model, the integrated design problem is formulated as an NLP problem where MOF descriptors and process operating conditions are simultaneously optimized. In the second step, multiple computational tools and in-house programs were used for *in silico* MOF synthesis and screening. Based on the optimal isotherm targets, the explicit relationships between adsorption properties and process feasibility were identified to develop model-based constraints for fast preliminary MOF screening. MOF building blocks were generated from 471 CoRE MOFs to create new hypothetical MOFs. The created MOF candidates were screened using the model-based constraints. In addition, GCMC simulations and PSA optimization were performed to further screen the MOFs. The best MOF candidate was identified, which shows a much better PE/PA separation performance than the benchmark CoRE MOF *SEYDUW*. The largest novelty is the use of multi-scale modeling approach to integrate the variations of MOF chemistry and structure into PSA process design. This provides a reliable and efficient way for computational adsorbent design to maximize the practical adsorption process performance.

References

1 Yao, Z., Sánchez-Lengeling, B., Bobbitt, N.S. et al. (2021). Inverse design of nanoporous crystalline reticular materials with deep generative models. *Nature Machine Intelligence* 3: 76–86.

2 Wang, T., Lin, E., Peng, Y.L. et al. (2020). Rational design and synthesis of ultramicroporous metal-organic frameworks for gas separation. *Coordination Chemistry Reviews* 423: 213485.

3 Bao, Z., Chang, G., Xing, H. et al. (2016). Potential of microporous metal–organic frameworks for separation of hydrocarbon mixtures. *Energy & Environmental Science* 9: 3612–3641.

4 Burns, T.D., Pai, K.N., Subraveti, S.G. et al. (2020). Prediction of MOF performance in vacuum swing adsorption systems for postcombustion CO_2 capture based on integrated molecular simulations, process optimizations, and machine learning models. *Environmental Science & Technology* 54: 4536–4544.

5 Zhang, X., Zhou, T., and Sundmacher, K. (2021). Integrated metal–organic framework and pressure/vacuum swing adsorption process design: descriptor optimization. *AIChE Journal* 68: e17524.

6 Asgari, M., Streb, A., van der Spek, M. et al. (2021). Synergistic material and process development: application of a metal-organic framework, Cu-TDPAT, in single-cycle hydrogen purification and CO_2 capture from synthesis gas. *Chemical Engineering Journal* 414: 128778.

7 Pullumbi, P., Brandani, F., and Brandani, S. (2019). Gas separation by adsorption: technological drivers and opportunities for improvement. *Current Opinion in Chemical Engineering* 24: 131–142.

8 Daglar, H. and Keskin, S. (2020). Recent advances, opportunities, and challenges in high-throughput computational screening of MOFs for gas separations. *Coordination Chemistry Reviews* 422: 213470.

9 Wilmer, C.E., Leaf, M., Lee, C.Y. et al. (2012). Large-scale screening of hypothetical metal–organic frameworks. *Nature Chemistry* 4: 83–89.

10 First, E.L., Hasan, M.M.F., and Floudas, C.A. (2014). Discovery of novel zeolites for natural gas purification through combined material screening and process optimization. *AIChE Journal* 60: 1767–1785.

11 Yeo, B.C., Kim, D., Kim, H., and Han, S.S. (2016). High-throughput screening to investigate the relationship between the selectivity and working capacity of porous materials for propylene/propane adsorptive separation. *Journal of Physical Chemistry C* 120: 24224–24230.

12 Li, Z., Zhang, Y., Liu, B. et al. (2021). Multilevel screening of computation-ready, experimental metal-organic frameworks for natural gas purification. *AIChE Journal* 67: e17279.

13 Yan, Y., Zhang, L., Li, S. et al. (2021). Adsorption behavior of metal-organic frameworks: from single simulation, high-throughput computational screening to machine learning. *Computational Materials Science* 193: 110383.

14 Rajagopalan, A.K., Avila, A.M., and Rajendran, A. (2016). Do adsorbent screening metrics predict process performance? A process optimisation based study for post-combustion capture of CO_2. *International Journal of Greenhouse Gas Control* 46: 76–85.

15 Khurana, M. and Farooq, S. (2016). Adsorbent screening for postcombustion CO_2 capture: a method relating equilibrium isotherm characteristics to an optimum vacuum swing adsorption process performance. *Industrial and Engineering Chemistry Research* 55: 2447–2460.

16 Ga, S., Jang, H., and Lee, J.H. (2017). New performance indicators for adsorbent evaluation derived from a reduced order model of an idealized PSA process for CO_2 capture. *Computers and Chemical Engineering* 102: 188–212.

17 Arora, A., Iyer, S.S., and Hasan, M.M.F. (2020). Computational material screening using artificial neural networks for adsorption gas separation. *Journal of Physical Chemistry C* 124: 21446–21460.

18 Pai, K.N., Prasad, V., and Rajendran, A. (2020). Generalized, adsorbent-agnostic, artificial neural network framework for rapid simulation, optimization, and adsorbent screening of adsorption processes. *Industrial and Engineering Chemistry Research* 59: 16730–16740.

19 Sen, T., Kawajiri, Y., and Realff, M.J. (2021). Integration of material and process design for kinetic adsorption separation. *Industrial and Engineering Chemistry Research* 60: 2536–2546.

20 Adjiman, C.S., Sahinidis, N.V., Vlachos, D.G. et al. (2021). Process systems engineering perspective on the design of materials and molecules. *Industrial and Engineering Chemistry Research* 60: 5194–5206.

21 Zhou, T., McBride, K., Zhang, X. et al. (2015). Integrated solvent and process design exemplified for a Diels–Alder reaction. *AIChE Journal* 61: 147–158.

22 Burger, J., Papaioannou, V., Gopinath, S. et al. (2015). A hierarchical method to integrated solvent and process design of physical CO_2 absorption using the SAFT-γ Mie approach. *AIChE Journal* 61: 3249–3269.

23 Bardow, A., Steur, K., and Gross, J. (2010). Continuous-molecular targeting for integrated solvent and process design. *Industrial and Engineering Chemistry Research* 49: 2834–2840.

24 Zhang, L., Pang, J., Zhuang, Y. et al. (2020). Integrated solvent-process design methodology based on COSMO-SAC and quantum mechanics for TMQ (2,2,4-trimethyl-1,2-*H*-dihydroquinoline) production. *Chemical Engineering Science* 226: 115894.

25 Khurana, M. and Farooq, S. (2017). Integrated adsorbent-process optimization for carbon capture and concentration using vacuum swing adsorption cycles. *AIChE Journal* 63: 2987–2995.

26 Khurana, M. and Farooq, S. (2019). Integrated adsorbent process optimization for minimum cost of electricity including carbon capture by a VSA process. *AIChE Journal* 65: 184–195.

27 Boyd, P.G., Lee, Y., and Smit, B. (2017). Computational development of the nanoporous materials genome. *Nature Reviews Materials* 2: 1–15.

28 Iyer, S.S. and Hasan, M.M.F. (2019). Mapping the material-property space for feasible process operation: application to combined natural-gas separation and storage. *Industrial and Engineering Chemistry Research* 58: 10455–10465.

29 Zhang, X., Zhou, T., and Sundmacher, K. (2022). Integrated metal-organic framework (MOF) and pressure/vacuum swing adsorption process design: MOF matching. *AIChE Journal* 68: e17788.

30 Jorge, M., Fischer, M., Gomes, J.R. et al. (2014). Accurate model for predicting adsorption of olefins and paraffins on MOFs with open metal sites. *Industrial and Engineering Chemistry Research* 53: 15475–15487.

31 Zhang, Y., Li, B., Krishna, R. et al. (2015). Highly selective adsorption of ethylene over ethane in a MOF featuring the combination of open metal site and π-complexation. *Chemical Communications* 51: 2714–2717.

32 Chong, S., Lee, S., Kim, B., and Kim, J. (2020). Applications of machine learning in metal-organic frameworks. *Coordination Chemistry Reviews* 423: 213487.

33 Krishnapriyan, A.S., Montoya, J., Haranczyk, M., Hummelshøj, J. and Morozov, D. (2020). Machine learning with persistent homology and chemical word embeddings improves prediction accuracy and interpretability in metal-organic frameworks. *arXiv preprint arXiv:2010.00532.*

34 Shevchenko, A.P., Alexandrov, E.V., Golov, A.A. et al. (2020). Topology versus porosity: what can reticular chemistry tell us about free space in metal–organic frameworks? *Chemical Communications* 56: 9616–9619.

35 Willems, T.F., Rycroft, C.H., Kazi, M. et al. (2012). Algorithms and tools for high-throughput geometry-based analysis of crystalline porous materials. *Microporous and Mesoporous Materials* 149: 134–141.

36 Gharagheizi, F., Tang, D., and Sholl, D.S. (2020). Selecting adsorbents to separate diverse near-azeotropic chemicals. *Journal of Physical Chemistry C* 124: 3664–3670.

37 Chung, Y.G., Haldoupis, E., Bucior, B.J. et al. (2019). Advances, updates, and analytics for the computation-ready, experimental metal–organic framework database: CoRE MOF 2019. *Journal of Chemical & Engineering Data* 64: 5985–5998.

38 Moosavi, S.M., Nandy, A., Jablonka, K.M. et al. (2020). Understanding the diversity of the metal-organic framework ecosystem. *Nature Communications* 11: 4068.

39 RDKit: Open-source cheminformatic. http://www.rdkit.org (accessed May 2021).

40 Tang, D., Wu, Y., Verploegh, R.J., and Sholl, D.S. (2018). Efficiently exploring adsorption space to identify privileged adsorbents for chemical separations of a diverse set of molecules. *ChemSusChem* 11: 1567–1575.

41 Lee, J.H., Shin, J., and Realff, M.J. (2018). Machine learning: overview of the recent progresses and implications for the process systems engineering field. *Computers & Chemical Engineering* 114: 111–121.

42 Jablonka, K.M., Ongari, D., Moosavi, S.M., and Smit, B. (2020). Big-data science in porous materials: Materials genomics and machine learning. *Chemical Reviews* 120: 8066–8129.

43 Ritter, J.A., Bhadra, S.J., and Ebner, A.D. (2011). On the use of the dual-process Langmuir model for correlating unary equilibria and predicting mixed-gas adsorption equilibria. *Langmuir* 27: 4700–4712.

44 Kahrs, O. and Marquardt, W. (2007). The validity domain of hybrid models and its application in process optimization. *Chemical Engineering and Processing* 46: 1054–1066.

45 Schweidtmann, A.M., Weber, J.M., Wende, C. et al. (2021). Obey validity limits of data-driven models through topological data analysis and one-class classification. *Optimization and Engineering* 23: 855–876.

46 Khalighi, M., Karimi, I.A., and Farooq, S. (2014). Comparing SiCHA and 4A zeolite for propylene/propane separation using a surrogate-based simulation/optimization approach. *Industrial and Engineering Chemistry Research* 53: 16973–16983.

47 Park, J.-H., Kim, J.-N., and Cho, S.-H. (2000). Performance analysis of four-bed H_2 PSA process using layered beds. *AIChE Journal* 46: 790–802.

48 Agarwal, A., Biegler, L.T., and Zitney, S.E. (2009). Simulation and optimization of pressure swing adsorption systems using reduced-order modeling. *Industrial and Engineering Chemistry Research* 48: 2327–2343.

49 Leperi, K.T., Snurr, R.Q., and You, F. (2016). Optimization of two-stage pressure/vacuum swing adsorption with variable dehydration level for postcombustion carbon capture. *Industrial and Engineering Chemistry Research* 55: 3338–3350.

50 Grande, C.A. and Rodrigues, A.E. (2005). Propane/propylene separation by pressure swing adsorption using zeolite 4A. *Industrial and Engineering Chemistry Research* 44: 8815–8829.

51 Wehring, M., Gascon, J., Dubbeldam, D. et al. (2010). Self-diffusion studies in CuBTC by PFG NMR and MD simulations. *Journal of Physical Chemistry C* 114: 10527–10534.

52 Ferreira, A.F.P., Santos, J.C., Plaza, M.G. et al. (2011). Suitability of Cu-BTC extrudates for propane–propylene separation by adsorption processes. *Chemical Engineering Journal* 167: 1–12.

53 Lamia, N., Jorge, M., Granato, M.A. et al. (2009). Adsorption of propane, propylene and isobutane on a metal–organic framework: molecular simulation and experiment. *Chemical Engineering Science* 64: 3246–3259.

54 Tang, S.-F., Song, J.-L., Li, X.-L., and Mao, J.-G. (2006). Luminescent lanthanide(III) carboxylate–phosphonates with helical tunnels. *Crystal Growth & Design* 6: 2322–2326.

55 Colón, Y.J., Gómez-Gualdrón, D.A., and Snurr, R.Q. (2017). Topologically guided, automated construction of metal–organic frameworks and their evaluation for energy-related applications. *Crystal Growth & Design* 17: 5801–5810.

56 Anderson, R. and Gómez-Gualdrón, D.A. (2019). Increasing topological diversity during computational "synthesis" of porous crystals: how and why. *CrystEngComm* 21: 1653–1665.

57 O'Boyle, N.M., Banck, M., James, C.A. et al. (2011). Open Babel: an open chemical toolbox. *Journal of Cheminformatics* 3: 33.

58 Sarkisov, L., Bueno-Perez, R., Sutharson, M., and Fairen-Jimenez, D. (2020). Materials informatics with PoreBlazer v4.0 and the CSD MOF database. *Chemistry of Materials* 32: 9849–9867.

59 Dubbeldam, D., Calero, S., Ellis, D.E., and Snurr, R.Q. (2016). RASPA: molecular simulation software for adsorption and diffusion in flexible nanoporous materials. *Molecular Simulation* 42: 81–101.

6

Reverse Design of Heat Exchange Systems Using Physics-Informed Machine Learning

Chang He[1,2] *and Yunquan Chen*[3]

[1] *Sun Yat-Sen University, School of Chemical Engineering and Technology, No. 135, Xingang Xi Road, Zhuhai, Guangdong 519082, China*
[2] *The Key Laboratory of Low-carbon Chemistry & Energy Conservation of Guangdong Province, Guangdong Engineering Center for Petrochemical Energy Conservation, 132 Waihuan East Road, University City, Panyu District, Guangzhou 510275, China*
[3] *Sun Yat-Sen University, School of Materials Science and Engineering, No. 135, Xingang Xi Road, Guangzhou 510275, China*

6.1 Introduction

Designing high-performance and low-cost heat exchange systems for energy savings and emissions reduction has been a long-lasting challenge in electronic devices, power plants, and petrochemical processes [1–3]. To address this challenge, traditional practices typically rely on a trial-and-error or empirical procedure [4–6], where successive candidate solutions are generated through space-filling experimental designs over the combined space of the inputs (including geometric variable $\mathcal{G}$, operating variable $\mathcal{O}$, and uncertain variable $\mathcal{U}$, Figure 6.1) until one that is best-performing is found by evaluating the outputs of all objectives $\{\text{obj}_1,\ldots\text{Obj}_k\ldots,\text{Obj}_K\}$. For each solution, to accurately capture the multi-phase fluid flow and interfacial heat and mass transfer, it is often required to build a high-fidelity numerical model by solving the first-principle governing equations via mesh-based methods [7–9] (e.g. finite volume method and finite element method) with the aid of computational fluid dynamics (CFDs) tools. In practice, however, it is well known that numerical modeling is time-consuming, especially for varying geometry design problems that require tedious preprocessing and calibration procedures such as mesh regeneration or calibration of initial and boundary conditions [10]. Hence, the resulting computational burden becomes a major bottleneck that slows down the research and development of cutting-edge heat exchange designs.

In contrast to intuition-based approaches, inverse design directly starts with the targeted functionality and sets it as an objective function to be optimized via partial differential equation (PDE)-constrained shape or topology optimization. When the geometry is represented by using the parameterized function of discrete geometric components, inverse design is known as *shape optimization*. On the other hand, when the geometry is parameterized by using a density function or level set so that

Applied AI Techniques in the Process Industry: From Molecular Design to Process Design and Optimization, First Edition. Edited by Chang He and Jingzheng Ren.

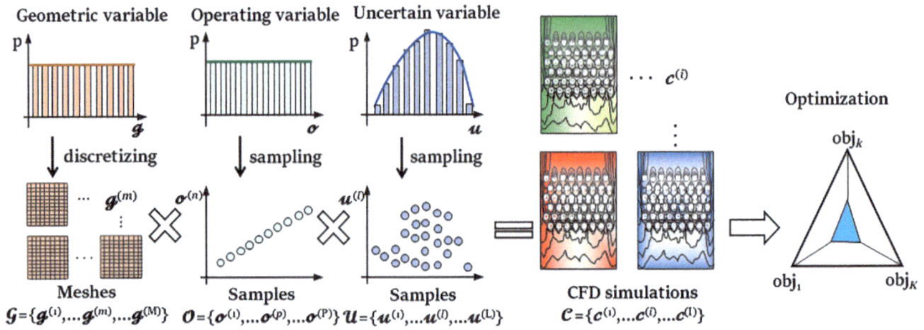

Figure 6.1 Traditional mesh-based numerical method for determining the optimal designs.

the connectivity/structure/layout is arbitrary, inverse design is known as *topology optimization* [11, 12]. In this case, PDE-constrained inverse design can be performed via gradient-based optimization algorithms, where the gradient is obtained using adjoint methods that compute the gradient of the objective function with respect to all involved parameters. These methods are commonly based on numerical PDE solvers built into commercial CFD software, which tracks the interactions and spatial movement of each particle characterized by computationally expensive PDEs [13]. Moreover, data interface and transfer issues would lead to inefficient integration of numerical models with equation-oriented optimization frameworks since a large number of recourses to the PDE solver have to be executed before converging to the optimum. The time and memory costs further increase exponentially, which may cause a combinatorial explosion in the computation burden if no effective approach is applied.

In recent times, a prevalent approach to addressing the aforementioned computational burden has been the use of data-driven surrogate-based optimization. This is typically achieved by extensively studying shortcut models or detailed full-order models (FOMs) [14]. Shortcut models, employed to parameterize the geometry, consist of computationally inexpensive analytical or lumped parameter equations (rather than governing PDEs) with numerous ideal assumptions [15, 16]. These assumptions include perfect mixing, plug flow, and equilibrium behavior, among more. Consequently, it becomes advantageous for researchers to develop automated procedures that can effectively explore the design space by coupling with optimization algorithms such as mathematical programming [16, 17] and metaheuristic methods [18]. However, such approaches do not scale up well with the system's size and would suffer from inaccurate and mismatch issues due to the insufficient description of the spatially distributed multiphysics phenomena inside a heat exchanger. By contrast, the FOM-based surrogate model can accurately capture the distributed phenomena. Neural networks, Kriging and radial basis functions, and specialized regression models are often employed in a supervised-learning paradigm to learn the nonlinear mapping from arbitrary designs to their associated objectives, i.e. neural networks are regarded as surrogate models of the original FOMs to accelerate the design and optimization [17, 18]. Nevertheless, it often requires a large number of datasets to form high-fidelity surrogate models, particularly for coping with complex varying geometry problems, where the generation of such datasets via conventional PDE solvers could become prohibitively expensive. As a result, the benefit of data-driven surrogate-based optimization must be weighed against model accuracy and time cost to resolve them, which is a major driver and could impede the development of better designs for heat exchange systems.

With the explosive growth of computing and data resources, deep learning has shown promise for simulation, modeling, and optimization due to its capability of handling strong nonlinearity and high dimensionality issues. Recently, the physics-informed neural network (PINN) developed by Karniadakis et al. and coworkers [11, 12, 19, 20] has opened up a deep learning, mesh-free route to solve governing PDEs by mathematically encoding the underlying physics (e.g. conservation laws, transfer laws, and kinetic equations) into the loss function of the neural network. The governing PDEs can take various forms, such as integer-order PDEs [21], integrodifferential equations [22], fractional or stochastic PDEs [23–25], and so on. One of the notable strengths of PINN is its ability to efficiently take the derivatives of a neural network by applying chain rules for differentiating the compositions of PDEs using automatic differentiation [26]. PINN has been successfully applied to diverse areas such as fluid mechanics [27–29], medical diagnosis [30–32], heat transfer analysis [33–35], and materials science [36–38]. Moreover, due to its capability of learning physics equations in rich representation as well as the use of automatic differentiation that removes the need for mesh generation, physics-informed learning is well-placed to become a natural bridge that integrates the numerical model with the optimization procedure for accelerating the discovery of optimal geometry designs.

Despite the recent success and promising prospects of PINN, it has been found that this approach is currently limited to tasks characterized by relatively simple and well-defined physics [10, 20]. Existing research on PINN generally aims to construct a high-fidelity surrogate model of the investigated system using a single deep neural network to control all variables and derivatives in the PDEs. It often struggles to accurately penalize residuals or even fails to train, particularly when tackling multiphysics and multiscale problems with high-frequency functions [10]. For example, heat sink systems typically involve a conjugate heat transfer process [39–41], which is a combination of heat transfer in fluids and heat transfer in solids. As a result, the involved physical variables or parameters (such as velocity, pressure, and temperature) have distinct meanings and may extend over several orders of magnitude in on spatial scales. However, most studies have applied encoding to the parametric inputs using a single variable or parameter in PDEs, without focusing on boundary conditions or geometric dimensions, as well as multivariable parameterization issues. When applying the stand PINN method to approximate this process, the trained neural networks and constructed surrogate models are not robust and stable enough, which are not conducive to subsequent tasks such as real-time prediction and design optimization. Therefore, the extension of the standard PINN to accomplish the parameterized system representation task is technically challenging and requires significant improvements in the structure and algorithm of the neural network for better accuracy, faster training, and improved generalization.

This chapter aims to not only use the PINN for modeling heat exchange systems with complex geometries but also propose an inverse design method that seamlessly combines the constructed surrogate model with multi-objective optimization and decision-making algorithms. For simulating the involved heat and mass transfer processes, specialized neural network structures are developed to decompose the standard PINN model into multiple interconnected subnetworks with identical architecture, which were functionally designed to approximate the latent solutions of the governing PDEs. In addition, the developed PINN applies the input encoding to the geometric and operating inputs in a fully decoupled setting and then concatenates them together in surrogate modeling. This surrogate model, combined with multi-objective optimization and decision-making methods, not only delivers the Pareto-optimal solutions directly but also allows real-time visualization of the distributions of the monitored state variables for better physical inspection. Results are presented for two illustrative examples: (i) In the first example, we demonstrate the proposed method's ability to describe the underlying behaviors of conjugate heat transfer inside the heat sink system. (ii) In the second example, we further leverage the capabilities of space decomposition, physics-informed deep learning, and transfer learning to accelerate the multi-objective stochastic optimization of a tubular air cooler system.

This chapter is derived from the work of He and coworkers [42, 43].

6.2 PINN-Based Inverse Design Method

This section begins by providing a brief introduction to the inverse design problem. Subsequently, the standard PINN method is presented, along with the PINN-based optimization and decision-making methods utilized to address this problem.

6.2.1 Overview of Inverse Design

Given a steady-state heat exchanger, the governing PDEs can be represented by a generic form defined on a bounded domain $\Omega \subset \mathbb{R}^3$

$$\mathcal{N}_i[\boldsymbol{\mu}(\mathbf{x}_\mathrm{f}); \lambda] = 0, \quad i \in [1, \dots, N_\mathcal{N}], \quad \mathbf{x}_\mathrm{f} = [x_\mathrm{f}, y_\mathrm{f}, z_\mathrm{f}] \in \Omega \tag{6.1}$$

with suitable boundary conditions

$$C_j[\boldsymbol{\mu}(\mathbf{x}_\mathrm{b}); \lambda] = 0, \quad j \in [1, \dots, N_C], \quad \mathbf{x}_\mathrm{b} = [x_\mathrm{b}, y_\mathrm{b}, z_\mathrm{b}] \in \partial\Omega \tag{6.2}$$

where $\mathcal{N}_i$ and C_j denote the general form of the PDE operator and boundary condition operator, respectively; $N_\mathcal{N}$ and N_C are the numbers of corresponding equations involved in the PDEs and boundary conditions. $\partial\Omega$ denotes the boundary of domain Ω that is required for defining the constraints. Generally, C_j may consist of differential, nonlinear, and identity terms, which could be subject to the following Dirichlet and Neumann boundary conditions [44], as given by

$$\text{Dirichlet:} \quad \boldsymbol{\mu}(\mathbf{x}_\mathrm{D}) - g(\mathbf{x}_\mathrm{D}) = 0, \quad \mathbf{x}_\mathrm{D} \in \partial_\mathrm{D}\Omega \tag{6.3}$$

$$\text{Neumann:} \quad \mathbf{n} \cdot \nabla \boldsymbol{\mu}(\mathbf{x}_\mathrm{N}) - h(\mathbf{x}_\mathrm{N}) = 0, \quad \mathbf{x}_\mathrm{N} \in \partial_\mathrm{N}\Omega \tag{6.4}$$

where $\mathbf{x} = [\mathbf{u}, p, T]^{\mathrm{T}}$, $\mathbf{x}_{\mathrm{D}}$, and $\mathbf{x}_{\mathrm{N}}$ are the coordinates of $\partial_{\mathrm{D}}\Omega$ and $\partial_{\mathrm{N}}\Omega$, $\partial\Omega = \partial_{\mathrm{D}}\Omega \cup \partial_{\mathrm{N}}\Omega$, $\partial_{\mathrm{D}}\Omega \cap \partial_{\mathrm{N}}\Omega = \emptyset$; $\mathbf{n}$ is the unit normal vector outward to the boundary at $\mathbf{x}_{\mathrm{N}}$. Here, the initial condition can be simply treated as a special type of Dirichlet boundary condition on the domain [45].

The latent solution of the PDEs, $\boldsymbol{\mu}(\mathbf{x}) = [\mu_1(\mathbf{x}), \ldots, \mu_{\mathrm{n}}(\mathbf{x})] \in \mathbb{R}^{\mathrm{n}}$, is determined by the independent variable set $\lambda = [\lambda^{(1)}, \lambda^{(2)}]$, which is our quantity of interest for the inverse design problem [11]. Here, it is noted that λ includes design variable $\lambda^{(1)}$ and control variable $\lambda^{(2)}$, which describe the specific shapes/structures to be manufactured and the corresponding operating conditions to be maintained for realizing an optimal design, respectively. To address this complex problem, traditional intuition-based approaches have to indirectly convert it into a large number of forward-design problems via experimental design or even brute-force search. This is done until a feasible solution is found by evaluating the outputs of all objective values of interest.

As presented in the Section 6.2.1.1, we utilize the advancements of PINN for inverse design optimization due to its natural capability of embedding physical models, as well as the use of automatic differentiation that removes the need for mesh generation. In particular, a deep neural network $\boldsymbol{\mu}_{\mathrm{net}}$ is developed as a parametric surrogate model of the solution $\boldsymbol{\mu}$, where $\lambda = [\lambda^{(1)}, \lambda^{(2)}]$ satisfies all the equality constraints enforced by the governing PDEs and the boundary conditions. With the aid of this surrogate model, we can directly search for the best independent variable set λ^* by minimizing objective functions of interest, $\boldsymbol{\Theta} = [\boldsymbol{\Theta}_1, \ldots, \boldsymbol{\Theta}_m]$. Besides, a set of equality/inequality constraints that stem from the multi-objective problems of the system, as well as the thermal and hydraulic requirements, are also incorporated in the inverse design. In all, the multi-objective inverse design problem can be formulated by

$$\mathbf{P1} \qquad \begin{aligned} &\lambda^* = \arg\min \boldsymbol{\Theta}(\lambda) = \{\boldsymbol{\Theta}_1, \ldots \boldsymbol{\Theta}_m\} \\ &\text{s.t.} \begin{cases} \mathcal{N}_i[\boldsymbol{\mu}(\mathbf{x}); \lambda] = 0 \\ \mathcal{C}_j[\boldsymbol{\mu}(\mathbf{x}_{\mathrm{b}}); \lambda] = 0 \\ \boldsymbol{h}(\boldsymbol{\mu}, \lambda) \leq 0 \end{cases} \end{aligned} \tag{6.5}$$

where $\boldsymbol{h}$ denotes the additional equality or inequality constraints.

6.2.1.1 Standard Physics-Informed Neural Networks

The PINN algorithm can parameterize a physical model by applying the input encoding to the temporal/spatial and parametric inputs and then concatenating these inputs together. It often uses a conventional feed-forward fullyconnected architecture, where a deep neural network represented by $\boldsymbol{\mu}_{\mathrm{net}}(\mathbf{x}, \lambda; \theta)$ is constructed as a surrogate of the solution $\boldsymbol{\mu}$. As shown in Figure 6.2, $\boldsymbol{\mu}_{\mathrm{net}}(\mathbf{x}, \lambda; \theta)$ takes the spatial coordinate $\mathbf{x} = [x, y, z]$ and parametric variables $\lambda = [\lambda_1, \lambda_2, \ldots \lambda_n]$ as the input and outputs a vector $\boldsymbol{\alpha}^{\mathrm{L}}$ (i.e. $\mathbf{u}, p, T$) that has the same dimension as $\boldsymbol{\mu}$. The training of the PINN aims to obtain the set of best parameters $\theta^* = [\boldsymbol{w}^*, \boldsymbol{b}^*]$ by minimizing loss terms ($\mathcal{L}_{\mathrm{f}}$ and $\mathcal{L}_{\mathrm{b}}$) such that the neural network can approximate the solution of the original PDEs.

$$\boldsymbol{\mu}_{\mathrm{net}}(\mathbf{x}, \lambda; \theta) \triangleq \boldsymbol{\alpha}^{\mathrm{L}}(\mathbf{x}, \lambda; \boldsymbol{w}, \boldsymbol{b}) \rightarrow \boldsymbol{\mu}(\mathbf{x}, \lambda) \tag{6.6}$$

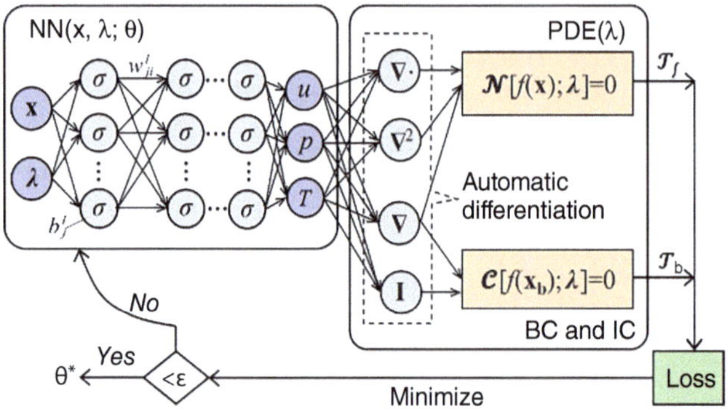

Figure 6.2 Schematic of the standard PINN algorithm for parametric surrogate modeling. Source: [19]/with permission of Elsevier. The derivative terms (i.e. $\nabla$and ∇^2) are handled via automatic differentiation. $a = \sigma[\cdot]$ denotes a nonlinear activation function.

where $\boldsymbol{w}$ and $\boldsymbol{b}$ are weight matrices and bias vectors of the neural network, respectively, which will be tuned at the training stage.

In order to train PINN, the derivative terms of $\boldsymbol{\mu}_{\text{net}}$ with respect to its inputs are computed by applying the chain rule for differentiating compositions of the functions using automatic differentiation. Next, the neural network $\boldsymbol{\mu}_{\text{net}}$ is restricted to the training points $\mathcal{T} = [\mathbf{x}_1,\ldots,\mathbf{x}_{|\mathcal{T}|}]$ to satisfy the physics imposed by the PDEs and boundary conditions. For this purpose, a composite loss function that penalizes the divergence of the neural network from the PDEs and boundary conditions is considered.

$$\mathcal{L}_{\text{tot}}(\theta; \lambda, \mathcal{T}) = \omega_{\text{f}}\mathcal{L}_{\text{f}}(\theta; \lambda, \mathcal{T}_{\text{f}}) + \omega_{\text{b}}\mathcal{L}_{\text{b}}(\theta; \lambda, \mathcal{T}_{\text{b}}) \tag{6.7}$$

where

$$\mathcal{L}_{\text{f}}(\theta; , \mathcal{T}_{\text{f}}) = \frac{1}{|\mathcal{T}_{\text{f}}|}\sum_{\mathbf{x}_{\text{f}}\in\mathcal{T}_{\text{f}}} \|\mathcal{N}[\boldsymbol{\mu}_{\text{net}}(\mathbf{x}_{\text{f}},\ \ \lambda; \theta)]\|_2^2 \tag{6.8}$$

$$\mathcal{L}_{\text{b}}(\theta; \lambda, \mathcal{T}_{\text{b}}) = \frac{1}{|\mathcal{T}_{\text{b}}|}\sum_{\mathbf{x}_{\text{b}}\in\mathcal{T}_{\text{b}}} \|\mathcal{C}[\boldsymbol{\mu}_{\text{net}}(\mathbf{x}_{\text{b}},\ \ \lambda; \theta)]\|_2^2 \tag{6.9}$$

and $\boldsymbol{\omega}_{\text{f}}$ and $\boldsymbol{\omega}_{\text{b}}$ are the weights that balance the interplay between the two loss terms $\mathcal{L}_{\text{f}}$ and $\mathcal{L}_{\text{b}}$; $\|\ \|_2$ denotes the L^2 norm of residuals. Two mutually independent sets, $\mathcal{T}_{\text{f}}$ and $\mathcal{T}_{\text{b}}$, are the sampled points in the domain and on the boundary, respectively, $\mathcal{T}_{\text{f}} \subset \Omega$, $\mathcal{T}_{\text{b}} \subset \partial\Omega$, $\mathcal{T} = [\mathcal{T}_{\text{b}}, \mathcal{T}_{\text{f}}]$. Since the loss function is often highly nonlinear and non-convex, the network parameters $\boldsymbol{\theta}$ of PINN are iteratively optimized by gradient-based optimizers, such as gradient descent, Adam, and L-BFGS [45].

$$\theta^* = \arg\min_{\theta} \mathcal{L}'_{\text{PINN}}(\theta; \lambda, \mathcal{T}) \tag{6.10}$$

To improve the training performance, the loss function of PINN can take the form of Monte Carlo integration approximation [46] since it can ensure the consistency of the loss per volume across the domain. Additionally, it is worth noting that the weighting coefficient in the loss function plays a pivotal role in enhancing the accuracy of PINN, which can be user-defined or tuned automatically. It is often defined as a fixed value that does not vary with the spatial position of the sampled points in the flow field or on the boundary. Nevertheless, the gradient profile at several positions of the flow field varies greatly (e.g. sampled points nearby the sharp corners or discontinuous areas), which makes a fixed weighting coefficient unreasonable. Here, the signed distance function (SDF) [47] is introduced in the loss function to weigh the loss terms, by which each weighting coefficient can be a function of the spatially distributed points in the spatial domain. The SDF is defined as follows [48]:

$$\omega(\mathbf{x}) = \begin{cases} d(\mathbf{x}), & \mathbf{x} \in \Omega \\ 0, & \mathbf{x} \in \partial\Omega \\ -d(\mathbf{x}), & \mathbf{x} \in \Omega \end{cases} \tag{6.11}$$

where

$$d(\mathbf{x}) = \underbrace{\min}_{\mathbf{x}_p \in \partial\Omega} \|\mathbf{x} - \mathbf{x}_p\|, \quad \mathbf{x} \in \Omega \tag{6.12}$$

where $d(\mathbf{x})$ represents the minimum distance of a given point from $\partial\Omega$. According to this definition, the absolute value of SDF often decreases close to zero near the boundary of the domain, where sharp gradients occur. Hence, weighting by the SDF tends to reduce the impact of sharp gradients on the convergence, which contributes to speeding up the learning rate and sometimes also improves prediction accuracy.

Overall, by combining the SDF and Monte Carlo integration approximation, the combined loss function for training the parameters of each subnetwork is as follows:

$$\begin{aligned} \mathcal{L}'_{\mathrm{PINN}}(\theta; \lambda, \mathcal{T}) &= \omega(\mathbf{x}_f)\mathcal{L}'_f(\theta; \lambda, \mathcal{T}_f) + \omega(\mathbf{x}_b)\mathcal{L}_b(\theta; \lambda, \mathcal{T}_b) \\ &= \omega(\mathbf{x}_f)\left(\int_\Omega \mathrm{d}\mathbf{x}\right)\frac{1}{|\mathcal{T}_f|}\sum_{\mathbf{x}_f \in \mathcal{T}_f} \|\mathcal{N}[\mu_{\mathrm{net}}(\mathbf{x}_f, \lambda; \theta)]\|_2^2 \\ &\quad + \omega(\mathbf{x}_b)\left(\int_\Omega \mathrm{d}\mathbf{x}\right)\frac{1}{|\mathcal{T}_b|}\sum_{\mathbf{x}_b \in \mathcal{T}_b} \|\mathcal{C}[\mu_{\mathrm{net}}(\mathbf{x}_b, \lambda; \theta)]\|_2^2 \end{aligned} \tag{6.13}$$

Another similar training strategy, namely integral continuity plane [49], can be used for enforcing the conservation constraints (e.g. mass, volume, and energy equations) of a given control volume in the domain. For example, we can specify that the volumetric flow of coolant passing through the heat exchanger system must be equal to that entering the system in the flow channel, as listed in Eq. (6.14). This conservation constraint helps to accelerate the solution of the continuity PDEs. Similarly, the heat flux of coolant entering the flow channel, along with the heat flux generated by the working heat source, must be equal to that leaving the channel. Such heat flux conservation is also expressed in an integral form, as given by Eq. (6.15).

$$\int (\mathbf{n} \cdot \mathbf{u})\mathrm{d}S_{\mathrm{plane}} = U_{\mathrm{inlet}}A_{\mathrm{inlet}} \tag{6.14}$$

$$\int \rho_{\mathrm{fluid}}C_{\mathrm{p,fluid}}[(\mathbf{n} \cdot \mathbf{u}) \cdot T_{\mathrm{fluid}}]\mathrm{d}S_{\mathrm{outlet}} = \rho_{\mathrm{fluid}}C_{\mathrm{p,fluid}}U_{\mathrm{inlet}}T_{\mathrm{inlet}} + q_s A_{\mathrm{source}} \tag{6.15}$$

where A_{inlet} and A_{source} are the areas of the channel inlet plane and heat source surface, respectively.

The training hyperparameters of the neural network, e.g. network sizes, learning rates, optimizers, initializations, and regularizations, also need to be fine-tuned to achieve a good level of accuracy. More details of the training strategy can be found in Refs [45, 49].

Finally, it should be pointed out that the standard PINN uses a single neural network to construct a surrogate of the solution μ_{net}. While this approach has demonstrated promising results in certain applications, its limitations become apparent in more complex multiphysics and multiscale tasks that involve high-dimensional and intricate geometric domains. For example, heat sink systems typically involve a convective heat transfer process, which entails a two-way coupling of fluid flow and heat transfer. In practice, however, it is challenging to precisely penalize the residuals in training the neural network of this process in a fully coupled setting. This is because the terms in the combined loss function have distinct physical meanings, such as velocity, pressure, and temperature, and may span several orders of magnitude in spatial scales, making the training process less robust and stable.

6.2.1.2 Design Optimization and Decision-making Methods

In the field of multi-objective optimization, previous studies [50–52] have often relied on evolutionary methods, which mainly include decomposition-based and Pareto-based approaches. The former typically uses a scalarizing function to aggregate all the objectives into a single scalar objective function or retain only one objective while enforcing the others as constraints. A major issue with this approach resides in the extensive knowledge of the problem structure, and not all scalarizing functions can guarantee that all Pareto-optimal solutions are obtainable [53]. The latter can make full use of the Pareto-dominance relations to induce partial ordering in the objective space. It provides a more convenient way to obtain a set of solutions toward the Pareto front and covers the entire Pareto front. One of the most widely used algorithms is the Non-dominated Sorting Genetic Algorithm II (NSGA-II) [54], which is employed in the present study. It is noted that the alternative solutions of the multi-objective optimization on the Pareto-optimal front have the same worthiness based on optimization objectives in the absence of additional preference information, which entails a need to identify the most preferred solution for decision-makers in practice. To this end, a classic decision-making method, namely technique for order preference by similarity to ideal solution (TOPSIS) [55, 56], is subsequently used to identify the most preferred solution closest to the positive-ideal solution and furthest away from the negative-ideal solution as the compromise scheme.

Assuming the number of alternative solutions is M, the procedure of the TOPSIS method can be divided into the following steps:

Step I: All objectives of interest are converted into extremely large ones using positive management

$$Y_{mn} = \left(\max_{m} \Theta^*_{mn}\right) - \Theta^*_{mn}, \quad \begin{array}{l} \forall m \in [1, 2, \ldots M] \\ \forall n \in [1, 2, \ldots N] \end{array} \tag{6.16}$$

where Θ^*_{mn} denotes the mth alternative solution in terms of the nth objective, and $\max_{m} \Theta^*_{mn}$ denotes the maximum value of all the alternative solutions in terms of the nth objective. Y_{mn} denotes the element of the positive matrix **Y.**

Step II: The weighted normalized decision matrix **Z** is constructed by normalizing and weighting the positive matrix **Y** according to the following equation:

$$Z_{mn} = \frac{Y_{mn}}{\sqrt{\sum_{m=1}^{M} Y_{mn}^2}} \times \omega_n, \quad \forall n \in [1, 2, \dots N] \tag{6.17}$$

where ω_n is the weighting factor of the nth criterion, $\sum \omega_n = 1$.

Step III: Determining the positive-ideal solution Z^+ and the negative-ideal solution Z^-:

$$Z_n^+ = \left(\max_{m} Z_{mn}\right), \quad \forall n \in [1, 2, \dots N] \tag{6.18}$$

$$Z_n^- = \left(\min_{m} Z_{mn}\right), \quad \forall n \in [1, 2, \dots N] \tag{6.19}$$

Step IV: The Euclidean distances for each alternative solution from the positive-ideal solution and the negative-ideal solution are calculated from Eq. (6.16).

$$R_n^+ = \sqrt{\sum_{n=1}^{N} \left(Z_n^+ - Z_{mn}\right)^2}, \quad \forall m \in [1, 2, \dots M] \tag{6.20}$$

$$R_m^- = \sqrt{\sum_{n=1}^{N} \left(Z_n^- - Z_{mn}\right)^2}, \quad \forall m \in [1, 2, \dots M] \tag{6.21}$$

Step V: Measuring the relative closeness to the positive-ideal solution by

$$S_m = \frac{R_m^-}{R_m^+ + R_m^-}, \quad \forall m \in [1, 2, \dots M] \tag{6.22}$$

Finally, the Pareto solution with the largest S_i is recommended as the desired solution for the bi-objective optimization problem.

6.3 Example 1: Finned Heat Sink Model

6.3.1 System Description and Objectives

A 3D finned heat sink system with forced air cooling [57] is used as a motivating example in this chapter. As shown in Figure 6.3, the base plate with a built-in chip is placed on the bottom center of a rectangular channel, which is directly in contact with the flat base of the heat sink system through thermal conductive adhesive. In this way, we can consider that almost all heat flux generated by the hot chip can be transferred to the fins on the flat base. Meanwhile, the coolant medium (air) introduced from the channel inlet flows to the surface of the fins, which aims to promptly

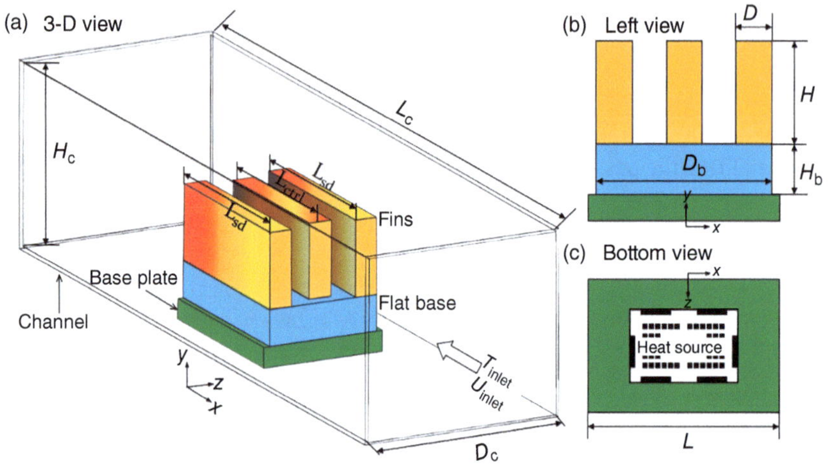

Figure 6.3 A schematic view of the geometry of the finned heat sink system. (a) 3D view, (b) left view, and (c) bottom view. In 3D modeling, the center of the channel is used as the origin of the Cartesian coordinate system. Source: [42]/with permission of Elsevier.

take away the transferred heat and prevent the junction temperature of the chip from reaching the threshold value. In summary, the heat sink system involves a 3D, steady-state conjugate heat transfer process.

The heat sink system involves a 3D, steady-state conjugate heat transfer process, which combines heat transfer in fluids with heat transfer in solids [58]. In fluids, heat convection dominates the heat transfer process. We assume that fluid passes through the system subject to incompressible and laminar flows, the governing continuity and Navier–Stokes equations are as follows [59]:

$$\rho_{\text{fluid}}(\nabla \cdot \mathbf{u}) = 0 \tag{6.23}$$

$$\rho_{\text{fluid}}(\mathbf{u} \cdot \nabla)\mathbf{u} - \nabla p + \mu \nabla^2 \mathbf{u} = 0 \tag{6.24}$$

where $\mathbf{u} = [u_x, u_y, u_z]$ denotes the fluid velocity vector; ∇ denotes the Hamiltonian operator; p, ρ_{fluid}, and μ denote the static pressure, density, and dynamic viscosity of the fluid, respectively. Here, it is assumed that the thermophysical properties of the fluid and solid, as listed in Table 6.1, are constant and insensitive to temperature change in the flow field.

Table 6.1 Thermophysical properties of fluid and solid.

Property	ρ (kg m^{-3})	κ (W (m K)$^{-1}$)	Cp (J (kg K)$^{-1}$)	μ (kg (m s)$^{-1}$)
Fluid	1.0	1.0	50	0.02
Solid	1.0	5.0	80	—

The occurrence of fluid heat transfer is based on the premise of fluid motion. Thereby, without internal heat sources, the governing equation of fluid heat transfer is given by

$$\rho_{\text{fluid}} C_{\text{p,fluid}} \mathbf{u} \cdot \nabla T_{\text{fluid}} - \kappa_{\text{fluid}} \nabla^2 T_{\text{fluid}} = 0 \tag{6.25}$$

where T_{fluid}, $C_{\text{p,fluid}}$, and κ_{fluid} are the temperature, specific heat, and thermal conductivity of the fluid. The first and second terms act as the convection term and diffusion term, which describe the thermal convection and heat conduction of fluid, respectively.

While heat conduction plays a major role in solids, the temperature distribution can be directly determined by solving Laplace's equation.

$$\frac{\kappa_{\text{solid}}}{\rho_{\text{solid}} C_{\text{p,solid}}} \nabla^2 T_{\text{solid}} = 0 \tag{6.26}$$

where T_{solid}, $C_{\text{p,solid}}$, and κ_{solid} are the temperature, specific heat, and thermal conductivity of the solid, respectively.

An essential condition for solving the aforementioned conjugate heat transfer problem is to provide well-defined boundary and initial conditions as model constraints, especially at the fluid–solid interface. First, for simulating the fluid flow process, a non-slip velocity condition ($\mathbf{u} = 0$) is applied to both the channel wall and the fluid–solid interface. The initial air velocity at the channel inlet is known ($u_x = U_{\text{inlet}}$, $u_y = 0$, $u_z = 0$), while the air pressure at the channel outlet is set to zero. For simulating the heat transfer process, the channel wall is modeled using an adiabatic condition ($\mathbf{n} \cdot (\kappa_{\text{fluid}} \cdot \nabla T_{\text{fluid}}) = 0$) with a known temperature at the channel inlet ($T_{\text{fluid}} = T_{\text{inlet}}$). It is assumed that the hot heat source could create an even heat flow at the bottom of the flat base. With this assumption, a constant heat flux q_s can be specified as a thermal boundary constraint for the heat sink model, as given by

$$-\mathbf{n} \cdot (\kappa_{\text{solid}} \cdot \nabla T_{\text{sb}}) = q_{\text{s}} \tag{6.27}$$

where ∇T_{sb} denotes the temperature gradient at the bottom of the flat base; $\mathbf{n}$ denotes the unit normal vector outward to the surface. Herein, to guarantee the continuity of heat flux and temperature at the fluid–solid interface, the following constraints are enforced on boundary conditions [60]:

$$T_{\text{sw}} - T_{\text{fw}} = 0 \tag{6.28}$$

$$\mathbf{n} \cdot (\kappa_{\text{solid}} \cdot \nabla T_{\text{sw}}) - \mathbf{n} \cdot (\kappa_{\text{fluid}} \cdot \nabla T_{\text{fw}}) = 0 \tag{6.29}$$

where the subscripts sw and fw denote the solid wall and fluid wall at the fluid–solid interface, respectively.

The heat sink system generally aims at cost-effectively cooling down the running heat source to prevent thermal shutdown by exchanging sensible heat with air flowing through the surface of this system. Based on this premise, the first objective of interest is termed the mean surface temperature of the heat source (T_{mean}), which is often used as a key indicator to measure the cooling capacity of a given heat sink system. In addition, decision-makers are concerned about the operating cost of achieving a higher cooling capacity. For example, using complex fins with a larger heat exchange area would help to reduce the working temperature of the heat source, but it also incurs an additional pressure drop (Δp) in the airflow. In this case, increased power consumption by the intake fan is required proportionally. These two conflicting objectives can be calculated through the solid heat transfer network and fluid flow network, combined with the Monte Carlo integral approximation, as given by.

$$\begin{aligned} T_{\text{mean}} &= \frac{1}{A_{\text{source}}} \int_{\text{source}} \boldsymbol{\mu}_{\text{heat\_net\_solid}}(\mathbf{x}_{\text{source}}, \lambda; \boldsymbol{\theta}^*) \mathrm{d}\mathbf{x} \\ &\approx \frac{1}{N_{\text{source}}} \sum_{k=1}^{N_{\text{source}}} \boldsymbol{\mu}_{\text{heat\_net\_solid}} \left(\mathbf{x}_{\text{source}}^k, \lambda; \boldsymbol{\theta}^* \right) \end{aligned} \tag{6.30}$$

$$\begin{aligned} \Delta p &= p_{\text{in}} - p_{\text{out}} \\ &= \frac{1}{A_{\text{in}}} \int_{\text{in}} \boldsymbol{\mu}_{\text{flow\_net}}(\mathbf{x}_{\text{in}}, \lambda; \boldsymbol{\theta}^*) \mathrm{d}\mathbf{x} - \frac{1}{A_{\text{out}}} \int_{\text{out}} \boldsymbol{\mu}_{\text{flow\_net}}(\mathbf{x}_{\text{out}}, \lambda; \boldsymbol{\theta}^*) \mathrm{d}\mathbf{x} \\ &\approx \frac{1}{N_{\text{in}}} \sum_{k=1}^{N_{\text{in}}} \boldsymbol{\mu}_{\text{flow\_net}} \left(\mathbf{x}_{\text{in}}^k, \lambda; \boldsymbol{\theta}^* \right) - \frac{1}{N_{\text{out}}} \sum_{k=1}^{N_{\text{out}}} \boldsymbol{\mu}_{\text{flow\_net}} \left(\mathbf{x}_{\text{out}}^k, \lambda; \boldsymbol{\theta}^* \right) \end{aligned} \tag{6.31}$$

where A_{out} denotes the cross-sectional area at the outlet of the channel. $\mathbf{x}_{\text{source}}^k$, $\mathbf{x}_{\text{in}}^k$, and $\mathbf{x}_{\text{out}}^k$ are the coordinates of samples collected on the heat source surface, channel inlet, and outlet sections, respectively; N_{source}, N_{in}, and N_{out} are the corresponding numbers of these samples.

The peak surface temperature of the heat source (T_{peak}) must be lower than the maximum allowable junction temperature (T_{max}) to ensure the safety and reliability of the system [61].

$$T_{\text{peak}} \begin{aligned} &= \max \boldsymbol{\mu}_{\text{heat\_net\_solid}}(\mathbf{x}_{\text{source}}, \lambda; \boldsymbol{\theta}^*) \\ &\approx \max \boldsymbol{\mu}_{\text{heat\_net\_solid}} \left(\mathbf{x}_{\text{source}}^k, \lambda; \boldsymbol{\theta}^* \right) \end{aligned} \tag{6.32}$$

A bi-objective optimization model is formulated to minimize the mean surface temperature and the pressure drop of the system, as given by:

$$\begin{aligned} \lambda^* &= \arg \min_{\lambda^{(1)}, \lambda^{(2)}} \boldsymbol{\Theta} = \{T_{\text{mean}}, \Delta p\} \\ &\text{s.t.} \begin{cases} \mathcal{N}_i[\boldsymbol{\mu}(\mathbf{x}); \lambda] = 0 \\ C_j[\boldsymbol{\mu}(\mathbf{x}_b); \lambda] = 0 \\ \lambda \in [\lambda^{\text{lb}}; \lambda^{\text{ub}}] \\ T_{\text{peak}} \leq T_{\text{max}} \end{cases} \end{aligned} \tag{6.33}$$

where Eqs. (6.29)–(6.31) are embedded as a surrogate model to predict the temperature and pressure drop. The superscripts lb and ub are the lower and upper bounds for the decision variables of interest for the present study, as listed in Table 6.2. For simplicity, the height (H) and thickness (D) of all fins of the heat sink system are consistent, while the length of the central fin (L_{ctrl}) and the length of two side fins (L_{sd}) are independent of each other. As a result, there will be four design variables in total, $\lambda^{(1)} = [H, D, L_{\text{ctrl}}, L_{\text{sd}}]$. Besides, only the inlet velocity of the coolant is considered the operating variable, $\lambda^{(2)} = [U_{\text{inlet}}]$.

Table 6.2 Summary of key variables and constraints.

Variable/constraint		Unit	Lower bound	Upper bound
Operating variable, $\lambda^{(2)}$				
Inlet air velocity	U_{inlet}	$m\,s^{-1}$	0	3.0
Design variable, $\lambda^{(1)}$				
Fin height	H	m	0	0.6
Fin thickness	D	m	0.05	0.15
Central fin length	L_{ctrl}	m	0.50	1.0
Side fin length	L_{sd}	m	0.50	1.0
Process constraint				
Inlet air temperature	T_{inlet}	°C	20	
Heat flux	q_s	$W\,m^{-2}$	1800	
Max allowable junction temperature	T_{max}	°C	90	

6.3.2 Improved PINN Structure

In this example, we develop a specialized neural network structure, namely hybrid PINN, to decompose the standard PINN into multiple small-size sub-networks, which can distinguish the difference between the state variables of interest in the combined loss function. According to the characteristics of fluid flow and heat transfer in the conjugate process, the neural network of this system representation is decomposed into three interconnected sub-networks with identical architecture, namely *flow_net*, *heat_net_fluid*, and *heat_net_solid*, which are functionally designed in this hybrid PINN model to approximate the latent solutions of Navier–Stokes, heat transfer in fluid, and heat transfer in solid, respectively. In Figure 6.4, each sub-network applies the input encoding to the same spatial and parametric inputs $[\mathbf{x}, \lambda]$. According to the relationship between the governing PDEs involved in these sub-networks, it is assumed that there is a one-way coupling between the fluid flow and heat transfer processes. Meanwhile, the process of heat transfer between fluid and solid is mutually independent and coupled in a parallel correlation. On this basis, it is possible to compute the temperature field once the training of the flow field has converged. To be more specific, once the training of *flow_net* is completed and can satisfy convergence criteria, the resulting field distribution of velocity $\mathbf{u}^*$ can be used as an intermediate for training *heat_net_fluid* and *heat_net_solid* simultaneously. Besides, the continuity condition at the fluid–solid interface is encoded into the loss function as a boundary condition, which can lead to a significant speedup of the multiphysics learning task.

6.3.3 Results

The neural network training and CFD simulation are implemented on TensorFlow-based NVIDIA Modulus [46] and COMSOL Multiphysics@5.5 [62], respectively.

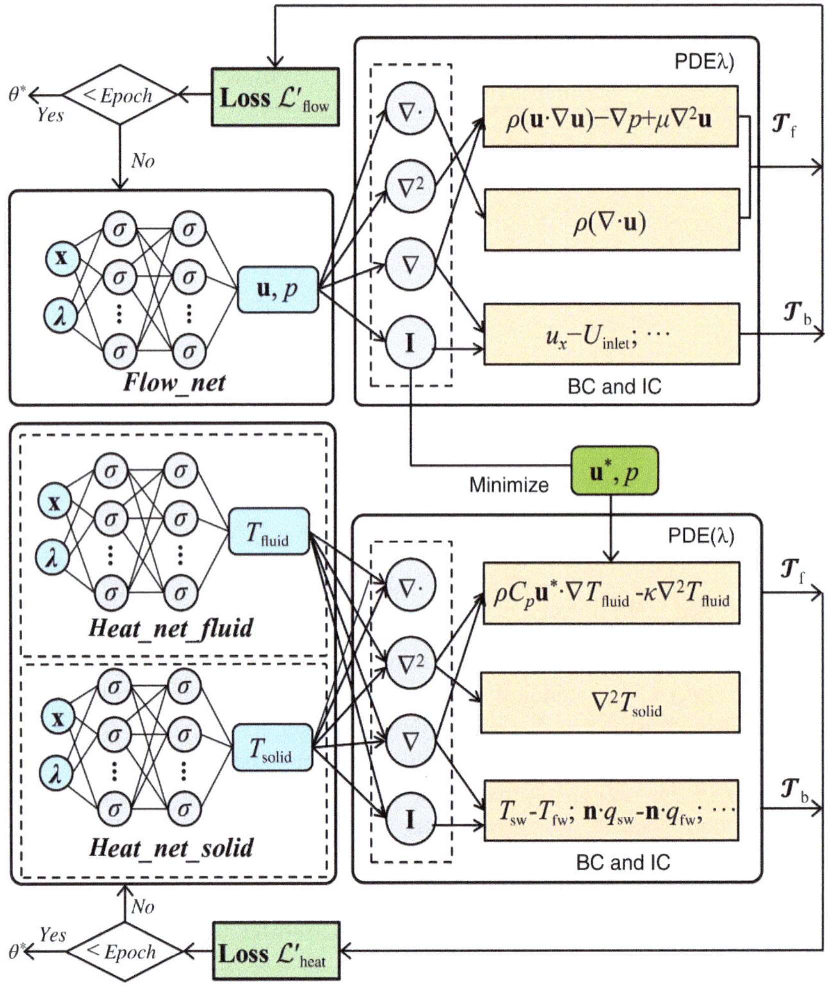

Figure 6.4 Schematic of the hybrid PINN strategy for simulating the conjugate heat transfer processes. Source: [42]/with permission of Elsevier.

During the neural network training, each sub-network of the hybrid PINN model that we have developed consists of six hidden layers, with 256 neurons in each layer. A minimum of 1×10^6 training steps for the *flow_net*, and 1×10^6 for both the *heat_net_fluid* and *heat_net_solid* are required to guarantee the convergence of the neural networks. The corresponding times for training are 38.3 and 75.6 hours. After training, the obtained networks are transferred to Matlab 2021b platform for implementing optimization and decision-making algorithms. These computational models were solved on a workstation with two Intel Xeon E5-2695v4 CPUs@ 2.1 GHz, 96GB RAM, and two NVIDIA GeForce RTX 3070 GPUs, using the Linux Ubuntu operating system.

The performance of the PINN-derived surrogate model for predicting the behavior of the conjugate heat transfer process is evaluated by comparing the results obtained from CFD simulations. To facilitate the model validation, a benchmark design is used with representative conditions ($H = 0.4$ m, $D = 0.1$ m, $L_{ctrl} = 1$ m, $L_{sd} = 1$ m, and $U_{inlet} = 1$ m s^{-1}). To quantitatively measure the divergence of the field distributions, the normalized mean absolute percentage error (NMAPE [63]) is employed as an evaluation index

$$\text{NMAPE} = 100\% \times \frac{1}{N_i} \sum_{1}^{N_i} \frac{|\phi_{i,\text{CFD}} - \phi_{i,\text{PINN}}|}{\max(\phi_{i,\text{CFD}}) - \min(\phi_{i,\text{CFD}})} \tag{6.34}$$

where ϕ denotes the evaluated state variable, and the subscripts CFD and PINN denote the results predicted by CFD simulation and PINN-derived surrogate model.

The statistics of NMAPE for the distributed velocity, pressure, and temperature are listed in Table 6.3. The PINN-derived surrogate model can lead to acceptable prediction accuracy, which is 0.40–1.14% for velocity, 1.53% for pressure, and 0.38–4.17% for temperature. Besides, the thermal and hydraulic behaviors of the heat sink system in terms of the pressure drop, the surface temperature, and the peak temperature on the heat source are also used to validate the PINN-derived surrogate model. The PINN-derived results for these indicators are 9.178 Pa, 70.34 °C, and 79.38 °C, which are only 0.45%, 2.66%, and 1.64% diverged from the reference CFD results, respectively. From the comparison, it is concluded that the PINN-derived surrogate model can offer good quantitative agreement and sufficient confidence to describe the underlying behaviors of the conjugate heat transfer process inside heat sink system.

Based on the surrogate model, the design optimization of the heat sink system was performed using the NSGA-II algorithm, with parameters for the initial population and maximum evolution generation set at 40 and 30, respectively. Figure 6.5 presents the Pareto-optimal solutions of the system under fixed inlet air velocities, which consist of passive cooling mode ($U_{inlet} = 0$ m s^{-1}) and active cooling mode ($U_{inlet} > 0$ m s^{-1}). The corresponding optimal geometric dimensions and operating conditions are provided in Figure 6.6. Note that, the solutions on the Pareto-optimal curve are all non-dominated and feasible, indicating that the heat source surface temperature is minimized relative to the specified pressure drop limit. Specifically, these solutions are all optimal options with balanced performance between cooling

Table 6.3 Error evaluation between PINN and CFD solutions.

NMAPE	u_x (m s^{-1})	u_y (m s^{-1})	u_z (m s^{-1})
	0.88%	1.14%	0.40%
	p (Pa)	T_{fluid} (°C)	T_{solid} (°C)
	1.53%	4.17%	0.38%
	Δp (Pa)	T_{mean} (°C)	T_{peak} (°C)
CFD	9.137	68.51	78.10
PINN	9.178	70.34	79.38
Relative error	0.45%	2.66%	1.64%

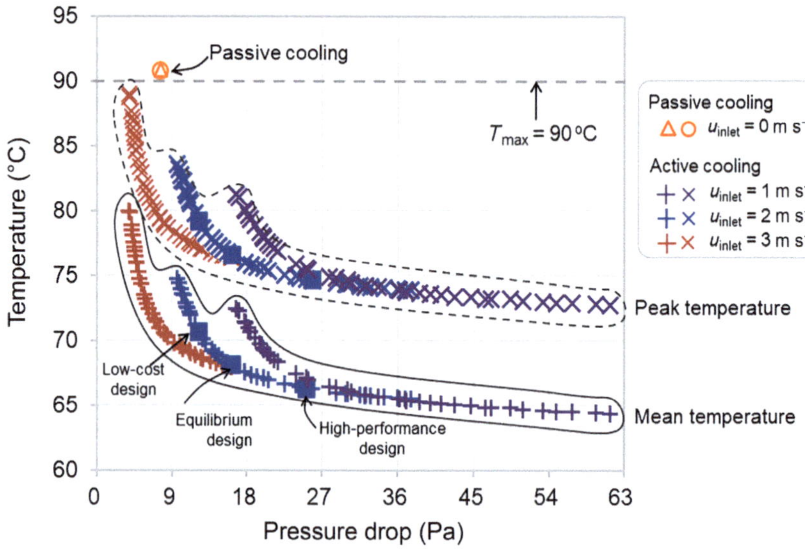

Figure 6.5 The Pareto front of multi-objective optimization under fixed inlet air velocities. Source: [42]/with permission of Elsevier.

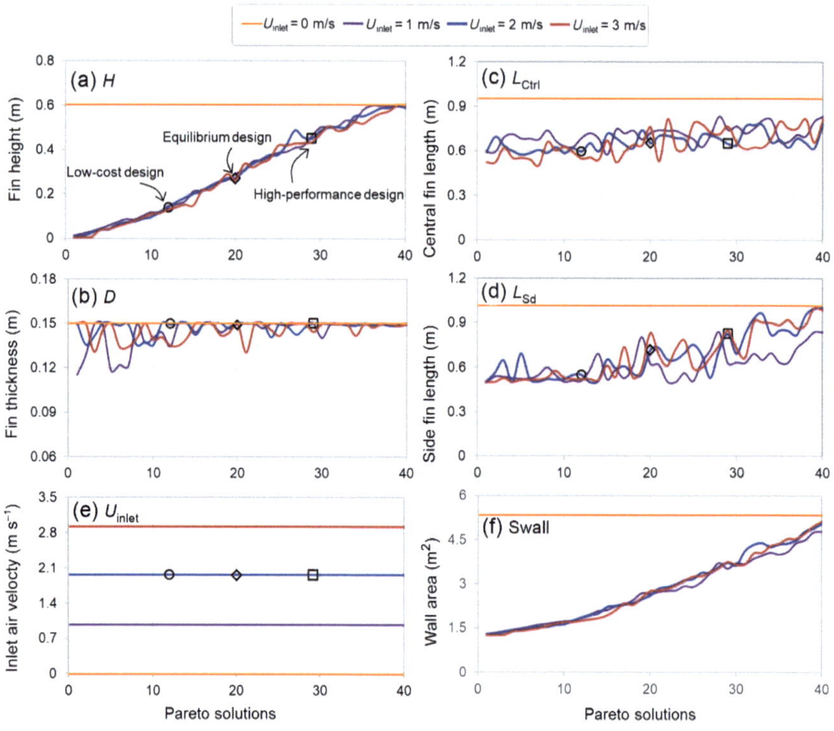

Figure 6.6 The optimal results of geometric dimensions and operating conditions under fixed inlet air velocities. (a) fin height, (b) fin trickiness, (c) central fin length, (d) side fin lengthen, (e) inlet air velocity, (f) wall area.

capacity and energy cost for decision-making. Therefore, selecting the most preferred solution from these optimal options is critical for designing the heat sink system. Here, we consider three representative scenarios using the TOPSIS method: high-performance design, equilibrium design, and low-cost design. These scenarios correspond to weighting factors of 0.8 (0.2), 0.5 (0.5), and 0.2 (0.8) allocated to the objective of T_{mean} (Δp) in the process of decision-making.

Though it requires $38.3 + 75.6 = 113.9$ hours to train the networks, note that the time cost for the subsequent multi-objective optimization and decision-making can be almost negligible, as they only take less than 10 seconds. This brings a huge advantage over the traditional trial-and-error or empirical methods. For example, if we want to obtain similar optimal solutions (e.g. active cooling, U_{inlet} = variable), at least $6^4 = 1296$ (four variables, with each variable having six evenly distributed values in its range) CFD runs are required to perform simultaneously. For a single CFD run, it would take about 10 minutes considering the time for mesh regeneration, communication, and computation. The total computational time required by CFD simulation is up to 1296×10 minutes = 216 hours. Besides, the traditional methods require additional time to screen the desired solutions from these 1296 candidates by analyzing the CFD simulation results. In contrast, the proposed inverse design method starts directly with the desired thermal and hydraulic objectives and works backward to identify the optimal geometry and corresponding operating conditions. It should *be pointed out that, a further increase in geometric variables or their values would amplify the advantages of the proposed method in terms of time cost and computational* efficiency.

In the passive cooling mode, the heat generated by the heat source can only dissipate through thermal conduction since there is no forced airflow. In Figure 6.5, the mean temperature of the heat source surface exactly coincides with its peak temperature at $T_{peak} = 91.53\,°\text{C}$ and $\Delta p = 8.90$ Pa, which is beyond the maximum allowable junction temperature ($T_{max} = 90\,°\text{C}$). As a result, the passive cooling mode would render the operation of the heat source inefficient and short-lived due to long-term overheating. Unlike the passive cooling mode, in the active cooling mode, the co-presence of thermal convection and conduction effectively reduces the thermal resistance between the fins and the surrounding airflow, which improves the cooling capability of the heat sink system. All generated Pareto-optimal solutions are distributed in the safe region, and even the peak temperatures are below 90 °C. The mean temperature of the heat source surface is lower than its peak temperature by 8–10 °C, while they exhibit similar trends of change. As the pressure drop increases from 0 to 27 Pa, both the mean temperature and the peak temperature have an obvious drop, particularly for high inlet velocity of air (e.g. $U_{inlet} = 3\,\text{m s}^{-1}$). As the pressure drop exceeds 27 Pa, the temperature decline is getting very slow, and almost all the Pareto-optimal curves overlap, indicating that the change in air inlet velocity has a negligible impact on the thermal performance of the heat sink system.

From Figure 6.6a–d, it is clear that in the passive cooling mode, the geometric dimensions remain unchanged and reach the upper limit of the allowable range ($H = 0.6$ m, $D = 0.15$ m, $L_{ctrl} = 1.0$ m, $L_{sd} = 1.0$ m). The resulting wall area is up to $5.38\,\text{m}^2$ for maximizing heat removal. In the active cooling mode, it can be observed that, though the fin thickness remains almost unchanged at $D = 0.15$ m, the fin height exhibits a strong linear growth trend from 0 to 0.6 m in the Pareto solutions. This growth trend suggests that the evolutionary algorithm focuses on optimizing conflicting objectives primarily by adjusting the fin height of the heat sink system. Followed by the fin height, it appears that fin length has a secondary role in the multi-objective optimal design of the heat sink system; e.g. in general, the total fin length ($L_{tot} = 2L_{sd} + L_{ctrl}$) increases from 1.6 to 2.8 m along the Pareto-optimal curve. Due to the growth of fin height and length, as seen in Figure 6.6f, the wall area of the fins in the active mode also has a linear growth from 0 to 3.90 m^2.

Figure 6.7 visualizes the distributed flow and pressure for the representative scenarios at $U_{inlet} = 2.0\,\mathrm{m\,s^{-1}}$. Due to the different serrated shapes of the fins, it is evident that the flow field exhibits distinct flow patterns as it flows through the heat sink system. For example, the height and total length of the spaced fins have an obvious rise from 0.14 and 1.70 m for the low-cost design to 0.45 and 2.30 m for the high-performance design, respectively. The increase in fin size can reduce the flow gap between the heat sink system and the channel wall. Note that, as the air flows past the heat sink system, it must change direction and create different hydrodynamic zones in front of and behind the system. A stagnation zone is formed as air encounters the heat sink system, where zero velocity (Figure 6.7a–c) and maximum pressure (Figure 6.7d) are observable. Similarly, the blockage in the flow direction causes a decrease in air velocity and an increase in air static pressure, forming a

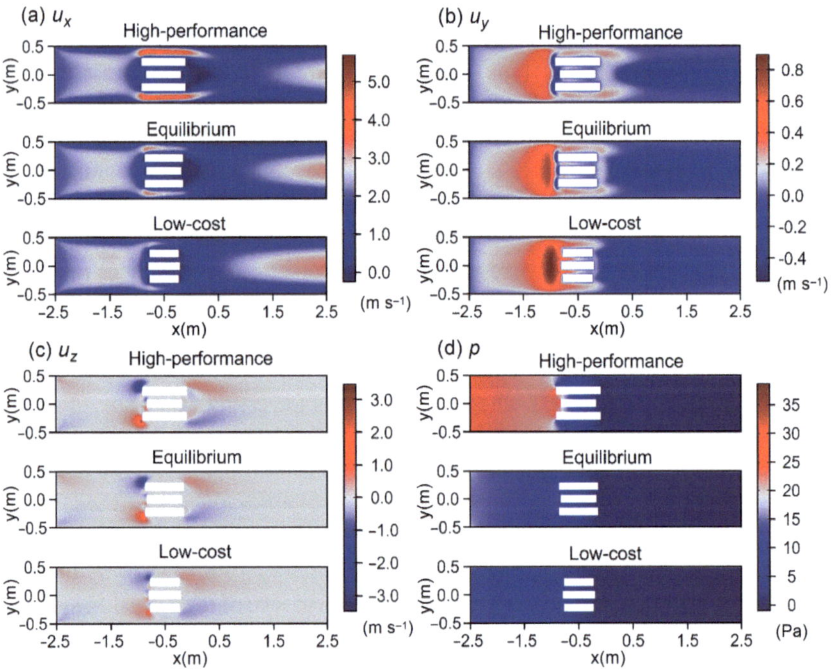

Figure 6.7 Comparison of flow and pressure fields among the representative scenarios at $U_{inlet} = 2.0\,\mathrm{m\,s^{-1}}$, in the cross-section ($z = -0.2$ m) of the coordinate system.

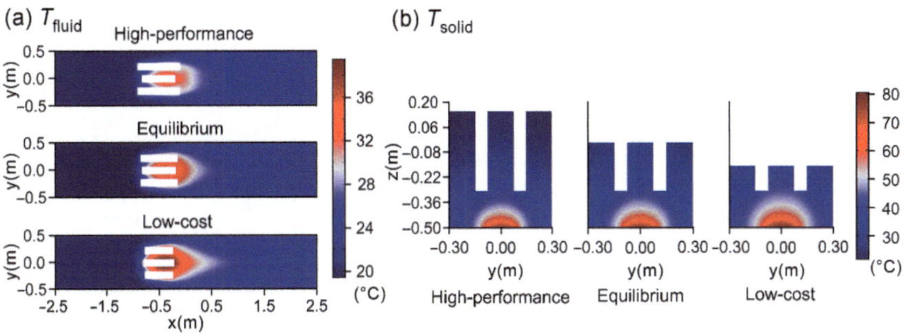

Figure 6.8 The optimal temperature field of the heat sink system at $U_{inlet} = 2.0\,m\,s^{-1}$, in the cross-section ($z = -0.2$ m; $x = -0.5$ m) of the coordinate system. (a) fluid temperature, (b) solid temperature.

wake zone behind the heat sink system. After that, the vortex flow is created by the separation and reattachment of airflow. In all, the high-performance design has significantly larger stagnation and wake zones than those in the low-cost design, with sizes ranging from 1.5 to 3.0 times larger.

As mentioned above, increasing the size of the fins can enlarge the stagnation and wake zones, which increases thermal resistance as the air contacts the heat sink system. Despite this, the increased height and length of the fins can increase heat transfer areas, thereby facilitating the inlet air to remove more heat from the heat source. Furthermore, at the same inlet velocity, the increase in the fin size causes the majority of the inlet air to pass through the reduced gap between the heat sink system and the channel wall. As shown in Figure 6.7a, this in turn enhances the velocity component of $\boldsymbol{u}_x$ according to the Bernoulli equation. The combined increase in heat transfer areas and flow velocity contributes to an overall improvement in heat transfer efficiency. At $U_{inlet} = 2.0\,m\,s^{-1}$, the total heat transfer areas and coefficient increase from $1.84\,m^2$ and $5.91\,W\,m^{-2}\,K^{-1}$ for the low-cost design to $3.72\,m^2$ and $6.29\,W\,m^{-2}\,K^{-1}$ for the high-performance design, respectively. Correspondingly, the mean temperatures of the heat source surface and fin wall declined from 70.68 to 66.33 °C, and 34.13 to 26.75 °C. These results are consistent with the change in temperature fields seen in Figure 6.8. On the other hand, note that the improvement in heat removed from the heat source comes at the cost of a greater pressure drop. The resulting pressure drop has increased by 2.1 times, from 12.19 Pa for the low-cost design to 25.16 Pa for the high-performance design. Relatively, the equilibrium design offers a more moderate option for the design of the heat sink system, where the mean temperature and pressure drop are 68.12 °C and 16.27 Pa, respectively.

6.4 Illustrative Example 2: Tubular Air Cooler Model

6.4.1 System Description and Objectives

In the second example, we utilize a tubular air cooling system to demonstrate the potential of PINN in conjunction with transfer learning. In Figure 6.9a, it is seen

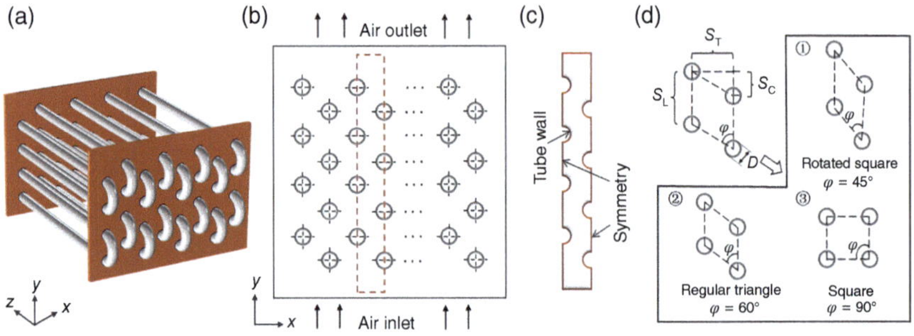

Figure 6.9 Illustrations of the geometry of the air-cooled heat exchanger system. (a) Schematic view. (b) Cross-section of the tube bundle. (c) Computational domain with assigned boundary conditions. (d) Geometric variable definition and three representative topology designs that are termed rotated square, regular triangle, and standard square, here φ is the intersection angle between tube centers.

that the air cooler system consists of eight-row and eight-column plain-type tubes with a staggered configuration. Figure 6.9b presents the corresponding cross-section of these tube bundles, where the region enclosed by the red dashed line is defined as the computational domain due to the symmetry of the geometry. The tube bundles can be divided into multiple heat exchange units, and each unit is composed of four copper tubes that have the same centerline spacing (Figure 6.9c). To accurately characterize this specific topology, in Figure 6.9d, we define the key geometric variables by using three tube pitch parameters, namely transversal tube pitch (S_T), longitudinal tube pitch (S_L), and cross tube pitch (S_C). S_C is a newly introduced geometric parameter and is defined as a multiple of S_L, i.e., $S_C = F \times S_L$, $F \in [-1, 1]$, where F is a ratio between them. Given the same diameter of the tube (D), the random combination of these geometric variables results in countless tube bundle arrangements, which is our quantity of interest since they inherently determine the thermal and hydraulic behaviors of the system.

The uncertain variables related to the inlet air can complicate the operation and management of the heat exchanger system in practice. In Figure 6.9a, the ambient air flows upwards and traverses the outer wall of the tube bundles, while the hot fluid pumped from the main tube is evenly divided and enters the tube bundles. A forced convective heat transfer process occurs inside the heat exchanger, where the sensible heat of the hot fluid is indirectly absorbed by upward airflow outside the tube bundles. In this forced convective heat transfer process, the temperature drop of the hot fluid only relies on sensible heat exchange with inlet air to achieve the required cooling capacity. This means that the thermal properties of the inlet air can strongly impact the characteristics of fluid flow and heat transfer inside the system. Specifically, the dry-bulb temperature of inlet air that depends on fluctuating weather conditions is considered an uncertain variable, while the velocity of inlet air is the major controllable variable that could mitigate the impact of the uncertainty and improve the system's behavior. In all, to obtain the optimal geometric and operating variables while fully accounting for the impacts of weather uncertainty, a

Table 6.4 The input space of different variables.

Input variables	Symbol	Unit	Lower bound	Upper bound
Geometric variables	$\mathcal{G}$			
Transverse tube pitch	S_T	m	$2D$	$4D$
Longitudinal tube pitch	S_L	m	$3D$	$5D$
Cross tube pitch	S_C	m	$-5D$	$5D$
Tube diameter	D	m	0.01	
Operating variables	$\mathcal{O}$			
Velocity of inlet air	U_{in}	m s^{-1}	1	2
Uncertain variables	$\mathcal{U}$			
Dry-bulb temperature of inlet air	T_{in}	K	259.15	306.15

stochastic optimization is conducted in this example. The ranges of variation for the aforesaid variables $\mathcal{G}$, $\mathcal{O}$, and $\mathcal{U}$ are listed in Table 6.4.

As shown in Figure 6.10, the reverse design starts with a decomposition of the geometric space, generating a batch of discrete geometric designs through the use of the Halton sequence sampling method, $^{(m)}\in\mathcal{G}$. That is, the entire design space is represented using M discrete geometric designs, and each design g represents a

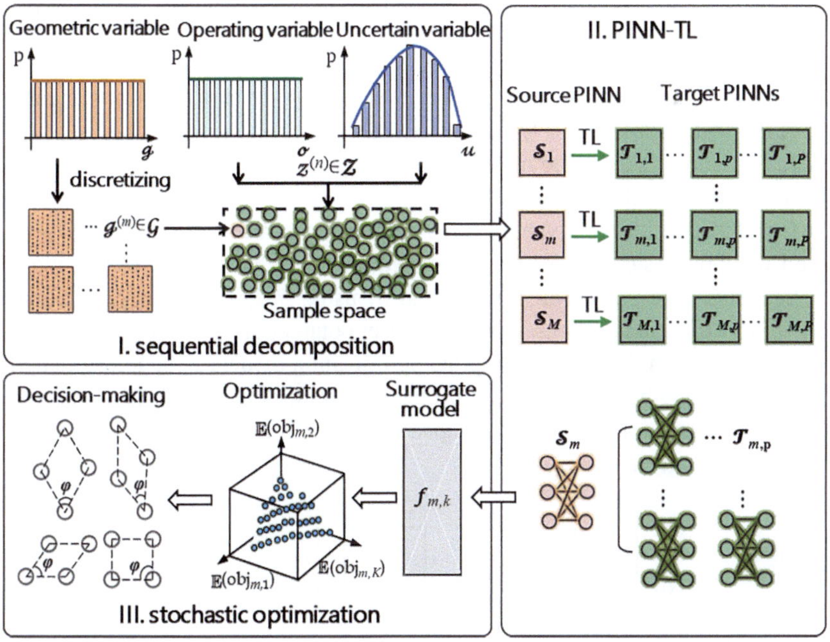

Figure 6.10 The proposed framework combines sequential decomposition, PINN, and TL for reverse design. (a) sequential decomposition; (b) PINN-TL; (c) stochastic optimization.

combined set of geometric variables. Meanwhile, the uncertainty of weather conditions is realized by extracting N uncertain samples that can represent the actual environment, and these samples are fed into each optimization model. We employ a non-intrusive sampling approach called the stochastic reduced-order model (SROM, see details in Chapter 8) to generate a finite set of stochastic samples with varying probabilities, $\mathscr{z}^{(n)} \in \mathcal{Z}$, $\mathscr{z}^{(n)} = \{\mathscr{o}^{(n)}, \mathscr{u}^{(n)}\}$.

In this example, if we construct a separate PINN model for each sample, there will be a total of $M \times N$ PINN models to be trained. For each geometric design, we randomly select a source model from N PINN models and then leverage it to train the rest $P = N - 1$ target models via transfer learning. The N PINN models obtained in total can be concatenated to construct the surrogate models that map all input and output variables of interest. Here, we encounter the following two conflicting objectives [64] that are often used in engineering design:

$$\text{Pressure drop:}\ \ \Delta p = \overline{p}_{\text{in}} - \overline{p}_{\text{out}} \tag{6.35}$$

$$\text{Nusselt number:}\ \ \text{Nu} = \frac{h_{\text{a}} D}{\lambda} \tag{6.36}$$

where $\Delta p = \overline{p}_{\text{in}} - \overline{p}_{\text{out}}$ is the pressure drop between the inlet and outlet of airflow in the computational domain, and h_{a} is the air-side convection heat transfer coefficient that can be calculated from:

$$h_{\text{a}} = \frac{Q_{\text{a}}}{A \Delta T_{\text{m}}} = \frac{\dot{m}_{\text{a}} C_{\text{pa}} \left(\overline{T}_{\text{a,out}} - \overline{T}_{\text{a,in}}\right)}{A\left\{\left(\overline{T}_{\text{a,out}} - \overline{T}_{\text{a,in}}\right) / \ln\left[\left(T_{\text{wall}} - \overline{T}_{\text{a,in}}\right) / \left(T_{\text{wall}} - \overline{T}_{\text{a,out}}\right)\right]\right\}} \tag{6.37}$$

where A and Q_{a} are the total heat transfer area and the air-side heat transfer rate, ΔT_{m} is the logarithmic mean temperature difference between the tube wall and the air, $\overline{T}_{\text{a,in}}$ and $\overline{T}_{\text{a,out}}$ are the mean temperatures of air at inlet and outlet, T_{wall} is the temperature of the tube wall.

We therefore consider two types of surrogate models regarding the bi-objectives of Nu and Δp of the system, which are embedded in the stochastic optimization models. These surrogate models are constructed by using the generalized polynomial chaos (gPC) expansion [43] based on the dataset obtained from the source and target PINN models. In this way, the original optimization model can be reformulated as a set of sampling-based stochastic nonlinear programming models. The goal is to obtain optimal solutions through the expected maximization of the distribution of the objectives, as given by:

$$\max_{\mathcal{O},\mathcal{U}} \left\{ \begin{bmatrix} \mathbb{E}(\Delta p) \\ \cdots \\ \mathbb{E}(\text{Nu}) \end{bmatrix}_1, \ldots, \begin{bmatrix} \mathbb{E}(\Delta p) \\ \cdots \\ \mathbb{E}(\text{Nu}) \end{bmatrix}_m, \ldots, \begin{bmatrix} \mathbb{E}(\Delta p) \\ \cdots \\ \mathbb{E}(\text{Nu}) \end{bmatrix}_M \right\} \tag{6.38}$$

$$\text{s.t.}\ \ \mathbb{E}(\text{obj}_{m,k}) = \sum_{n}^{N} \text{prob}_{m,n} \times f_{m,k}(\mathcal{O}_n, \mathcal{U}_n)$$

$$\mathcal{O}_n \in [\mathcal{O}^{\text{lb}}, \mathcal{O}^{\text{ub}}]$$

$$\mathcal{U}_n \in [\mathcal{U}^{\text{lb}}, \mathcal{U}^{\text{ub}}]$$

$$\boldsymbol{\Psi} \in [\boldsymbol{\Psi}^{\text{lb}}, \boldsymbol{\Psi}^{\text{ub}}]$$

$$\forall m \in \{1, \dots, M\}, \quad \forall n \in \{1, \dots, N\}$$

where $\mathcal{O}_n$ and $\mathcal{U}_n$ are the realizations of operating and uncertain variables in a specific sample n; lb and ub are the corresponding lower and upper bounds for these variables; prob_n is the probability related to the occurrence of a specific sample n; Ψ is the design constraint that defines the outlet temperature of the hot fluid to meet the cooling requirement. $f_{m,k}(\cdot)$ is the surrogate model corresponding to the kth objective under mth geometric design, which is used as the objective function depending on the operating and uncertain variables.

The Pareto solutions $\{\Delta p^*_{m,n}, Nu^*_{m,n}\}$ are generated by repeatedly solving this optimization model for $M \times N$ times according to the numbers of uncertain samples and geometric designs. Similar to example 1, the NSGA-II [54] is also adopted to handle each optimization problem since it can efficiently search for the Pareto solutions. After that, the TOPSIS method is applied to identify the optimal operating conditions for each uncertain sample and determine the expected value of the optimal objective function $\left\{\mathbb{E}\left(\Delta\overline{p}^*_m\right), \mathbb{E}\left(\overline{\text{Nu}}^*_m\right)\right\}$ for each geometric design. Finally, the most preferred solutions regarding the best-performing geometric design $\{\mathbb{E}(\Delta\overline{p}^*), \mathbb{E}(\overline{\text{Nu}}^*)\}$ are obtained by re-applying with the TOPSIS method.

6.4.2 Improved PINN Structure

Herein, we propose a specialized segregated-network architecture to effectively decompose the standard PINN architecture into multiple small-size sub-networks. This segregation allows for a better distinction between the state variables of interest within the combined loss function. Figure 6.11 illustrates the segregated-network architecture for surrogate modeling of multi-physical fields in the heat exchanger system. As shown in this figure, the neural network of this system representation is decomposed into two serial sub-networks (namely the flow sub-network and the heat sub-network) that take the spatial coordinate $[\mathbf{x}] = [x, y]$ as inputs and are designed to successively approximate the latent solutions of Navier–Stokes and energy equations. According to the correlation between these two governing equations for incompressible flow, it is assumed that there is a one-way coupling between the fluid flow process and the heat transfer process. This assumption implies that once the fluid flow sub-network has been trained and can satisfy the convergence criteria, the resulting field distribution of velocity $\{\boldsymbol{u}^*, \boldsymbol{v}^*\}$ could be used as an intermediate input for training the heat transfer sub-network. Meanwhile, we disable the gradient computation for velocity components, which effectively prevents the update of parameters in the flow sub-network. In this way, it is possible to substantially speed up the learning rate of multiphysics regarding field distributions of key state variables including velocity, pressure, and temperature, i.e. $[\mathbf{x}] \rightarrow [u, v, p]$, $[\mathbf{x}] \rightarrow [T]$.

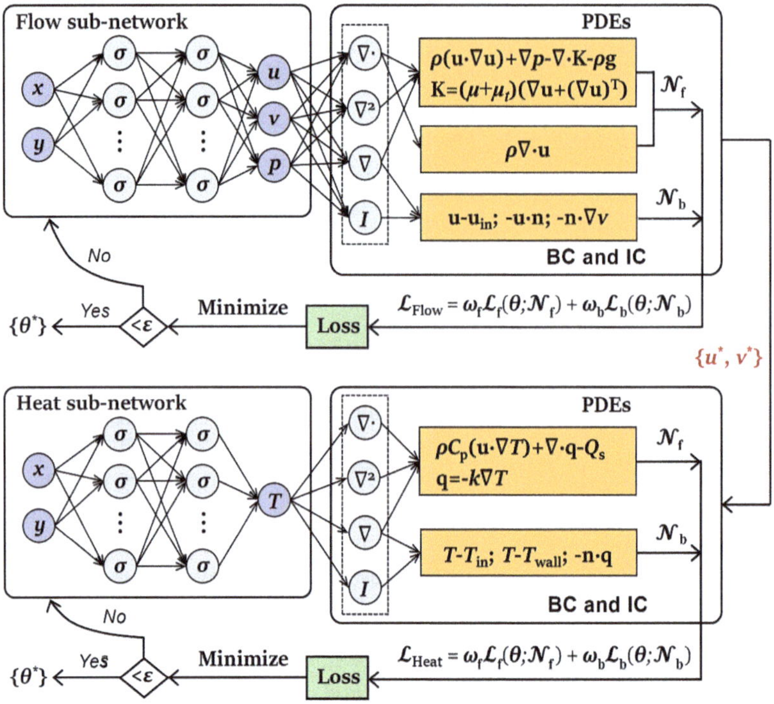

Figure 6.11 A schematic diagram of the segregated-network PINN architecture for surrogate modeling of fluid flow and heat transfer behavior. For the governing equations of the heat exchanger system, the boundary conditions include differential, nonlinear, and identity terms, which could be subject to Dirichlet or Neumann conditions.

It is well known that the network architecture plays a crucial role in improving the prediction ability of PINN, particularly for multiphysics learning tasks. Here, we address the convective heat transfer modeling problem for the heat exchanger system by comparing three representative types of architecture, i.e. fully-connected network (FCN), Fourier network (FN), and modified Fourier network (MFN). The theories for the three network architectures are detailed in Ref. [43] In order to compute arbitrary-order derivatives, we need to use a smooth and differentiable activation function. Note that the commonly used activation functions, such as ReLU, may fail to satisfy the continuous second-order derivatives. To address this issue, the Swish function with continuous derivatives, $\sigma = \mathbf{x} \times \text{sigmoid}(\beta \mathbf{x})$, with a fixed parameter $\beta = 1$, is employed in each layer except the last one [65]. To improve the convergence performance, the loss function of each PINN can take the form of a Monte Carlo integration approximation since it can maintain the consistency of the loss per area across the domain and ensure that the loss is minimal throughout the entire spatial domain.

6.4.3 Transfer Learning

A generic description of transfer learning is as follows [66]: Given a source domain $\mathcal{D}_S = \{\mathcal{X}_S, P(X_S)\}$ and learning task $\mathcal{T}_S = \{\mathcal{Y}_S, f_S(\cdot)\}$, a target domain $\mathcal{D}_T = \{\mathcal{X}_T,$

$P(X_T)\}$ and learning task $\mathcal{T}_T = \{\mathcal{Y}_T, f_T(\cdot)\}$, it aims to improve the learning of the target predictive function $f_T(\cdot)$ using the knowledge in $\mathcal{D}_S$ and $\mathcal{T}_S$, where $\mathcal{D}_S \neq \mathcal{D}_T$ or $\mathcal{T}_S \neq \mathcal{T}_T$; $\mathcal{X}$ and $\mathcal{Y}$ are the feature and label spaces; $P(X)$ is a marginal probability distribution, where $X \in \mathcal{X}$. In this study, we perform design space exploration and operating parameter optimization tasks, which are based on constructing a surrogate model using the solutions obtained from a batch of PINN models. In theory, neural network solvers can transfer knowledge across these PINN models via transfer learning since both source PINN and target PINN originate from the decomposition of the same design space of the heat exchanger system. Here, a parameter-transfer learning approach that assumes $\mathcal{T}_S$ and $\mathcal{T}_T$ share partial parameters or prior distributions of the hyper-parameter of the source model is employed, where the transferred knowledge acquired by a source PINN for a specific geometry is encoded into the shared parameters or priors. By discovering the shared parameters or priors, the acquired knowledge can be applied to the construction of target PINN models that have similar geometry and slightly different operating and uncertain conditions.

A schematic of the parameter-transfer learning approach is shown in Figure 6.12. Firstly, the source PINN model is pre-trained with two serial sub-networks regarding fluid flow and heat transfer using the sample set, namely, $\mathcal{N}_S = [\mathcal{N}_{S,f}, \mathcal{N}_{S,b}]$. The parameters of these two sub-networks are tuned by minimizing the following combined losses $\mathcal{L}_{S,\text{Flow}} = \omega_{S,f}^{\text{Flow}}\mathcal{L}_{S,f}^{\text{Flow}} + \omega_{S,b}^{\text{Flow}}\mathcal{L}_{S,b}^{\text{Flow}}$ and $\mathcal{L}_{S,\text{Heat}} = \omega_{S,f}^{\text{Heat}}\mathcal{L}_{S,f}^{\text{Heat}} + \omega_{S,b}^{\text{Heat}}\mathcal{L}_{S,b}^{\text{Heat}}$. The trained optimal network parameter set $\theta_S^* = \left[\theta_{S,\text{Flow}}^*, \theta_{S,\text{Heat}}^*\right]$ is further used for the initialization of network parameters for each target PINN model. Based on the sample set $\mathcal{N}_T = [\mathcal{N}_{T,f}, \mathcal{N}_{T,b}]$, the initialized

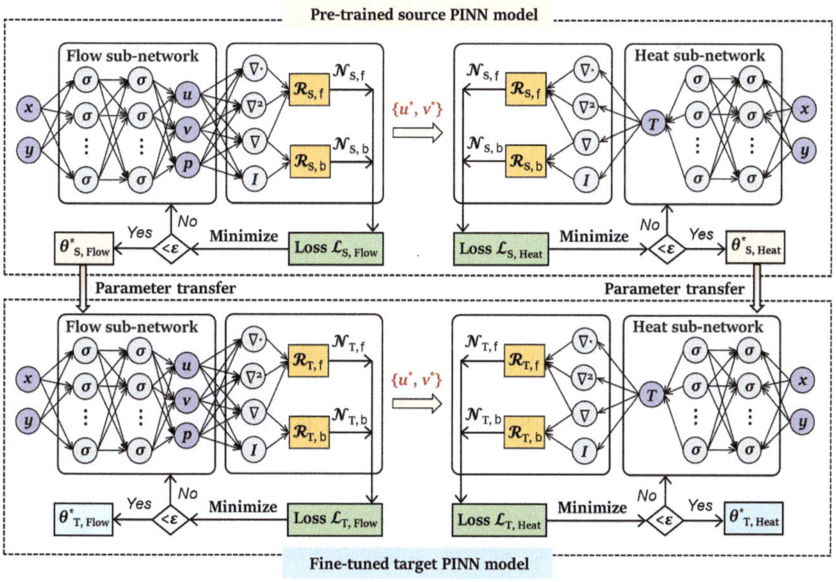

Figure 6.12 Schematic representation of the proposed parameter-transfer learning between source PINN and target PINN models.

Table 6.5 Hyperparameters setting of source and target PINN models.

	Source PINN		
Parameters	**Flow**	**Heat**	**Target PINN**
No. neurons	256	256	256
No. layers	6	6	6
Learn rate schedule	Exponential decay	Exponential decay	Exponential decay
Learning rate	5×10^{-4}	5×10^{-4}	1×10^{-4}
Decay step	15 000	8000	3000
No. iterations	1.5×10^{6}	0.8×10^{6}	0.3×10^{6}
Activation function	Swish	Swish	Swish
Optimizer	Adam	Adam	Adam

target PINN model is then re-trained to cope with the changes in operating and uncertain conditions without having the NN model trained from scratch. In this stage, the parameters of two similar sub-networks in the target PINN models are fine-tuned by minimizing the combined losses, $\mathcal{L}_{\text{T,Flow}} = \omega_{\text{T,f}}^{\text{Flow}}\mathcal{L}_{\text{T,f}}^{\text{Flow}} + \omega_{\text{T,b}}^{\text{Flow}}\mathcal{L}_{\text{T,b}}^{\text{Flow}}$ and $\mathcal{L}_{\text{T,Heat}} = \omega_{\text{T,f}}^{\text{Heat}}\mathcal{L}_{\text{T,f}}^{\text{Heat}} + \omega_{\text{T,b}}^{\text{Heat}}\mathcal{L}_{\text{T,b}}^{\text{Heat}}$, with a limited number of iterations, smaller learning rate, and a faster learning rate decay, as listed in Table 6.5. Iteratively, $\boldsymbol{\theta}_{\text{S}}^{*}$ is updated and $\boldsymbol{\theta}_{\text{T}}^{*} = \left[\boldsymbol{\theta}_{\text{T,Flow}}^{*}, \boldsymbol{\theta}_{\text{T,Heat}}^{*}\right]$ is obtained at the end of this stage.

Although the space decomposition and transfer learning make the multiphysics problem trainable, the final computational time may still far exceed the allowable specification for engineering design. In the PINN framework, Monte Carlo sampling of spatial coordinates is employed to generate residual points, which serve as the input dataset for the neural network. These residual points are similar to the meshes used in traditional numerical methods. Increasing the number of residual points can improve computational accuracy, but it also leads to longer training times and increased computational load. In transfer learning, the data size of the source domain is generally much larger than that of the target domain, $|X_{\text{S}}| \gg |X_{\text{T}}|$. Therefore, it can be inferred that the number of residual points provided to the target model can be reduced to shorten the computational time. However, it is necessary to investigate the appropriate size of residual points $|\mathcal{N}_{\text{T}}|$ for training the target PINN models. Herein, we propose a point density adjustment (PDA) strategy that can quickly identify an optimal ratio of sample size in the domain to that on the boundary by comparing different sampling schemes. For example, the size of residual points on the boundary remains constant ($|\mathcal{N}_{\text{T,b}}| = |\mathcal{N}_{\text{S,b}}|$), while the proportion of interior residual points drops gradually.

Note that, the presence of numerous pointwise operations in network learning could put huge pressure on the memory sub-system of a GPU. To streamline the training process, we further propose the use of a domain-specific compiler called accelerated linear algebra (XLA) [67, 68] in conjunction with transfer learning. XLA can enable kernel fusion and just-in-time compilation of TensorFlow graphs. With

XLA, a batch of pointwise operations can be executed simultaneously in a single kernel, reducing the number of memory transfers from GPU memory to the computation units.

6.4.4 Results

The results of NMAPEs for the velocity, pressure, and temperature predicted by the PINN model with different network architectures are listed in Table 6.6. These results clearly show that both FN and MFN architectures outperform FCN architecture. For example, the y-velocity and temperature NMAPEs for the MFN architecture have a significant increase from 4.60% to 32.99% and 3.15% to 64.14% compared with the FCN case, respectively. This increase reflects that Fourier-derived architectures lead to significant improvement in dealing with the increased nonlinearity problems over the regular FCN architecture because they can effectively reduce the impact of spectral bias [69] and have the ability to capture steep gradient variation (i.e. near the tube wall). Compared with the FN architecture, the MFN architecture further improves the prediction accuracy, e.g. the NMAPEs of x-velocity, y-velocity, pressure, and temperature drop from 2.57% to 1.54%, 9.47% to 4.60%, 10.60% to 6.80%, and 6.22% to 3.15%, respectively. From these results, it is concluded that the PINN model with MFN architecture is a more accurate solution to describe the underlying behaviors of fluid flow and heat transfer inside the heat exchanger system under consideration.

The aforesaid best-performing PINN trained with MFN architecture is used as the source model for implementing parameter-transfer learning. Herein, for each geometric design, a total of 12 groups of stochastic samples from the combined operating and uncertain spaces are generated according to the requirement for constructing high-fidelity surrogate models. Based on these samples, we only randomly select a single one as the source PINN model, while the rest are the target PINN models. As previously mentioned, the target PINN models are fine-tuned using parameter-transfer learning, as opposed to training them from scratch. Taking the rotated square design, for example, the source PINN model is fully trained (full run) with a total of 2.3×10^6 iterations (fluid flow 1.5×10^6, heat transfer 0.8×10^6), as shown in Figure 6.13. The training loss of target PINN models in transfer learning runs converged at a much faster speed of 3×10^5 for both fluid flow and heat transfer sub-networks, which leads to a time reduction by a factor of 3.73 (the average total training time of a single target PINN model is 20.35 hours).

Table 6.6 NMAPEs for the trained PINN models using different network architectures.

Outputs	FCN	FN	MFN
$\boldsymbol{u}$ (m s^{-1})	6.67%	2.57%	1.54%
$\boldsymbol{v}$ (m s^{-1})	32.99%	9.47%	4.60%
$\boldsymbol{p}$ (Pa)	16.95%	10.60%	6.80%
$\boldsymbol{T}$ (K)	64.14%	6.22%	3.15%

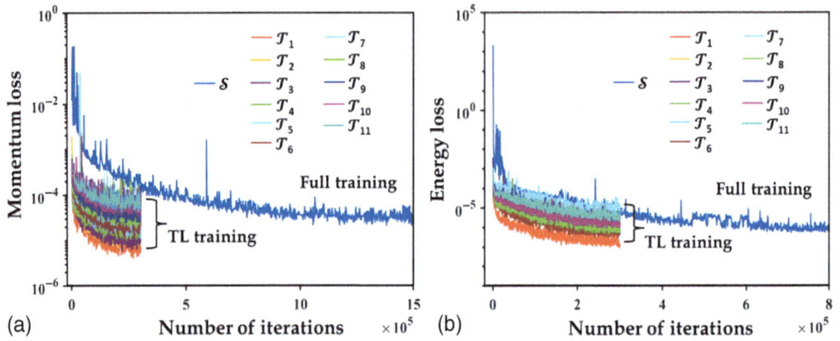

Figure 6.13 Training losses at different air velocities and temperatures with/without transfer learning. (a) Fluid flow and (b) heat transfer. The model training was performed using an HP Precision T7920 workstation, where the basic configuration is as follows: NVIDIA 3090 GPU, Intel(R) Xeon E5-6230 CPU@ 2.30 GHz, 128GB RAM, 4TB SSD.

Although transfer learning performs well in improving convergence rates, the total training time is still unacceptable for a complex system since the simultaneous training of hundreds of target PINN models is required in transfer learning. In this example, the decomposition strategy sequentially generates 30 geometric designs and 12 stochastic samples; the total training time for the decomposed PINN models is expected to be up to 75.82 hours × 30 × 12 = 27 295 hours (3.16 years). As shown in Figure 6.14b, by combining with parameter-transfer learning, the expected total training time can be preliminarily reduced by 3.12 times from 27 295 to 8 748 hours. On this basis, we can further cut down the abovementioned training time to 3133 hours by using simultaneous PDA and XLA approach in transfer learning, as shown in Figure 6.14c,d. Despite the significant time reduction, the total computational cost is still unbearable in practice. With the support of the NVIDIA TensorFlow container, parallel processing using multi-GPU configuration has been introduced to the training process of this example. In Figure 6.14e, taking the use of two NVIDIA 3090 GPUs as an example, the training time can be remarkably reduced by about 3.6 times from 3133 to 871 hours. Furthermore, it is noted that one great advantage of transfer learning is that the source and target models are independent of each other, and thus they allow for being deployed to parallelly run on a multi-workstation. In Figure 6.14f, when these target models are evenly assigned to four nodes, the resulting training time is proportionately reduced to 326 hours (13.6 days, without considering the overhead time between nodes).

Figure 6.15a presents the optimal solutions regarding the expected values of the Nu and Δp obtained from multi-objective stochastic optimization. By applying the TOPSIS method, the most preferred solution is $\mathscr{g}^{(27)}$, where $\mathbb{E}(\overline{\mathrm{Nu}}^*) = 25.38$ and $\mathbb{E}(\overline{\Delta p}^*) = 2.81$ Pa. For a better comparison, three representative topology designs (see Figure 6.9d) often used in actual engineering design and manufacturing are also evaluated, namely rotated square ($\mathscr{g}^{\mathrm{RS}}$, $S_\mathrm{T} = 0.5S_\mathrm{L}$, $S_\mathrm{C} = 0.5S_\mathrm{L}$, $\varphi = 45°$), regular triangle ($\mathscr{g}^{\mathrm{RT}}$, $S_\mathrm{T} = \sqrt{3}/2S_\mathrm{L}$, $S_\mathrm{C} = 0.5S_\mathrm{L}$, $\varphi = 60°$), and standard square ($\mathscr{g}^{\mathrm{SS}}$, $S_\mathrm{T} = S_\mathrm{L}$, $S_\mathrm{C} = 0$, $\varphi = 90°$). The expected Nusselt numbers for $\mathscr{g}^{\mathrm{RS}}$, $\mathscr{g}^{\mathrm{RT}}$, and $\mathscr{g}^{\mathrm{SS}}$ are 26.92,

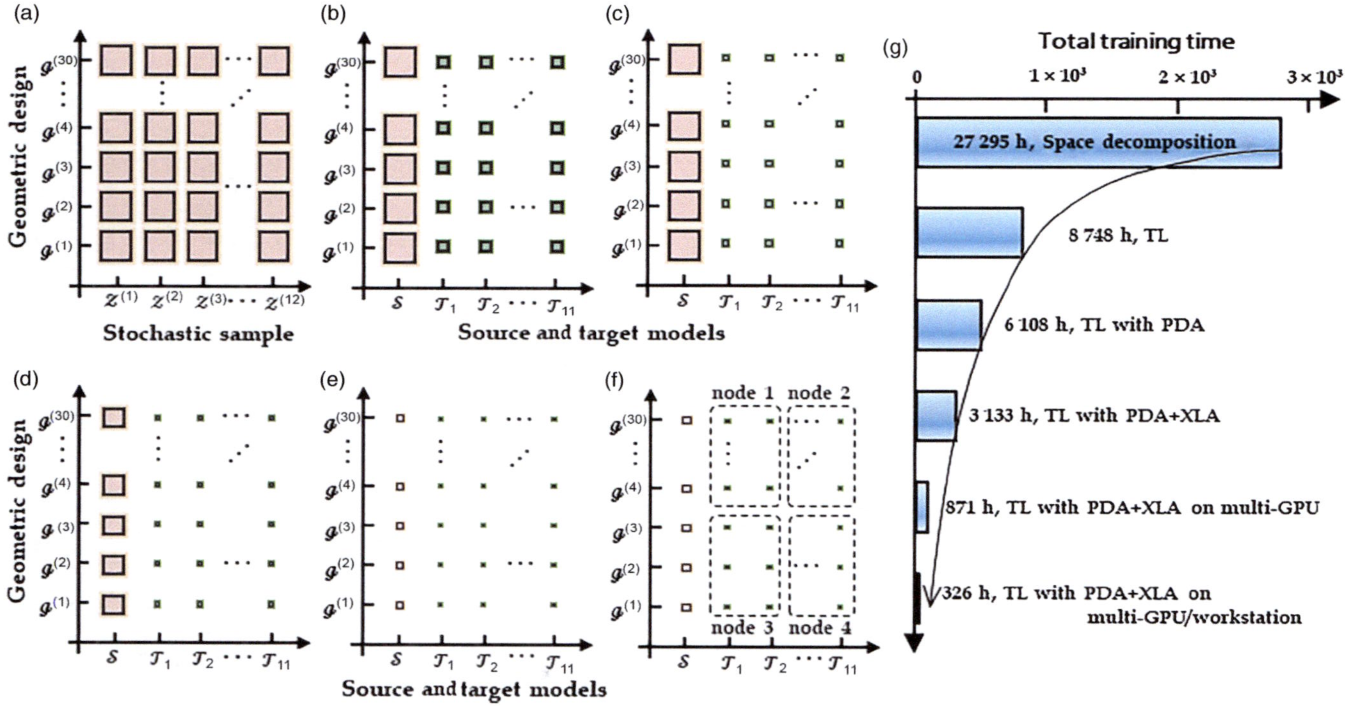

Figure 6.14 Review of training time for PINN models of the heat exchanger system with (a) standard PINN with space decomposition, (b) PINN-TL without acceleration, (c) PINN-TL with PDA, (d) PINN-TL with PDA and XLA, (e) PINN-TL with PDA and XLA on multi-GPU configuration, (f) PINN-TL with PDA and XLA on multi-GPU and multi-workstation configuration, and (g) the corresponding total training time for the aforesaid methods. Note that, the area of each square in (a) corresponds to a training time of 75.82 hours, while the training times for the squares in (a–f) are proportional to their areas.

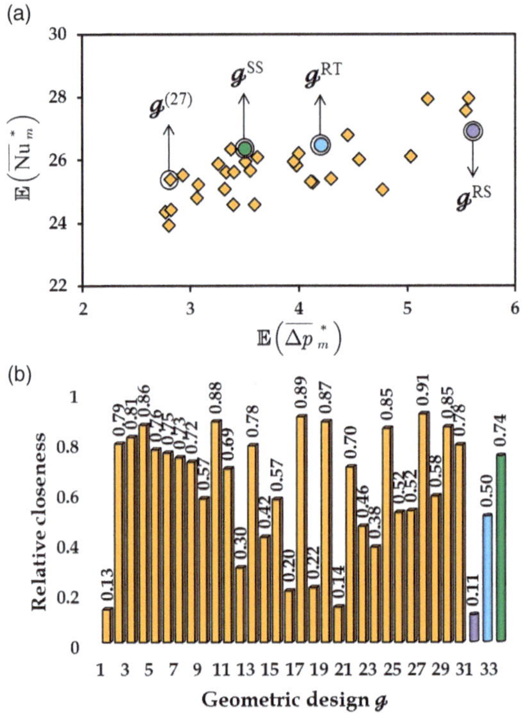

Figure 6.15 (a) The expected values of pressure drop and Nusselt number, (b) the relative closeness from TOPSIS analysis for each geometric design.

26.47, and 26.37, which is 6.07%, 4.29%, and 3.90% higher than that for $^{(27)}$, respectively. Meanwhile, the expected pressure drop for $^{(27)}$ has a dramatic rise from 2.81 to 5.61 Pa for g^{RS}, 4.20 Pa for g^{RT}, and 3.51 Pa for g^{SS}. Figure 6.13b shows the relative closeness of each design point with their corresponding values. Herein, a larger value of relative closeness means a closer distance to the ideal solution and a further distance away from the nadir solution. The relative closeness of the $^{(27)}$ has the largest relative closeness (0.91) as compared to the other geometric designs; therefore it is recommended as the most preferred solution for the case study.

Figure 6.16 depicts the detailed geometry for $^{(27)}$, g^{RS}, g^{RT}, and g^{SS} with respect to the optimal tube arrangement. As depicted, the value of transversal tube pitch S_T in $^{(27)}$ is $S_T = 0.0369$ m, which is greater than that in g^{RS} ($S_T = 0.02$ m), g^{RT} ($S_T = 0.026$ m), and g^{SS} ($S_T = 0.03$ m). Notably, the $^{(27)}$ has nearly the same longitudinal tube pitch $S_L = 0.0302$ m as compared to g^{RT} ($S_L = 0.03$ m) and g^{SS} ($S_L = 0.03$ m). Among them, the g^{RS} has the longest longitudinal tube pitch $S_L = 0.04$ m. The different included angles between tubes φ can be formed by changing the cross tube pitch S_C. To be more specific, the value of S_C is $S_C = -0.0056$ m in $^{(27)}$, and an included angle of $\varphi = 98.6°$ is formed, which is greater than that in g^{RS} ($S_C = 0.02$ m, $\varphi = 45°$), g^{RT} ($S_C = 0.015$ m, $\varphi = 60°$) and g^{SS} ($S_C = 0$ m, $\varphi = 90°$).

In Figure 6.17a, we present the probability of occurrence for all 50 uncertain samples used in the stochastic optimization. As shown, they have a scattered and wide distribution that ranges from 0.0012 to 0.0627. According to these samples, Figure 6.17b presents the optimal value of the expected inlet air velocity. The

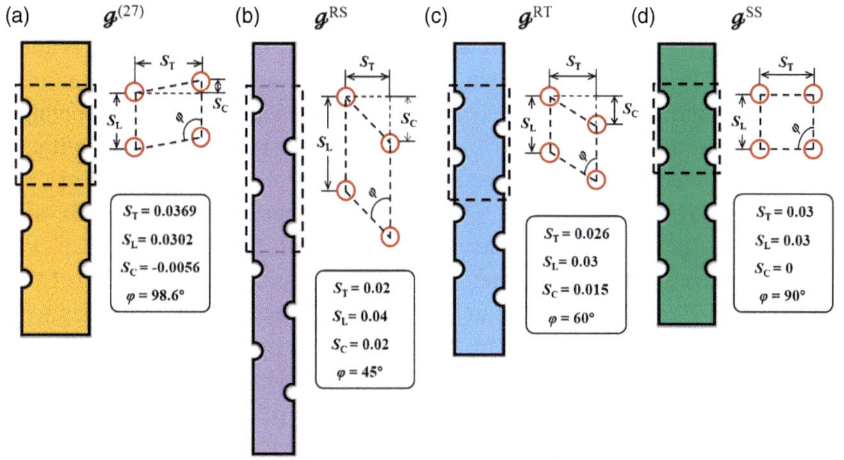

Figure 6.16 Optimal geometric details of (a) the most preferred solution ($^{(27)}$), (b) rotated square (g^{RS}), (c) regular triangle (g^{RT}), and (d) standard square (g^{SS}).

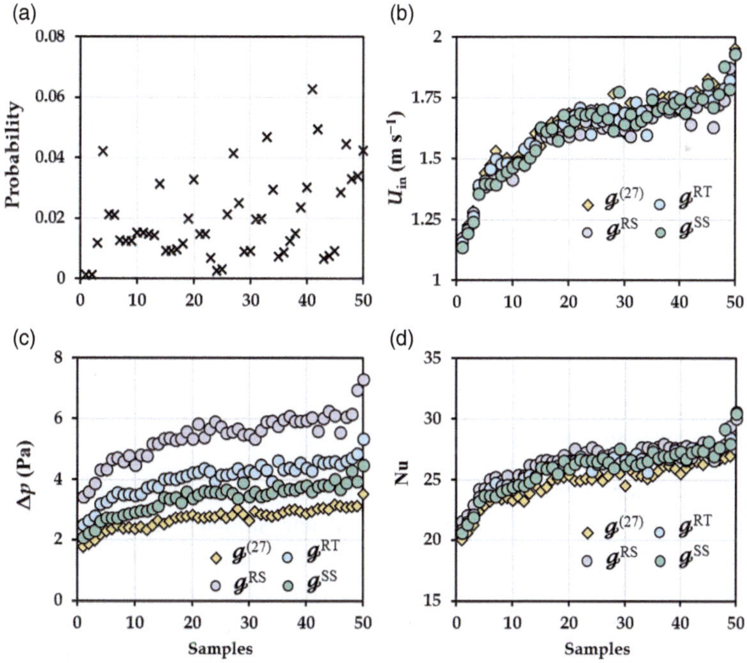

Figure 6.17 The optimal results for all uncertain samples in $^{(27)}$, g^{RS}, g^{RT}, and g^{SS}. (a) The probability of occurrence, (b) pressure drop, (c) Nusselt number, and (d) inlet air velocity. All samples are rearranged in ascending order of the dry-bulb temperature of inlet air.

velocity of inlet air under each geometric design shows an almost rising trend with the increase in the dry-bulb temperature of inlet air. This phenomenon is due to the decrease in heat transfer driving force when the air dry-bulb temperature is increased, resulting in the air velocity having to increase to meet the cooling target of the system. Besides, $^{(27)}$, $\mathscr{g}^{\mathrm{RS}}$, $\mathscr{g}^{\mathrm{RT}}$, and $\mathscr{g}^{\mathrm{SS}}$ have nearly the same optimal solutions, and the corresponding expected values are 1.67, 1.64, 1.66, and 1.65 m s^{-1}, respectively. It is seen from Figure 6.15c that, the optimal pressure drops for all samples in $\mathscr{g}^{\mathrm{RS}}$, $\mathscr{g}^{\mathrm{RT}}$, and $\mathscr{g}^{\mathrm{SS}}$ increase from 3.43 to 7.30, 2.49 to 5.34, and 2.07 to 4.49 Pa, respectively, which are higher than the case of $\mathscr{g}^{(27)}$ that grows from 1.79 to 3.52 Pa. This is mainly attributed to the fact that $^{(27)}$ has a wider flow channel, leading to a smaller pressure drop. In Figure 6.15d, the optimal Nusselt numbers for all samples in the $^{(27)}$ increase from 20.01 to 30.03, which are slightly lower than that in the $\mathscr{g}^{\mathrm{RS}}$ (21.55–30.05), $\mathscr{g}^{\mathrm{RT}}$ (20.82–30.58), and $\mathscr{g}^{\mathrm{SS}}$ (20.61–30.44). This is because a larger S_{T} leads to larger fully developed regions with a smaller Nusselt number. It should be noted that the effect of S_{T} on the Nusselt number is less than that on pressure drop.

6.5 Conclusion

Geometry optimization is essential for heat exchanger design, but it is often limited by the high computational time and costs associated with traditional trial-and-error or empirical methods. By utilizing PINN to parameterize the geometric and operating inputs of a heat sink system, this study proposed a new inverse design method that started with the desired objectives and worked backward to find the optimal designs. For simulating heat and mass transfer processes, specialized neural network structures have been developed. It efficiently decomposed the standard PINN model into multiple interconnected subnetworks with identical architecture, which were functionally designed to approximate the latent solutions of governing PDEs, i.e. Navier–Stokes, heat transfer in fluid, and heat transfer in solid, respectively. Based on this PINN model, a parametric surrogate model was developed according to the specific input and targets, which can be further coupled with multi-objective optimization and decision-making algorithms. The performance is verified by discovering the best-performing geometric design and the corresponding optimal operating conditions in the following illustrative examples.

(1) A 3D finned heat sink system was used to demonstrate the proposed method's ability to provide good quantitative agreement and sufficient confidence in describing the underlying behaviors of fluid flow and heat transfer inside the heat sink system. It not only significantly accelerated the design optimization process from more than 216 hours for the traditional methods to about 113.9 hours for the proposed method. For better physical inspection, three representative scenarios, namely high-performance design, equilibrium design, and low-cost design, were further considered according to the weighting factors allocated to each objective in decision-making.

(2) We leveraged the capabilities of physics-informed deep learning and transfer learning to develop a novel computational framework that accelerates the design of a heat exchanger system. We developed a sequential space decomposition strategy that generated a batch of discrete geometric designs, followed by finite stochastic samples from the combined space of operating and uncertain variables. The total training time can be sharply reduced from 27 295 to 326 hours, indicating that the developed method is effective in promoting the application of the deep learning method in practical engineering design and optimization.

In the foreseeable future, the training costs for such PINN models can be further reduced to a more advantageous level, e.g. a few days or even hours, with the advances in computational resources and acceleration algorithms or strategies. Some potential avenues for achieving this include: (i) developing new differential and optimization algorithms specifically designed for the loss functions, such as coupled automatic and numerical differential methods [70, 71]; (ii) exploring more efficient training techniques, like reinforcement learning, meta-learning, and multi-task learning; and (iii) devising next-generation computational paradigms by synergistically combining the strengths of classical numerical methods and PINN techniques, such as the random feature method [72].

References

1 Wu, Z., Lu, Z., Zhang, B. et al. (2022). Stochastic bi-objective optimization for closed wet cooling tower systems based on a simplified analytical model. *Energy* 250: 123703.

2 Khattak, Z. and Ali, H.M. (2019). Air cooled heat sink geometries subjected to forced flow: a critical review. *International Journal of Heat and Mass Transfer* 130: 141–161.

3 Alihosseini, Y., Zabetian Targhi, M., Heyhat, M.M., and Ghorbani, N. (2020). Effect of a micro heat sink geometric design on thermo-hydraulic performance: a review. *Applied Thermal Engineering* 170: 114974.

4 Zobeiry, N. and Humfeld, K.D. (2021). A physics-informed machine learning approach for solving heat transfer equation in advanced manufacturing and engineering applications. *Engineering Applications of Artificial Intelligence* 101: 104232.

5 Xie, G., Wang, Q., and Sunden, B. (2010). Application of a genetic algorithm for thermal design of fin-and-tube heat exchangers. *Heat Transfer Engineering* 29 (7): 597–607.

6 Cho, D.-H., Seo, S.-K., Lee, C.-J., and Lim, Y. (2017). Optimization of layer patterning on a plate fin heat exchanger considering abnormal operating conditions. *Applied Thermal Engineering* 127: 1036–1048.

7 Chu, W.-X., Tsai, M.-K., Jan, S.-Y. et al. (2020). CFD analysis and experimental verification on a new type of air-cooled heat sink for reducing maximum junction temperature. *International Journal of Heat and Mass Transfer* 148: 119094.

8 Wang, C.-C., Hung, C.-I., and Chen, W.-H. (2012). Design of heat sink for improving the performance of thermoelectric generator using two-stage optimization. *Energy* 39 (1): 236–245.

9 Nguyen, N.P., Maghsoudi, E., Roberts, S.N., and Kwon, B. (2023). Shape optimization of pin fin array in a cooling channel using genetic algorithm and machine learning. *International Journal of Heat and Mass Transfer* 202: 123769.

10 Karniadakis, G.E., Kevrekidis, I.G., Lu, L. et al. (2021). Physics-informed machine learning. *Nature Reviews Physics* 3 (6): 422–440.

11 Lu, L., Pestourie, R., Yao, W. et al. (2021). Physics-informed neural networks with hard constraints for inverse design. *SIAM Journal on Scientific Computing* 43 (6): B1105–B1132.

12 Raissi, M., Perdikaris, P., and Karniadakis, G.E. (2019). Physics-informed neural networks: a deep learning framework for solving forward and inverse problems involving nonlinear partial differential equations. *Journal of Computational Physics* 378: 686–707.

13 Xie, X., Liu, H., He, C. et al. (2019). Deciphering the heat and mass transfer behaviors of staggered tube bundles in a closed wet cooling tower using a 3-D VOF model. *Applied Thermal Engineering* 161: 114202.

14 Biegler, L.T., Lang, Y.-D., and Lin, W. (2014). Multi-scale optimization for process systems engineering. *Computers and Chemical Engineering* 60: 17–30.

15 Lang, Y., Zitney, S.E., and Biegler, L.T. (2011). Optimization of IGCC processes with reduced order CFD models. *Computers and Chemical Engineering* 35 (9): 1705–1717.

16 Lemos, J.C., Costa, A.L.H., and Bagajewicz, M.J. (2018). Globally optimal linear approach to the design of heat exchangers using threshold fouling modeling. *AIChE Journal* 64 (6): 2089–2102.

17 Souza, P.A., Costa, A.L.H., and Bagajewicz, M.J. (2018). Globally optimal linear approach for the design of process equipment: the case of air coolers. *AIChE Journal* 64 (3): 886–903.

18 Doodman, A.R., Fesanghary, M., and Hosseini, R. (2009). A robust stochastic approach for design optimization of air cooled heat exchangers. *Applied Energy* 86 (7): 1240–1245.

19 Raissi, M., Yazdani, A., and Karniadakis, G.E. (2020). Hidden fluid mechanics: learning velocity and pressure fields from flow visualizations. *Science* 367 (6481): 1026–1030.

20 Cai, S., Mao, Z., Wang, Z. et al. (2021). Physics-informed neural networks (PINNs) for fluid mechanics: a review. *Acta Mechanica Sinica* 37 (12): 1727–1738.

21 Kharazmi, E., Cai, M., Zheng, X. et al. (2021). Identifiability and predictability of integer- and fractional-order epidemiological models using physics-informed neural networks. *Nature Computational Science* 1 (11): 744–753.

22 Yuan, L., Ni, Y.-Q., Deng, X.-Y., and Hao, S. (2022). A-PINN: auxiliary physics informed neural networks for forward and inverse problems of nonlinear integro-differential equations. *Journal of Computational Physics* 462: 111260.

23 Pang, G., Lu, L., and Karniadakis, G.E. (2019). fPINNs: fractional physics-informed neural networks. *SIAM Journal on Scientific Computing* 41 (4): A2603–A2626.

24 Yang, L., Zhang, D., and Karniadakis, G.E. (2020). Physics-informed generative adversarial networks for stochastic differential equations. *SIAM Journal on Scientific Computing* 42 (1): A292–A317.

25 Zhang, D., Lu, L., Guo, L., and Karniadakis, G.E. (2019). Quantifying total uncertainty in physics-informed neural networks for solving forward and inverse stochastic problems. *Journal of Computational Physics* 397: 108850.

26 Baydin, A.G., Pearlmutter, B.A., Radul, A.A., and Siskind, J.M. (2015). Automatic differentiation in machine learning: a survey. *The Journal of Machine Learning Research*, 18(1): 5595–5637.

27 Cai, S., Mao, Z., Wang, Z. et al. (2022). Physics-informed neural networks (PINNs) for fluid mechanics: a review. *Acta Mechanica Sinica* 37: 1–12.

28 He, Q., Barajas-Solano, D., Tartakovsky, G., and Tartakovsky, A.M. (2020). Physics-informed neural networks for multiphysics data assimilation with application to subsurface transport. *Advances in Water Resources* 141: 103610.

29 Cao, Y., Xu, R., and Jiang, P. (2023). Physics-informed machine learning based RANS turbulence modeling convection heat transfer of supercritical pressure fluid. *International Journal of Heat and Mass Transfer* 201: 123622.

30 Kissas, G., Yang, Y., Hwuang, E. et al. (2020). Machine learning in cardiovascular flows modeling: predicting arterial blood pressure from non-invasive 4D flow MRI data using physics-informed neural networks. *Computer Methods in Applied Mechanics and Engineering* 358: 112623.

31 Yin, M., Zheng, X., Humphrey, J.D., and Em, K.G. (2021). Non-invasive inference of thrombus material properties with physics-informed neural networks. *Computer Methods in Applied Mechanics and Engineering* 375: 113603.

32 Arzani, A., Wang, J.-X., and D'Souza, R.M. (2021). Uncovering near-wall blood flow from sparse data with physics-informed neural networks. *Physics of Fluids* 33 (7): 071905.

33 Kim, K.M., Hurley, P., and Duarte, J.P. (2022). Physics-informed machine learning-aided framework for prediction of minimum film boiling temperature. *International Journal of Heat and Mass Transfer* 191: 122839.

34 Lu, Z., Li, Y., He, C. et al. (2022). Integrating physics-informed neural networks with partitioned coupling strategy for modeling conjugate heat transfer. *CIESC Journal* 73 (12): 5483–5493.

35 Lu, Z., Qu, J., Liu, H. et al. (2021). Surrogate modeling for physical fields of heat transfer processes based on physics-informed neural network. *CIESC Journal* 72 (3): 1496–1503.

36 Xie, J., Chai, Z., Xu, L. et al. (2022). 3D temperature field prediction in direct energy deposition of metals using physics informed neural network. *The International Journal of Advanced Manufacturing Technology* 119 (5–6): 3449–3468.

37 Raabe, D., Mianroodi, J.R., and Neugebauer, J. (2023). Accelerating the design of compositionally complex materials via physics-informed artificial intelligence. *Nature Computational Science* 3 (3): 198–209.

38 Zhu, Q., Liu, Z., and Yan, J. (2021). Machine learning for metal additive manufacturing: predicting temperature and melt pool fluid dynamics using physics-informed neural networks. *Computational Mechanics* 67 (2): 619–635.

39 Chamoli, S., Lu, R., Chen, H. et al. (2019). Numerical optimization of design parameters for a modified double-layer microchannel heat sink. *International Journal of Heat and Mass Transfer* 138: 373–389.

40 Kumar, S., Sarkar, M., Singh, P.K., and Lee, P.S. (2019). Study of thermal and hydraulic performance of air cooled minichannel heatsink with novel geometries. *International Communications in Heat and Mass Transfer* 103: 31–42.

41 Rao, R.V. and Waghmare, G.G. (2015). Multi-objective design optimization of a plate-fin heat sink using a teaching-learning-based optimization algorithm. *Applied Thermal Engineering* 76: 521–529.

42 Lu, Z., Li, Y., He, C. et al. (2024). Multi-objective inverse design of finned heat sink system with physics-informed neural networks. *Computers and Chemical Engineering* 180: 108500.

43 Wu, Z., Zhang, B., Yu, H. et al. (2023). Accelerating heat exchanger design by combining physics-informed deep learning and transfer learning. *Chemical Engineering Science* 282: 119285.

44 Zhu, Y., Zabaras, N., Koutsourelakis, P.-S., and Perdikaris, P. (2019). Physics-constrained deep learning for high-dimensional surrogate modeling and uncertainty quantification without labeled data. *Journal of Computational Physics* 394: 56–81.

45 Lu, L., Meng, X., Mao, Z., and Karniadakis, G.E. (2021). DeepXDE: a deep learning library for solving differential equations. *SIAM Review* 63 (1): 208–228.

46 NVIDIA Corporation. (2021). Modulus user guide (release v21.06). https://developer.nvidia.com/modulus (accessed 03 April 2023).

47 Sukumar, N. and Srivastava, A. (2022). Exact imposition of boundary conditions with distance functions in physics-informed deep neural networks. *Computer Methods in Applied Mechanics and Engineering* 389: 114333.

48 Zhao, Z., Zhu, Q., and Yan, J. (2021). A thermal multi-phase flow model for directed energy deposition processes via a moving signed distance function. *Computer Methods in Applied Mechanics and Engineering* 373: 113518.

49 Hennigh, O., Narasimhan, S., Nabian, M.A. et al. (2021). NVIDIA SimNet™: an AI-accelerated multi-physics simulation framework. *International Conference on Computational Science* (16–18 June 2021). pp. 447–461. Krakow, Poland: Springer.

50 Laszczyk, M. and Myszkowski, P.B. (2019). Improved selection in evolutionary multi–objective optimization of multi–skill resource–constrained project scheduling problem. *Information Sciences* 481: 412–431.

51 Gutjahr, W.J. and Pichler, A. (2016). Stochastic multi-objective optimization: a survey on non-scalarizing methods. *Annals of Operations Research* 236 (2): 475–499.

52 Bonnel, H. and Collonge, J. (2014). Stochastic optimization over a Pareto set associated with a stochastic multi-objective optimization problem. *Journal of Optimization Theory and Applications* 162 (2): 405–427.

53 Giagkiozis, I. and Fleming, P.J. (2015). Methods for multi-objective optimization: an analysis. *Information Sciences* 293: 338–350.

54 Deb, K., Pratap, A., Agarwal, S., and Meyarivan, T. (2002). A fast and elitist multiobjective genetic algorithm: NSGA-II. *IEEE Transactions on Evolutionary Computation* 6 (2): 182–197.

55 Lai, Y.J., Liu, T.Y., and Hwang, C.L. (1994). TOPSIS for MODM. *European Journal of Operational Research* 76 (3): 486–500.

56 Wang, T.H., Wu, H.C., Meng, J.H., and Yan, W.M. (2020). Optimization of a double-layered microchannel heat sink with semi-porous-ribs by multi-objective genetic algorithm. *International Journal of Heat and Mass Transfer* 149: 119217.

57 Electronic chip cooling, COMSOL multiphysics application library, application ID: 47721, created in COMSOL multiphysics 5.5. https://cn.comsol.com/model/electronic-chip-cooling-47721 (accessed 16 February 2022).

58 Xiao, H., Liu, Z., and Liu, W. (2021). Conjugate heat transfer enhancement in the mini-channel heat sink by realizing the optimized flow pattern. *Applied Thermal Engineering* 182: 116131.

59 Yang, X.-H., Tan, S.-C., Ding, Y.-J., and Liu, J. (2017). Flow and thermal modeling and optimization of micro/mini-channel heat sink. *Applied Thermal Engineering* 117: 289–296.

60 Ermagan, H. and Rafee, R. (2018). Geometric optimization of an enhanced microchannel heat sink with superhydrophobic walls. *Applied Thermal Engineering* 130: 384–394.

61 Li, J., Zhou, G., Tian, T., and Li, X. (2021). A new cooling strategy for edge computing servers using compact looped heat pipe. *Applied Thermal Engineering* 187: 116599.

62 Electronic chip cooling, application ID: 47721, created in COMSOL multiphysics 5.5. https://cn.comsol.com/model/electronic-chip-cooling-47721 (accessed 03 April 2023).

63 Laubscher, R. (2021). Simulation of multi-species flow and heat transfer using physics-informed neural networks. *Physics of Fluids* 33 (8): 087101.

64 Kong, Y., Yang, L., Du, X., and Yang, Y. (2016). Air-side flow and heat transfer characteristics of flat and slotted finned tube bundles with various tube pitches. *International Journal of Heat and Mass Transfer* 99: 357–371.

65 Ramachandran, P., Zoph, B., and Le, Q.V. (2017). Searching for activation functions. *arXiv preprint arXiv:1710.05941*.

66 Pan, S.J. and Yang, Q. (2009). A survey on transfer learning. *IEEE Transactions on Knowledge and Data Engineering* 22 (10): 1345–1359.

67 Li, M., Liu, Y., Liu, X. et al. (2020). The deep learning compiler: a comprehensive survey. *IEEE Transactions on Parallel and Distributed Systems* 32 (3): 708–727.

68 Leary, C. and Wang, T. (2017). XLA: TensorFlow, compiled. *TensorFlow Dev Summit* 2 (3).

69 Tancik, M., Srinivasan, P., Mildenhall, B. et al. (2020). Fourier features let networks learn high frequency functions in low dimensional domains. *Advances in Neural Information Processing Systems* 33: 7537–7547.

70 Chiu, P.-H., Wong, J.C., Ooi, C. et al. (2022). CAN-PINN: a fast physics-informed neural network based on coupled-automatic–numerical differentiation method. *Computer Methods in Applied Mechanics and Engineering* 395: 114909.

71 Wu, Z., Wang, H., He, C. et al. (2023). The application of physics-informed machine learning in multiphysics modeling in chemical engineering. *Industrial and Engineering Chemistry Research* 62 (44): 18178–18204.

72 Chen, J., Chi, X., Weinan, E., and Yang, Z. (2022). Bridging traditional and machine learning-based algorithms for solving PDEs: the random feature method. *Journal of Machine Learning* 1 (3): 268–298.

7

Integrating Incomplete Prior Knowledge into Data-Driven Inferential Sensor Models Under Variational Bayesian Framework

Zhichao Chen[1], Hao Wang[1], Yiran Ma[1], Cheng Qiu[1], Le Yao[2], Xinmin Zhang[1], and Zhihuan Song[1]

[1]*Zhejiang University, College of Control Science and Engineering, State Key Laboratory of Industrial Control Technology, 38 Zheda Rd. (Yuquan Campus), Hangzhou, Zhejiang Province, 310027, P. R. China*
[2]*Hangzhou Normal University, School of Mathematics, 2318 Yuhangtang Rd., Hangzhou, Zhejiang Province, 311121, P. R. China*

7.1 Introduction

Inferential sensors are pivotal in process automation, significantly contributing to areas such as modeling, control, and optimization [1, 2]. Their relevance extends across various dimensions of the process industry. One critical aspect involves addressing the demanding challenges of boosting profitability, reducing material usage, enhancing safety, and safeguarding the environment, which has spurred the extensive adoption of artificial intelligence methods, especially deep learning (DL). Concurrently, the escalation in the scale of process manufacturing has intensified the necessity for efficient management of substantial industrial big data volumes [3]. Consequently, the integration of DL techniques into inferential sensors is imperative, aimed at facilitating precise control and optimal decision-making in the realm of process industries.

In recent decades, the field of data-driven inferential sensor modeling, especially those methods driven by DL, has seen substantial research and development, fueled by advancements in machine learning algorithms, distributed control systems, and database technologies [1]. DL-based inferential sensor techniques can be classified into several distinct model architectures: (stacked-)autoencoders (SAEs)/feedforward neural networks (FNNs) [4], convolutional neural networks (CNNs) [5], recurrent neural networks (RNNs), and transformers. Each architecture presents its own set of unique advantages and disadvantages. For example, SAEs/FNNs excel in handling process nonlinearity but are less effective with dynamic data. CNNs, with their convolutional kernel structure, offer local memory capabilities but are limited by the predefined size of their receptive field. RNNs adeptly manage dynamic processes [6], yet they are prone to issues like gradient explosion or vanishing during the training phase [7]. Transformers, employing self-attention mechanisms (SAMs), greatly enhance parallelization in sequence

Applied AI Techniques in the Process Industry: From Molecular Design to Process Design and Optimization, First Edition. Edited by Chang He and Jingzheng Ren.

modeling, but their inherent "position-insensitivity" due to this mechanism can be a drawback [8]. This diversity among models ensures a well-rounded strategy in addressing the various challenges encountered in inferential sensor modeling.

Despite the significant strides made by current DL-based inferential sensor modeling approaches, a critical question persists: How can we effectively integrate established knowledge of industrial processes with these DL models? Industrial processes are subject to stringent evaluation before operation, leading to data that often conforms to empirical rules specific to areas like unit operations and reaction engineering. Incorporating this well-established knowledge into DL models holds the potential to greatly improve their efficacy and relevance. However, unlike physically informed neural networks (PINNs) [9], which model industrial processes at the transport process scale, current inferential sensor-based models, which model processes at unit operation scale, often lack the capability to capture all the necessary coefficients in semi-empirical equations. For instance, the accurate computation in applications such as an absorber column critically depends on the knowledge of mass transport coefficients. However, these coefficients are typically not obtainable through standard instrumentation, resulting in a gap in the necessary data. This chapter categorizes this gap as "incomplete knowledge". One of the primary challenges we face is the effective articulation and incorporation of this incomplete knowledge into data-driven methodologies. Doing so is essential for improving the performance of inferential sensors. This integration demands not only a deep understanding of the underlying industrial processes but also innovative strategies to compensate for the missing information in a way that enhances the overall efficacy of the inferential sensor models.

Recent research increasingly adopts graph neural networks (GNNs) to encapsulate prior knowledge in industrial processes [10–13]. This approach leverages the concept that industrial processes can be graphically represented, with instruments functioning as nodes within this framework. Such a configuration significantly enhances modeling efficacy by enforcing interconnections among various instruments. This methodology mirrors the principles used in recommender systems, where modeling user-item pairs leads to performance improvements [14]. However, this direct adoption of predefined graphs can encounter the following specific challenges:

1. **Data Shifting:** The training dataset may span various operational periods, leading to spatial and temporal variations in the relevant knowledge required for modeling, as indicated by Ma et al. [15].
2. **Knowledge Selection:** The physical properties of materials can vary with changing operating conditions, suggesting the need for a model capable of dynamically adjusting its underlying knowledge base.

To address the first issue effectively, it is advised to incorporate prior knowledge dynamically within the prior term of the loss function, and subsequently exclude it during the model's inference stage. This approach contrasts with the conventional method of statically embedding prior knowledge into the model's input graph and maintaining it throughout the inference stage. Addressing the second issue involves designing a model that inherently and seamlessly adjusts to the evolving prior knowledge. In response to these challenges, this chapter introduces the "variational inference over graph" module, a novel approach designed to adeptly handle both data shifting and knowledge selection in dynamic modeling environments.

This chapter is organized as follows: We begin with a comprehensive review of the relevant literature, identifying and discussing the existing technical gaps in the field. Then, in Section 7.3, we delve into the derivation of our proposed methodology, thoroughly detailing the architecture of the innovative module we introduce. Following this, Section 7.4 is dedicated to the empirical validation of our approach, showcasing its application and effectiveness in a real-world catalytic shift unit. The chapter concludes with Section 7.5, where we summarize our key findings and offer insights derived from our research.

This chapter is derived from the works of Chen et al. [16, 17].

7.2 Literature Review

The integration of prior knowledge into industrial process modeling to enhance model performance is a well-explored domain in data-driven approaches specific to industrial processes. Unlike other fields, industrial processes are subject to rigorous design and rating protocols before they begin operation. This results in process measurement data that generally conforms to established (semi)empirical guidelines, facilitating the seamless integration of prior knowledge into inferential models for industrial process data. In Sections 7.2.1 and 7.2.2, this chapter will be divided into two main areas for detailed discussion: the transport process scale and the unit operation scale, each offering unique insights into the application of prior knowledge in industrial process modeling.

7.2.1 Transport Process Scale

In the realm of transport processes, industrial operations are governed by well-established conservation laws of momentum, heat, and mass. Reflecting this foundation, PINNs, which replace nonlinear terms in partial differential equations (PDEs) or ordinary differential equations (ODEs) with neural networks, have gained substantial attention in recent research, as highlighted in works by Raissi et al. [18] and Zhiyong et al. [19]. A prominent example is in the prediction of turbulent flows, where Wang et al. [20] developed the turbulent-flow net, integrating trainable spectral filters with Reynolds-averaged Navier–Stokes equations and large Eddy simulation in a specialized U-net structure. Addressing few-shot learning challenges, Zheng et al. [21] proposed a physics-informed RNN within a model predictive control framework, ensuring compliance with physical laws while substituting key equations with an RNN.

While PINN-based models have found considerable application in process control, as noted by Alhajeri et al. [22] and Zheng et al. [21], their direct application using process measurement data in PINNs faces challenges due to the complex transport phenomena in unit operations. This complexity limits the broader application of PINN-based methods in industrial process inferential sensors.

7.2.2 Unit Operation Scale

At the unit operation scale, practitioners typically use graph structures rather than strict ODE/PDE formulations to represent process knowledge. In this context, Bayesian networks (BNs), as discussed by Zeng et al. [23] and Khosbayar et al. [24], are a natural choice for modeling the correlations between process variables. Khosbayar et al. [24], for instance, designed BN structures based on flowsheets, developed a corresponding parameter estimation algorithm using the expectation–maximization algorithm, and validated its effectiveness in a bitumen upgrading process.

However, the training and inference of BNs, typically based on the expectation propagation algorithm, are not easily integrated into the mini-batch stochastic gradient descent-based DL framework. To address this, recent studies have increasingly focused on GNN-based models [11, 12, 25]. Notably, Ren et al. [12] used normalized mutual information to extract the graph of process variables, applied a multi-level knowledge graph to model large-scale industrial process data, and demonstrated its effectiveness in a cobalt–nickel removal process.

7.2.3 Overall Summary and Technical Gap

Despite significant advancements in embedding prior knowledge in industrial process inferential sensors, the following key issues remain unresolved:

1. **Data Shifting Problem:** Current GNN-based methods fix prior knowledge without considering that the most suitable knowledge may vary spatially and temporally. Consequently, the knowledge embedded into models should not be in a rigid, unchanging form.
2. **Knowledge Selection Problem:** Existing approaches predominantly employ message passing mechanisms, overlooking the necessity of knowledge selection, which is critical for adapting to internal or external process changes.

To bridge these identified technical gaps, we introduce the variational inference over graph (VIOG) framework in Section 7.3.

7.3 Proposed Approach

In this section, our approach begins by reinterpreting the graph within a Bayesian framework and modifying the loss function to include graph constraints, thereby addressing the "data shifting problem" issue. Next, we consider the knowledge representation problem in detail. Specifically, we examine the process of information selection among various variables, leading to the development of a SAM aimed at resolving "knowledge selection problem" issue. Additionally, we analyze the parallels between this SAM and graph convolution operations to underscore its validity. The section culminates with the presentation of the overall architectural design of the VIOG framework.

7.3.1 Loss Function Derivation

De facto, the task of inferential sensing can be regarded as a specific case of a supervised learning task. Consequently, if we represent the process variables by x, the quality variable by y, and the parameters by θ, our learning objective can be articulated as follows:

$$\arg\max_{\theta} \log p_\theta(y \mid x) \tag{7.1}$$

Based on Eq. (7.1), we can reformulate the learning objective as follows:

$$\begin{aligned}
&\log p(y \mid x) \\
&= \log \int p(y, \mathcal{A} \mid x) \mathrm{d}\mathcal{A} \\
&= \underbrace{\log \int \frac{p(y, \mathcal{A} \mid x)}{q(\mathcal{A})} q(\mathcal{A}) \mathrm{d}\mathcal{A}}_{\mathbb{E}_{q(\mathcal{A})}\left[\frac{p(y,\mathcal{A}|x)}{q(\mathcal{A})}\right]} \\
&\geq \mathbb{E}_{q(\mathcal{A})}\left[\log \frac{p(y, \mathcal{A} \mid x)}{q(\mathcal{A})}\right] \\
&= \mathbb{E}_{q(\mathcal{A})}[\log p(y \mid \mathcal{A}, x)] - \mathbb{D}_{\mathrm{KL}}[q(\mathcal{A}) \parallel p(\mathcal{A})]
\end{aligned} \tag{7.2}$$

where the third line is built upon the Jessen's inequality [26, 27]. Besides, we can further divide the last line of Eq. (7.2) and formulate a novel optimization problem as follows:

$$\begin{aligned}
&\min \ \mathbb{E}_{q(\mathcal{A})}[\log p(y \mid \mathcal{A}, x)] \\
&\text{s.t.} \quad \mathbb{D}_{\mathrm{KL}}[q(\mathcal{A}) \parallel p(\mathcal{A})] = 0
\end{aligned} \tag{7.3}$$

Comparing Eqs. (7.2) and (7.3), it becomes evident that our learning objective transforms the conventional constraint term, typically a constant in most GNN-based models, into a penalty term. This transformation can be understood through the application of the celebrated Lagrangian multiplier method:

$$\begin{aligned}
&\min \ \mathbb{E}_{q(\mathcal{A})}[\log p(y \mid \mathcal{A}, x)] \\
&\text{s.t.} \quad \mathbb{D}_{\mathrm{KL}}[q(\mathcal{A}) \parallel p(\mathcal{A})] = 0 \\
&\Rightarrow \\
&\mathbb{E}_{q(\mathcal{A})}[\log p(y \mid \mathcal{A}, x)] - \beta \times \mathbb{D}_{\mathrm{KL}}[q(\mathcal{A}) \parallel p(\mathcal{A})]
\end{aligned} \tag{7.4}$$

where β functions as the Lagrangian multiplier, with the learning objective of the β-VAE outlined on the right-hand side of Eq. (7.4). A key innovation within our variational inference framework is the conversion of the standard predefined graph, a common feature in GNN-based models, into a penalty form. This adaptation enables our model to operate without reliance on a specific graph structure, enhancing its versatility. This approach effectively addresses the "data shifting" issue as discussed in Section 7.1. Section 7.3.2 will explore the representation and integration of prior knowledge in our model, further clarifying its significance and operational mechanism.

7.3.2 Knowledge Representation

In this subsection, we aim to elucidate the methodology for integrating incomplete knowledge into the prior term $p(\mathcal{A})$ (knowledge description) and to outline the framework for such a representation within our model (knowledge utilization).

7.3.2.1 Knowledge Description

Echoing approaches in the natural language processing domain, where sentences are delineated using graph-like grammar trees [28], our model in GNN-based industrial process modeling similarly employs graphs to represent knowledge regarding process covariates. However, as previously mentioned, this prior knowledge is characteristically uncertain. The essence of utilizing this uncertain knowledge effectively lies in its probabilistic representation. Consequently, this subsection is dedicated to introducing an approach for articulating prior knowledge from a probabilistic perspective.

According to previous works [29, 30], the graph adjacency matrix $\mathcal{A}$ can be decomposed using the mean-field assumption:

$$p(\mathcal{A}) = \prod_{i=1}^{v}\prod_{j=1}^{v} p(\mathcal{A}_{i,j}) \tag{7.5}$$

where v indicates the unique element number of the process variables. In light of this, the edges (denoted as α) in the graph may be present ($\alpha = 1$) or absent ($\alpha = 0$). This phenomenon indicates that the Bernoulli distribution can be adopted to describe the edges' state as Eq. (7.6) shown:

$$p(\mathcal{A}_{ij}) = \rho^{\alpha}(1-\rho)^{1-\alpha}, \quad \rho \in [0, 1] \tag{7.6}$$

where ρ represents the expected value of the Bernoulli distribution.

Given our comprehension of the process, knowledge can be classified into three discrete states: definitively present, ambiguously present, and definitively absent. To address this, we employ the 2σ bound within the Gaussian distribution, following the Pauta criterion, to determine the value of ρ. This method enables us to formulate the subsequent disjunctive expression:

$$\begin{bmatrix} S_1 \\ \rho_1 = 0.955 \end{bmatrix} \veebar \begin{bmatrix} S_2 \\ \rho_2 = 0.500 \end{bmatrix} \veebar \begin{bmatrix} S_3 \\ \rho_3 = 0.045 \end{bmatrix} \tag{7.7}$$

where $\veebar$ indicates that only one square bracket will take effect.

7.3.2.2 Knowledge Section via Self-Attention Mechanism

Having delineated our prior graph $p(\mathcal{A})$, our focus now shifts to the variational distribution $q(\mathcal{A})$. In line with existing literature, graph construction hinges on nodal similarity, which, in our case, translates to the similarity among different process variables. However, solely measuring node similarity is insufficient to tackle the problem of knowledge selection. For instance, consider a source node denoted by $i \in \{1, 2, \ldots, v\}$ and a target node represented as $j \in \{1, 2, \ldots, v\}$. Each target node j might receive input from v source nodes, potentially leading to redundant information. Therefore, it is imperative to implement a competitive mechanism. This mechanism should not only limit the information received by the target node j but also regulate the information emitted from the source node i.

In summary, based on our analysis, the following design requirements are essential to address the "knowledge selection" issue:

(1) Construct the graph based on similarity measurements.
(2) Implement restrictions on information emission and reception by the nodes.

To address requirement (1), we can conduct similarity measurement as follows:

$$\mathcal{S} = \mathbf{Q}\mathbf{K}^{\top} \tag{7.8}$$

where $\mathcal{S}$, $\mathbf{Q}$, and $\mathbf{K}$ represent the similarity matrix, the information from the source, and target nodes, respectively. To impose restrictions on this information, we employ a row-by-row normalization using the softmax operator:

$$s_{ij} = \frac{\exp(s_{ij})}{\sum_{j=1}^{v} \exp(s_{ij})} \quad i,j \in \{1, 2, \dots, v\} \tag{7.9}$$

Here, s_{ij} denotes an element of the matrix $\mathcal{S}$. It is important to note that directly applying a competitive approach to a matrix $\mathcal{A}$ could lead to a "winner takes all" scenario, where the rows of the matrix $\mathcal{S}$ transform into one-hot vectors. Such an occurrence could complicate the model training process and induce significant variance during the gradient estimation phase [31]. To mitigate these issues, the matrix $\mathcal{A}$ is derived from $\mathcal{S}$:

$$\mathcal{A} = \text{Softmax}\left(\frac{\mathcal{S}}{\sqrt{v}}\right) = \text{Softmax}\left(\frac{\mathbf{Q}\mathbf{K}^{\top}}{\sqrt{v}}\right) \tag{7.10}$$

The abovementioned graph construction is similar to the SAM [32] in the transformer model as depicted in Figure 7.1. In the next part, we want to discuss the similarity between the GNN and transformer model.

7.3.2.3 Similarity of GCN and SAM

In Section 7.3.2.3, our primary focus lies in the comprehensive examination of the similarities between GNN-based networks and transformer models. For simplicity, we consider graph CNNs, the most widely used in the process industry, as an analysis object. It is worth emphasizing, as articulated in the scholarly reference [33], that the GNN exhibits distinct characteristics within the realm of message passing neural networks, as explicitly elucidated in Eq. (7.11):

$$x_i^{(k)} = \gamma^{(k)}\left(x_i^{(k-1)}, \underset{j\in\mathbb{N}(i)}{\zeta}\, \phi^{(k)}\left(x_i^{(k-1)}, x_j^{(k-1)}, e_{i,j}\right)\right) \tag{7.11}$$

where the symbol ϕ represents an operator that transforms the features of node i and those from its neighboring nodes j into a single, unified message. Concurrently, ζ denotes a permutation-invariant operator responsible for aggregating all messages associated with node i. This aggregation can be executed through various methods, such as summation, mean calculation, or maximization. Finally, the γ operator is

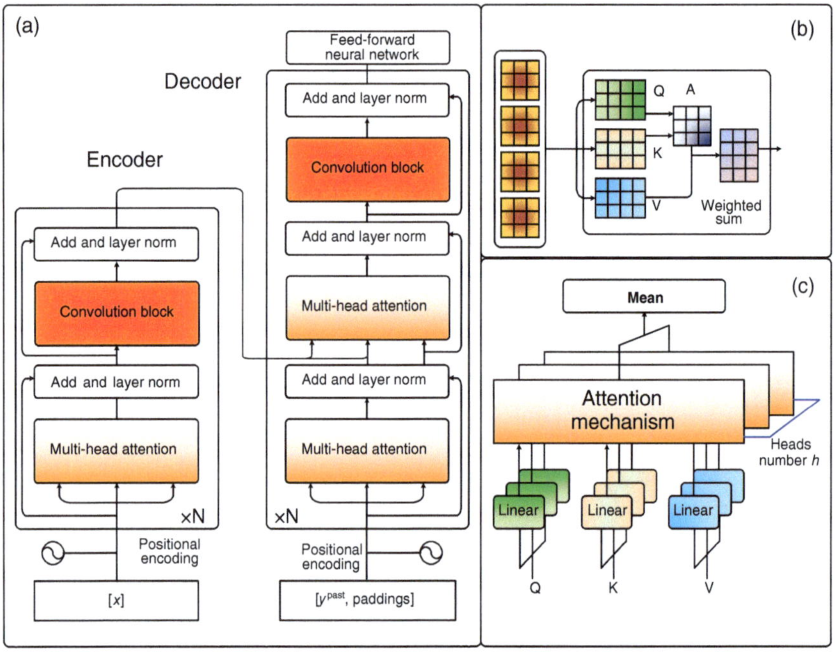

Figure 7.1 Model Structure of (a) transformer Model, (b) SAM, and (c) multi-head SAM.

employed to map the features of node i onto the aggregated message generated by the ζ operator, effectively integrating node-specific information with broader network insights. As such, the feature at node i in the kth GC layer is obtained:

$$x_i^{k+1} = \sigma\left(x_i^k U_i^k + \sum_{j \in \mathbb{N}(i)} x_j^k W_j^k\right) \tag{7.12}$$

Equation (7.11) can be rewritten as Eq. (7.12) for the GC operator [34], where the U and W are the weights that should be learned, the + is adopted as the γ operator in Eq. (7.11), the $\sum$ (sum) is chosen to be the ζ operator, and ϕ operator is linear transformation. Besides, in (7.12), the self-loop (the node i itself is included in the set of its neighboring node $\mathbb{N}(i)$) exists for all nodes. By analyzing Eq. (7.12), the corresponding structure can be drawn as Figure 7.2a shows.

Meanwhile, as shown in Figure 7.2b, the SAM [32] can be described as Eq. (7.13) to Eq. (7.15). Equation (7.13) shows that the input data are mapped into the vector space named **Q** (query), **K** (key), and **V** (value), respectively. The similarity for vectors in **Q** space and **K** space can be transformed into the weight on vectors in **V** space, considering the scaling factor D as shown in Eq. (7.14), where the weight is called attention value, and the matrix formed by the attention values is denoted as $\mathcal{A}$. Equation (7.15) stands for the residual block which means that the final output is the sum of the original input data x and the h vector obtained from Eq. (7.14).

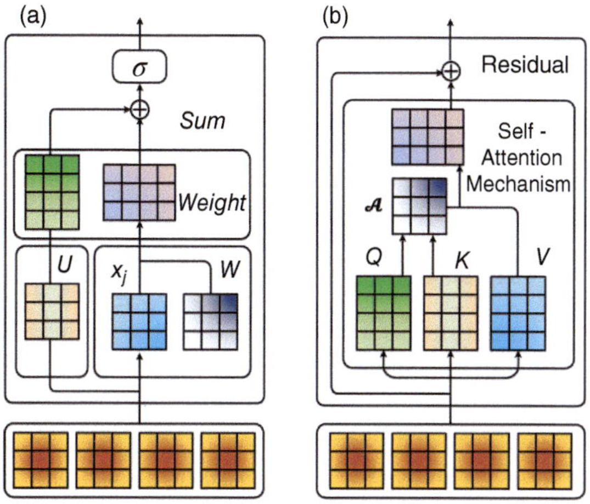

Figure 7.2 The structure of (a) the message passing of GNN, (b) the self-attention mechanism.

$$\begin{cases} q = xW_{\mathrm{Q}} + b_{\mathrm{Q}} \\ k = xW_{\mathrm{K}} + b_{\mathrm{K}} \\ v = xW_{\mathrm{V}} + b_{\mathrm{V}} \end{cases} \tag{7.13}$$

$$\begin{aligned} h_n &= \sum_{j=1}^{N} \alpha_{nj} v_j \\ &= \sum_{j=1}^{N} \mathrm{Softmax}\left(\frac{k^{\mathrm{T}} q}{\sqrt{D}}\right) v_j \end{aligned} \tag{7.14}$$

$$h_{\mathrm{final}} = h_n + x_n W_{\mathrm{res}} \tag{7.15}$$

This analysis leads to the conclusion that the similarity computation achieved through the inner product and the weighted summation operation, as described in Eq. (7.13), is essentially analogous to the operation of summing the neighboring nodes detailed in Eq. (7.12). In essence, this implies that the SAM can be equated to GC. This equivalence suggests that the attention matrix $\mathcal{A}$, derived through SAM, can represent the interconnections between different nodes in the network. The process of obtaining this attention matrix can thus be viewed as a data-driven approach to knowledge discovery. Based on these insights, it becomes feasible to design our model architecture with a foundation in SAM.

7.3.2.4 Sampling from Posterior

In Section 7.3.2.3, we conduct a detailed analysis of the reason for introducing SAM and the similarity of SAM and GC operator. Nevertheless, the model training issue remains a challenge; since the loss function is derived within the variational inference framework, where obtaining sample from the posterior distribution is vital to proceed the model training. To this end, the remain of this part will focus on how to obtain samples from the posterior distribution. First, let us review the normalization constraint introduced by SAM:

$$\sum_{j=1}^{v} p(\alpha_{ij}) = 1 \tag{7.16}$$

Building on this framework, the distribution of each row in $\mathcal{A}$ is subject to certain constraints, specifically:

- The distribution must exist on a simplex.
- Any sample drawn from this distribution should be nonnegative.

Although the Dirichlet distribution meets these criteria and promotes sparsity, making it an intuitive choice, it poses a challenge in that it is not parameterizable for backends based on gradient descent-driven DL techniques. Consequently, in alignment with the approach suggested in Ref. [35], a strategy is adopted wherein random variables are drawn from a *nonnegative distribution* and subsequently normalized to adhere to the simplex constraint. This process enables the derivation of the column vector $\vec{\alpha}_i$ for the attention matrix or graph $\mathcal{A}$, as delineated in Eq. (7.6):

$$\vec{\alpha}_i = \frac{\vec{\widetilde{a}}_i}{\sum_{j=1}^{v} \widetilde{a}_{ij}}, \quad i \in \{1, 2, \ldots, v\} \tag{7.17}$$

where the subscript i indicates the row vector at row i, and the $\widetilde{a}$ is a random sample drawn from a *nonnegative distribution*.

In this chapter, the log-normal distribution defined in Eq. (7.18) is chosen as the nonnegative distribution, and σ of the log-normal distribution is treated as a global hyperparameter:

$$p(\widetilde{\varepsilon} \mid \mu, \sigma) = \frac{1}{\widetilde{\varepsilon}\sigma\sqrt{2\pi}} \exp\left[-\frac{(\log \widetilde{\varepsilon} - \mu)^2}{2\sigma^2}\right] \tag{7.18}$$

Since sampling from a log-normal distribution is equivalent to sampling from a normal distribution:

$$\widetilde{a} \sim \textit{Log-}\text{Normal}(\mu, \sigma^2) \Leftrightarrow \widetilde{a} = \exp(\hat{\alpha}), \quad \hat{\alpha} \sim \mathcal{N}(\mu, \sigma^2) \tag{7.19}$$

Thereby, the unnormalized attention weight $\hat{\alpha}$ can be obtained via the expected value μ of the normal distribution, and the expected value μ can be obtained as Eq. (7.20):

$$\mu_{ij} = \left[\frac{\mathbf{QK}^{\top}}{\sqrt{v}} - \frac{\sigma^2}{2}\right]_{ij} \tag{7.20}$$

Note that the sampling operation ensures Eq. (7.21) set up, which means that the gradient estimation of the unnormalized graph $\widetilde{\mathcal{A}}$ (the entity is denoted as $\widetilde{a}_{ij}$) is unbiased:

$$\mathbb{E}(\widetilde{a}_{ij}) = \left[\exp\left(\frac{\boldsymbol{QK}^{\top}}{\sqrt{v}}\right)\right]_{ij} \tag{7.21}$$

7.3.3 Model Expressions

Based on our abovementioned approaches we name our module VIOG, where we use the variational inference framework, and our model can be treated as a special kind of GNN. Besides, we have not introduced the model structure assumption in our derivation process, which indicates that our proposed approach can be a plug-in module and can be adapted to most of the current model structure. On this basis, our proposed model architecture is given as follows:

Figure 7.3 presents the illustration of the VIOG, which consists of the VIOG module and downstream DL models. The Bayesian SAM parameterized by ϕ in the VIOG module serves as $q_\phi(\mathcal{A} \mid x)$, which undertakes the task of leveraging and reconciling the prior knowledge. While the DL models are parameterized by θ serve as $p_\theta(y \mid \mathcal{A}, x)$ support tasks like soft sensors and process monitoring.

Suppose the input tensors are in shape like [batch size, sequence length, v]. Before all operations, the reshape operation in Eq. (7.22) is first executed to avoid breaking their time–series structure.

$$x = \text{reshape}(x, [\text{batch\_size} \times \text{seq\_length}, v]) \tag{7.22}$$

After that, the linear projection for input tensor embedding on each dimension in shape [batch size × seq length, 1] is conducted as Eq. (7.23):

$$\begin{cases} q_i = x_i W_q^i + b_q^i \\ k_i = x_i W_k^i + b_k^i \\ v_i = x_i W_v^i + b_v^i \end{cases}; \quad i \in \{1, \ldots, v\} \tag{7.23}$$

Typically, instead of performing a single attention function, it is beneficial to linearly project the queries, keys, and values H times with different, learned linear transformations. Thereby, (7.23) will execute parallelly in H heads.

Thereafter, the similarity s_{ij} can be obtained as (7.24):

$$s_{ij} = \frac{q_i \times k_j^{\mathrm{T}}}{\sqrt{v}}; \quad i,j \in \{1, \ldots, v\} \tag{7.24}$$

On this basis, $\widehat{s}_{ij}$ is calculated row-by-row as (7.25):

$$\vec{\widehat{s}}_i = \log \text{Softmax}_j(s_{ij}); \quad i,j \in \{1, \ldots, v\} \tag{7.25}$$

where the subscript j signifies that the log-softmax operator (log-softmax) is applied column-wise, specifically targeting the column index j, and executed on a row-by-row basis. As such, the $\widehat{a}_{ij}$ is reparametrized with noise sampled from $\mathcal{N}(0, I)$ as per (7.20):

$$\widehat{a}_{ij} = \widehat{s}_{ij} + \varepsilon \times \sigma - \frac{\sigma^2}{2}, \quad \varepsilon \sim \mathcal{N}(0, I) \tag{7.26}$$

And thus, the nonnegative random variable α_{ij} can be obtained as per (7.27):

$$\widetilde{\alpha}_{ij} = \exp(\widehat{a}_{ij}) \tag{7.27}$$

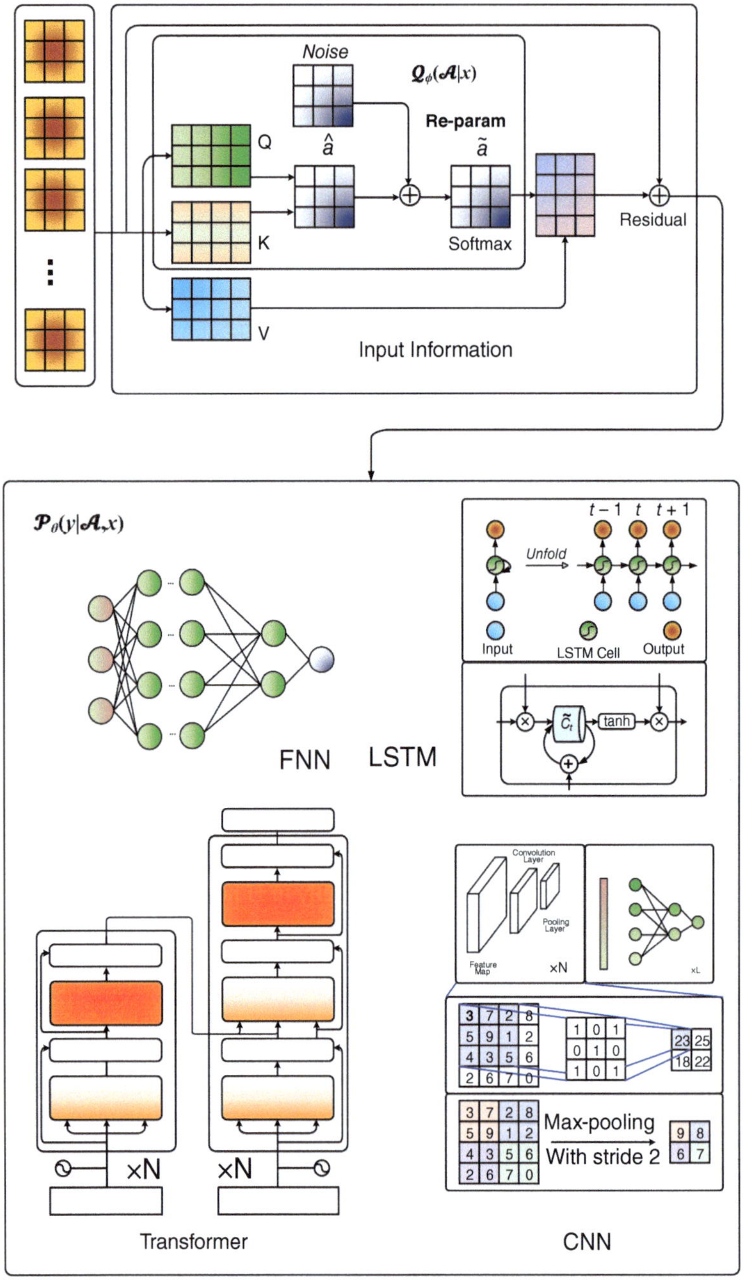

Figure 7.3 The illustrator of the VIOG module.

The entity α_{ij} of $\mathcal{A}$ is obtained via row-by-row normalization as per (7.17). It is worth noting that (7.27) and (7.17) are functionally equivalent to the softmax operation:

$$\vec{\alpha}_i = \text{Softmax}_j(\hat{\alpha}_{ij}) \tag{7.28}$$

In this procedure, the unnormalized attention weight, instrumental in calculating the KL divergence term, is provided as depicted in (7.29):

$$\widetilde{\mathcal{A}} = \log \text{Softmax}(\mathbf{KQ}^\top) \tag{7.29}$$

Note that the inference of the label $p_\theta(y \mid \mathcal{A}, x)$ is conditioned on A and x. While the abovementioned description mainly concentrates on$q_\phi(\mathcal{A} \mid x)$. To include the information of x, the weighted sum in (7.30) and residual operations (7.31) similar to the SAM is also indispensable:

$$h_v = \mathcal{A} \times V \tag{7.30}$$

$$h_{\text{final}} = h_v + xW_{\text{res}} + b_{\text{res}} \tag{7.31}$$

where h is the hidden feature, and subscript res stands for residual. Besides, to avoid the over-fitting of the VIOG module, it is suggested to adopt the dropout layer after obtaining the final feature h_{final} [32, 36].

And finally, the loss function for the DL models with the VIOG module can be rewritten as (7.32):

$$\begin{aligned}
\text{Loss} &= -\text{ELBO}_\beta \\
&= -\mathbb{E}_{q_\phi(\mathcal{A})}[\log p_\theta(y \mid x, \mathcal{A})] + \beta KL(q_\phi(\mathcal{A} \mid x) \parallel p(\mathcal{A})) \\
&= \|\hat{y} - y\|_2^2 + \sum_{i=1}^{V}\sum_{j=1}^{V}\frac{1}{2}\left[\frac{(\hat{a} - a)^2}{\sigma^2}\right]
\end{aligned} \tag{7.32}$$

where β acts as a balancing mechanism, harmonizing the trade-off between prediction error and knowledge-based regularization.

7.4 Experimental Results

In this section, the effectiveness of the VIOG module for industrial process inferential sensors will be demonstrated on a catalytic shift conversion (CSC) unit. We devise experiments on an industrial inferential sensor dataset, to verify the superiority of OC–NDPLVM and answer the research questions as follows:

- **Performance:** Does VIOG module work? The evaluation of VIOG's performance in comparison to various baseline methods is detailed in Section 7.4.5.
- **Uniqueness:** Does the knowledge regularization term have an effect? Section 7.4.6 is dedicated to comparing the regularization effect of the VIOG module with that of L1 and L2 regularization terms.
- **Sensitivity:** Is it sensitive to key hyperparameters? Section 7.4.7 elucidates the impact of different hyperparameters on prediction accuracy, analyzing the system's responsiveness to parameter alterations.

7.4.1 Evaluation Metrics

The root mean square error (RMSE), coefficient of determination (R^2), mean absolute error (MAE), and mean absolute percentage error (MAPE) are leveraged as the evaluation indices. The detailed expressions are given in Eqs. (7.33)–(7.36), respectively.

$$\mathrm{RMSE} = \sqrt{\frac{1}{N}\sum_{n=1}^{N}(y_n - \tilde{y}_n)^2} \tag{7.33}$$

$$R^2 = 1 - \frac{\sum_{n=1}^{N}(y_n - \tilde{y}_n)^2}{\sum_{n=1}^{N}(y_n - \bar{y})^2} \tag{7.34}$$

$$\mathrm{MAE} = \frac{1}{N}\sum_{n=1}^{N}|y_n - \tilde{y}_n| \tag{7.35}$$

$$\mathrm{MAPE} = \frac{100\%}{N}\sum_{n=1}^{N}\left|\frac{y_n - \tilde{y}_n}{y_n}\right| \tag{7.36}$$

where $\bar{y}$ is the mean value of the label, $\tilde{y}$ is the predicted value, y is the real value, and N is the data number of the testing dataset. For RMSE, MAE, and MAPE, the smaller value, the more accurate the model. On the contrary, the closer R^2 to 1, the better the model fits.

7.4.2 Process Description

To showcase the effectiveness of our proposed method, we conduct a case study focusing on quality prediction in a real CSC unit, which belongs to an ammonia synthesis process. In the following content, we offer an in-depth overview of the technical background pertinent to these industrial processes.

Figure 7.4 illustrates the process flow diagram of the CSC unit, as described in Ref. [37], which is integral to an actual ammonia synthesis process. The chemical reaction, as specified in Eq. (7.37), occurs within fixed-bed reactors. A critical aspect of ammonia synthesis is maintaining an optimal carbon–hydrogen ratio. This unit comprises two isothermal fixed-bed reactors arranged sequentially. The initial step involves compressing the reactant gas into the first high-temperature reactor, ensuring the achievement of thermodynamic equilibrium and an appropriate reaction rate. Subsequently, the gas is cooled in the second reactor, modifying the equilibrium phase to favor increased hydrogen yield. This process results in a final product gas, ready for further processing in subsequent downstream operations. Given that Eq. (7.37) outlines an exothermic reaction, as detailed in Ref. [38] (where $\delta\mathbf{H}$ indicates the enthalpy change):

$$CO + H_2O \leftrightarrow CO_2 + H_2, \quad \delta\mathbf{H}_{298K} = -41.4\ \mathrm{kJ\ mol^{-1}} \tag{7.37}$$

wherein elevated temperatures may shift the equilibrium toward the reverse reaction. This shift, in accordance with Le Chatelier's principle, typically results in a

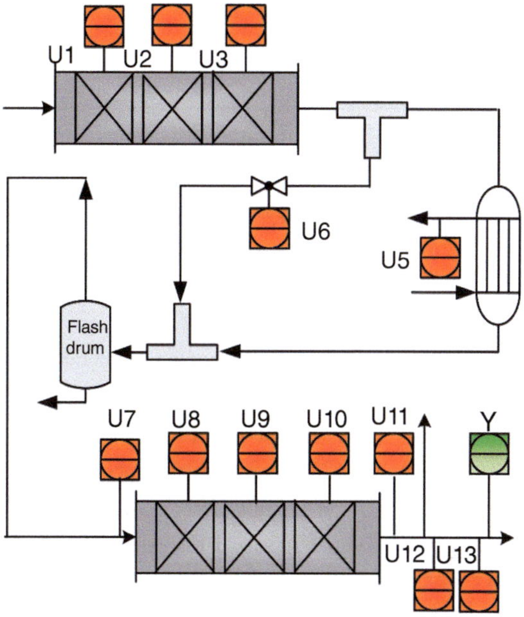

Figure 7.4 Flowsheet of CSC unit.

decreased conversion rate of carbon monoxide. Conversely, the reaction kinetics, guided by the Arrhenius equation, indicate that the reaction rate is significantly reduced at lower temperatures. Consequently, to optimize the conversion efficiency, it is practical to conduct the reaction using different catalysts at varied temperature ranges. This approach effectively balances the thermodynamic and kinetic considerations, ensuring optimal conversion from both perspectives. The technological specifications previously described necessitate accurate, real-time monitoring of carbon monoxide levels, specifically at the outlet of the low-temperature bed, which is highlighted in green in Figure 7.4. Traditional measurement techniques, such as gas chromatography, however, introduce significant delays in carbon monoxide content detection. To enhance the control of the reactor and enable immediate monitoring of carbon monoxide concentrations, hard sensors have been deployed to gather essential process variables. For the purpose of constructing a robust quality prediction model, 13 key variables have been selected. These variables and their respective roles in the process are elaborately detailed in Table 7.1.

7.4.3 Prior Knowledge Analysis

Based on the principles of chemical reaction engineering, the reactors are analyzed employing the plug-flow reactor model, as depicted in Figure 7.5. By analyzing the length-l infinitesimal ($\mathrm{d}V$) marked in the white zone, the mass conservation of reactant γ can be derived as shown in (7.38), where ζ is the bed voidage, $\mathcal{F}$ is the flow rate, and $\mathcal{R}$ is the reaction rate.

$$\mathcal{F}_\gamma - (\mathcal{F}_\gamma + \mathrm{d}\mathcal{F}_\gamma) = (-\mathcal{R}_\gamma)(1 - \zeta)\gamma \tag{7.38}$$

Table 7.1 Variable of CSC process.

Variable	Description
U1	High-temperature bed, temperature 1
U2	High-temperature bed, temperature 2
U3	High-temperature bed, temperature 3
U4	Outlet temperature of high-temperature bed
U5	Outlet temperature of cooling water
U6	Split-gas temperature
U7	Inlet temperature of low-temperature bed
U8	Low-temperature bed temperature 1
U9	Low-temperature bed temperature 2
U10	Low-temperature bed temperature 3
U11	Outlet temperature of low-temperature bed
U12	Outlet pressure of low-temperature bed
U13	Product gas pressure
Y	Carbon monoxide concentration

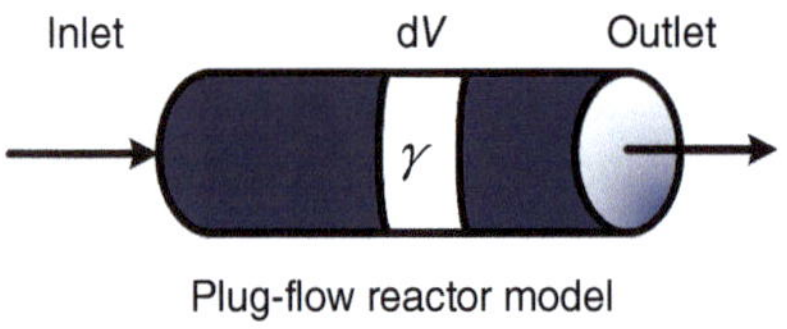

Figure 7.5 The plug-flow reactor model.

Meanwhile, the heat conservation [39] can be given in (7.39):

$$G_c\mathcal{T} - G_c(\mathcal{T} + d\mathcal{T}) = \mathcal{R}_\gamma(1-\zeta)(-\delta H)\gamma dl \tag{7.39}$$

where G_c indicates gas heat capacity, subscript c is the abbreviation of capacity, and $\mathcal{T}$ indicates temperature.

According to Ergun equation, the pressure drop ($\frac{dp}{dl}$) can be derived as follows:

$$\frac{dp}{dl} = \left(\frac{150}{Re} + 1.75\right)\left(\frac{1-\eta}{\zeta^3}\right)\left(\frac{\rho_g u_g}{d_s}\right) \tag{7.40}$$

where Re is Reynold's number, ρ_g is reactant gas density, u_g is gas velocity, and subscript g is the abbreviation of "gas." Note that the gas velocity determines the ratio of gas reactant diffusion rate and reaction rate (also known as Thiele modulus). And thus, in the designing stage of the reactor, the velocity and the voidage should be well designed to make the pressure drop as small as possible for higher reactant conversion [40].

Throughout the analysis of the CSC process, we can have the following prior knowledge:

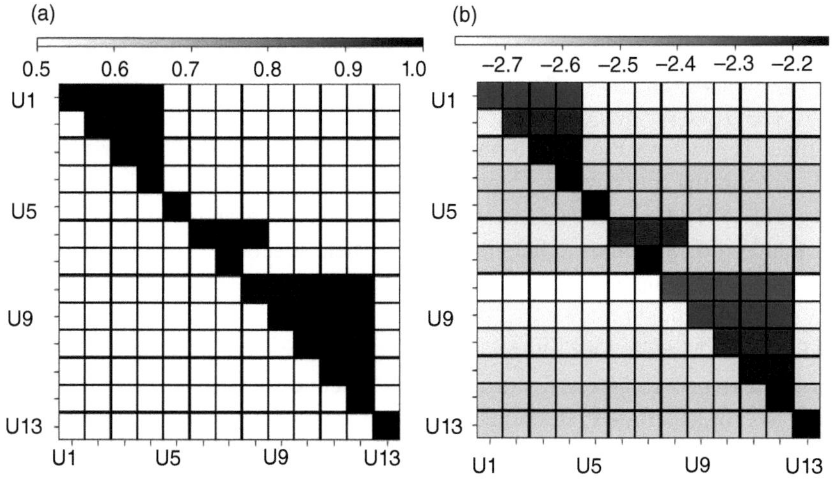

Figure 7.6 (a) The prior knowledge adjacency matrix of CSC; (b) The normalized prior knowledge adjacency matrix of CSC.

- U1, U2, and U3 in the high-temperature bed can derive temperature difference by minus along the reactant flow direction.
- U7, U8, and U9 along the gas flow direction can derive the temperature.
- U11, U12, and U13 in the low-temperature bed can derive pressure difference by minus along the reactant flow direction.
- U8, U9, U10, and U11 in the low-temperature bed can derive pressure difference by minus along the reactant flow direction.

Consequently, the prior knowledge of the CSC process can be obtained as Figure 7.6a shows, where the prior knowledge part and the diagonal are 0.955, and other elements are 0.045. The normalized graph by log-softmax function by rows is given in Figure 7.6b, which is adopted as the value of μ in the prior log-normal distribution.

7.4.4 Baseline Models

We choose the following kinds of baseline models to demonstrate the effectiveness of our VIOG module. The hyperparameter setting and other training protocols are reported in Appendix 7.A. To facilitate comprehension, the final column showcases the "win counts" as per Ref. [41], representing the metrics where the model featuring the VIOG module outperforms the baseline models:

- **GNN-Based Models:** Our VIOG can be treated as a special kind of GNN, and thus we select graph convolution network (GCN) [34] and graph attention networks (GATs) [42] as our baseline models.
- **Synthesizer-Based Models:** In the VIOG framework, our approach begins by leveraging the scaled-dot product attention mechanism to autonomously learn the graph structure. Progressing from this foundational step, we introduce

significant modifications: the scaled-dot product attention is replaced by two new elements – a multi-layer perceptron (MLP) and a learnable matrix. As elaborated by Tay et al. [43], these adaptations are identified as the dense synthesizer (SynDense) and the random synthesizer (SynRandom). The primary goal of these changes is to assess the effectiveness of the similarity function in our model. It is noteworthy that current adaptive GNN models, including Graph WaveNet [44], MTGNN [45], and FC–GAGA [46], can be viewed as variants of either SynRandom or SynDense. Therefore, due to their significance in the adaptive GNN field, SynRandom and SynDense are chosen as the baseline models for our comparative evaluation.

- **VAE-Based Model:** Our model, which is fundamentally based on variational inference techniques, employs two baseline models for comparative analysis – the supervised variational autoencoder (SVAE) and the attentive neural process (ANP). The SVAE, as discussed in references [47, 48], primarily focuses on variable space, while the ANP, described in Kim et al. [49], operates within functional space. These models were specifically selected due to their pertinent application of variational inference methods, aligning closely with the theoretical underpinnings of our model.

7.4.5 Model Performance Comparisons

Table 7.2 presents the evaluation metrics results for different models. The following observations can be obtained:

1. For most downstream models, our VIOG model can have the best or second-best performance.
2. The VIOG model does not improve the transformer downstream model as much as other downstream models.
3. Synthesizer-based models have more competitive performance than the scaled-dot product attention model.

Observation 1 highlights the efficacy and superiority of our proposed VIOG module across most downstream models in industrial data modeling tasks. However, as noted in Observation 2, the VIOG module demonstrates only marginal improvements when compared to other GNN models. This limited enhancement could be attributed to the complexity of the model architecture. For instance, the transformer model, characterized by its larger parameter count and more intricate encoder and layer structure, presents greater optimization challenges compared to other models. Consequently, the VIOG module's impact on improving the transformer model is less pronounced than it is on other downstream models. Further supporting this, Observation 3 points out the similarity between traffic flow prediction tasks and inferential sensor tasks. In traffic flow predictions, adaptive GNN models such as Graph WaveNet [44], MTGNN [45], and FC–GAGA [46], which can be considered variants of synthesizer-based models [43], align with the findings of Observation 3.

Table 7.2 Comparison to baseline models.

Structure	RMSE	R^2	MAPE	MAE
VIOG	0.144 ± 2.414 × E-6	0.939 ± 3.986 × E-6	0.058 ± 5.842 × E-7	0.114 ± 2.249 × E-6
GCN	0.161 ± 9.606 × E-5	0.904 ± 2.689 × E-4	0.065 ± 1.614 × E-5	0.127 ± 6.230 × E-5
GAT	0.179 ± 9.385 × E-4	0.877 ± 4.352 × E-3	0.069 ± 1.340 × E-4	0.137 ± 4.929 × E-4
SynDense	0.174 ± 4.127 × E-4	0.898 ± 7.115 × E-4	0.069 ± 5.540 × E-5	0.135 ± 2.136 × E-4
SynRandom	0.147 ± 3.648 × E-5	0.934 ± 6.423 × E-5	0.059 ± 7.612 × E-6	0.116 ± 2.933 × E-5
SVAE	0.159 ± 1.997 × E-5	0.928 ± 1.681 × E-5	0.064 ± 2.517 × E-6	0.125 ± 9.625 × E-6
ANP	0.239 ± 2.121 × E-4	0.836 ± 4.134 × E-4	0.096 ± 3.981 × E-5	0.188 ± 1.537 × E-4

7.4.6 Comparison with L1 and L2 Regularization Terms

The VIOG module incorporates a prior knowledge regularization term into its learning objective, a feature that inherently enhances model performance. In this subsection, we aim to conduct a comparative analysis between this prior knowledge-based regularization term and the conventional L1 and L2 regularization terms. This comparison is intended to further demonstrate our method's superiority. Following the approach used in Section 7.4.5, we implement a grid search for the L1 and L2 regularization terms and present the findings in Table 7.3.

From Table 7.3, we found that our VIOG module consistently outperforms conventional L1 and L2 regularization terms across various downstream models. This outcome highlights the effectiveness and superiority of the knowledge regularization term introduced by VIOG. However, it is noted that the prior knowledge regularization term underperforms when applied to LSTM and transformer downstream models within the context of the CSC dataset. This can likely be attributed to reasons analogous to those discussed in Section 7.4.5. Specifically, the CSC unit's direct connection to downstream production units, which maintains high reactant purity and stable production content at set points, can be equated to "conceptual drift" in machine learning terminology. As a result, the VIOG-augmented downstream model exhibits diminished performance compared to traditional L1 and L2 regularization when tested on the CSC dataset. Nonetheless, in most tested scenarios, the VIOG's regularization term demonstrably enhances the performance of downstream models beyond what is achieved with standard L1 and L2 terms. This finding emphasizes the broad-scale superiority of the VIOG module in enhancing model performance.

Table 7.3 Comparison to baseline models.

Downstream	Regularization	RMSE	R^2	MAPE	MAE
FNN	VIOG	0.144 ± 2.414 × E-6	0.939 ± 3.986 × E-6	0.058 ± 5.842 × E-7	0.114 ± 2.249 × E-6
	L1	0.177 ± 2.043 × E-4	0.907 ± 1.656 × E-4	0.070 ± 3.487 × E-5	0.128 ± 5.882 × E-4
	L2	0.188 ± 1.171 × E-4	0.908 ± 1.563 × E-4	0.075 ± 1.481 × E-5	0.147 ± 5.726 × E-5
CNN	VIOG	0.143 ± 3.583 × E-5	0.934 ± 8.610 × E-5	0.057 ± 5.353 × E-6	0.112 ± 2.048 × E-5
	L1	0.147 ± 4.017 × E-5	0.931 ± 8.462 × E-5	0.058 ± 5.514 × E-6	0.115 ± 2.122 × E-5
	L2	0.151 ± 1.104 × E-4	0.928 ± 9.043 × E-5	0.060 ± 1.616 × E-5	0.118 ± 6.235 × E-5
LSTM	VIOG	0.150 ± 5.858 × E-5	0.921 ± 1.965 × E-4	0.060 ± 9.164 × E-6	0.117 ± 3.531 × E-5
	L1	0.145 ± 3.775 × E-5	0.930 ± 9.769 × E-5	0.058 ± 5.670 × E-6	0.113 ± 2.194 × E-5
	L2	0.149 ± 4.249 × E-5	0.924 ± 1.376 × E-4	0.059 ± 5.456 × E-6	0.116 ± 2.102 × E-5
Transformer	VIOG	0.173 ± 8.454 × E-5	0.906 ± 1.676 × E-4	0.0692 ± 1.426 × E-5	0.1361 ± 5.497 × E-5
	L1	0.180 ± 3.464 × E-4	0.890 ± 1.211 × E-3	0.0737 ± 6.530 × E-5	0.1425 ± 2.508 × E-4
	L2	0.172 ± 3.250 × E-4	0.902 ± 8.558 × E-4	0.0694 ± 5.509 × E-5	0.1364 ± 2.117 × E-4

7.4.7 Sensitivity Analysis

In this subsection, we conduct a sensitivity analysis of the VIOG model paired with a transformer downstream model, utilizing the CSC dataset. The outcomes of this analysis are depicted in Figure 7.7. Here, we observe how the model's performance varies with adjustments in regularization strength and batch size. As evident from this figure, despite some fluctuations in performance due to changes in hyperparameters, the model consistently exhibits strong performance. This indicates that the VIOG module is effectively designed, demonstrating resilience across a broad range of hyperparameter settings.

7.5 Conclusions

In this study, we introduced the VIOG module, designed to automatically encode and select knowledge as DL features in data-driven industrial process modeling. The VIOG module enhances traditional DL models by incorporating a VIOG inference

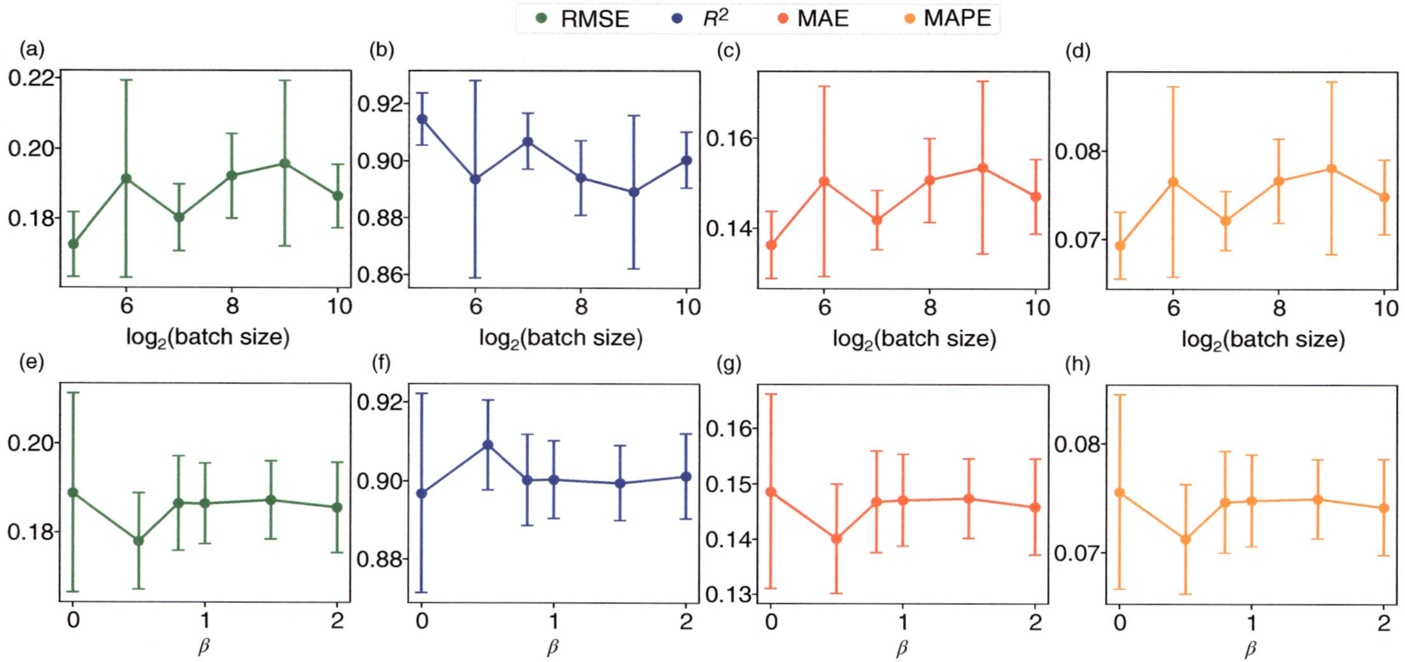

Figure 7.7 The sensitivity analysis results on the transformer downstream model of (a) RMSE, (b) R^2, (c) MAE, (d) MAPE along batch size for the CSC dataset. The sensitivity analysis results of (e) RMSE, (f) R^2, (g) MAE, and (h) MAPE along regularization strength β for the CSC dataset.

network and a novel prior knowledge regularization term. Our approach involved analyzing the probabilistic graph structures inherent in conventional GNNs and DL models. This analysis guided the development of the variational inference technique, which utilizes prior knowledge as a regulatory factor. Furthermore, we crafted the SAM as a unique spatial-based GNN variant, aiming to assimilate prior knowledge effectively. The loss function of the VIOG module was also meticulously derived to align with these objectives. To demonstrate its efficacy, the VIOG module was applied in inferential sensor experiments within the CSC process, where it exhibited superior performance.

While the VIOG module shows promising potential, it stands to benefit from advancements in two key areas. First, the computation of attention scores in the VIOG module is computationally demanding, requiring significant time and memory resources. A potential solution lies in adopting sampling techniques similar to those proposed in [41], which could significantly reduce the training costs associated with VIOG. Secondly, there is room for improvement in the precision of data probabilistic density estimation. The current reliance of the VIOG module on amortized variational inference techniques leads to what is known as an "amortization gap" – a discrepancy between the log-likelihood and the evidence lower bound (ELBO) as detailed in [50]. This gap might be narrowed by integrating variational inference with importance sampling methods, enhancing the tightness of the ELBO, a concept explored in [51].

7.A Experimental Settings

For simplicity, we set the regularization strength β of the VIOG module as 1.0. We conduct grid search on learning rate in [0.0001, 0.001, 0.01, 0.1, 1.0] and batch size in [32, 64, 128, 256, 512, 1024]. For L1 and L2 regularization terms, we select the model regularization strength as 0.001 by applying grid search in [0.001, 0.005, 0.010, 0.020]. The learning rate and batch size adopted in models are listed in Table 7.A.1.

Other hyperparameters are listed as follows:

- For the transformer model, the encoder layer is 3, decoder layer is 3. The head number of the attention is set as 8. The sequence length is set as 6 while the start token length is set as 0.
- The layer of the LSTM model is 1, where the number of hidden units is 32.
- The feature extraction layer of the CNN model is 3, where the convolution kernel size is [4, 4, 4] while the pooling kernel size is [2, 2, 2]. The stride of them is set as 1, and the hidden unit number of the FC layer is set as $30 \rightarrow 20$.
- The hidden unit of the FNN is set as $13 \rightarrow 10 \rightarrow 7 \rightarrow 5$.

In addition, both GCN and GAT necessitate a predefined graph structure during the model inference phase. To address this requirement, we consider the existence of edges in the prior graph as being governed by a Bernoulli distribution. Following this approach, the edges of the graph are independently sampled from their respective Bernoulli distributions, aligning with the mean-field assumption on graphs as

Table 7.A.1 Hyperparameters of the CSC dataset.

	batch				Lr			
Structure	**FNN**	**CNN**	**LSTM**	**Transformer**	**FNN**	**CNN**	**LSTM**	**Transformer**
VIOG	1024	64	64	32	0.01	0.001	0.001	0.005
GCN	1024	64	64	32	0.01	0.001	0.001	0.005
GAT	1024	64	64	32	0.01	0.001	0.001	0.005
SynDense	1024	64	256	64	0.01	0.001	0.001	0.001
SynRandom	1024	128	256	1024	0.01	0.01	0.001	0.001
SVAE	32	256	512	1024	0.01	0.01	0.01	0.01
ANP	128	64	64	32	0.01	0.01	0.01	0.01
L1	1024	64	64	32	0.01	0.001	0.001	0.005
L2	1024	64	64	32	0.01	0.001	0.001	0.005

outlined in Ref. [30]. We organize the dataset in ascending order based on timestamps. Subsequently, we allocated the first 60% of the data for training purposes, the subsequent 10% (from 60% to 70%) for validation, and the remaining portion for testing. For model optimization, we employ the Adam optimizer, as outlined in [52]. All experimental procedures are performed on a high-specification workstation equipped with four Intel Xeon E5 processors, eight NVIDIA GTX 1080 graphics cards, and 128 GB of RAM. The model training and inference processes are conducted using Python 3.8, with PyTorch 1.10 [53] serving as the DL backend. To ensure robustness and reduce randomness in our results, each experiment is replicated a minimum of three times, each time using one of seven distinct random seeds.

References

1 Sun, Q. and Ge, Z. (2021). A survey on deep learning for data-driven soft sensors. *IEEE Transactions on Industrial Informatics* 17 (9): 5853–5866. https://doi.org/10.1109/TII.2021. 3053128.

2 Luo, Y., Zhang, X., Kano, M. et al. (2023). Data-driven soft sensors in blast furnace ironmaking: a survey. *Frontiers of Information Technology & Electronic Engineering* 24 (3): 327–354.

3 Qian, F., Yaochu Jin, S., Qin, J., and Sundmacher, K. (2021). Guest editorial special issue on deep integration of artificial intelligence and data science for process manufacturing. *IEEE Transactions on Neural Networks and Learning Systems* 32 (8): 3294–3295. https://doi.org/10.1109/TNNLS.2021.3092896.

4 Qian, J., Song, Z., Yao, Y. et al. (2022). A review on autoencoder based representation learning for fault detection and diagnosis in industrial processes. *Chemometrics and Intelligent Laboratory Systems* 231: 104711.

5 Wang, K., Shang, C., Liu, L. et al. (2019). Dynamic soft sensor development based on convolutional neural networks. *Industrial & Engineering Chemistry Research* 58 (26): 11521–11531.

6 Yan, F., Yang, C., and Zhang, X. (2022). Dsted: a denoising spatial–temporal encoder–decoder framework for multistep prediction of burn-through point in sintering process. *IEEE Transactions on Industrial Electronics* 69 (10): 10735–10744. https://doi.org/10.1109/TIE. 2022.3151960.

7 Goodfellow, I., Bengio, Y., and Courville, A. (2016). *Deep Learning*, vol. 1. Cambridge: MIT Press.

8 Zeng, A., Chen, M., Zhang, L., and Xu, Q. (2023). Are transformers effective for time series forecasting? *Proceedings of the AAAI Conference on Artificial Intelligence* 37 (9): 11121–11128. https://doi.org/10.1609/aaai.v37i9.26317.

9 Karniadakis, G.E., Kevrekidis, I.G., Lu, L. et al. (2021). Physics-informed machine learning. *Nature Reviews Physics* 3 (6): 422–440.

10 Chen, Z., Xu, J., Alippi, C. et al. (2021). Graph neural network-based fault diagnosis: a review. *arXiv preprint arXiv:2111.08185.*

11 Chen, Z., Jiamin, X., Peng, T., and Yang, C. (2022). Graph convolutional network-based method for fault diagnosis using a hybrid of measurement and prior knowledge. *IEEE Transactions on Cybernetics* 52 (9): 9157–9169. https://doi.org/10.1109/TCYB.2021.3059002.

12 Ren, H., Chen, Z., Liang, X. et al. (2023). Association hierarchical representation learning for plant-wide process monitoring by using multilevel knowledge graph. *IEEE Transactions on Artificial Intelligence* 4 (4): 636–649. https://doi.org/10.1109/TAI.2022.3161860.

13 Ren, H., Liang, X., Yang, C. et al. (2023). Spatial-temporal associations representation and application for process monitoring using graph convolution neural network. *Process Safety and Environmental Protection* 180: 35–47.

14 Guo, H., Tang, R., Ye, Y. et al. (2017). Deepfm: a factorization-machine based neural network for CTR prediction. *Proceedings of the Twenty-Sixth International Joint Conference on Artificial Intelligence, IJCAI-17*. pp. 1725–1731. https://doi.org/10.24963/ijcai.2017/239.

15 Ma, Y. and Huang, B. (2018). Extracting dynamic features with switching models for process data analytics and application in soft sensing. *AIChE Journal* 64 (6): 2037–2051.

16 Chen, Z. and Ge, Z. (2022). Knowledge automation through graph mining, convolution, and explanation framework: a soft sensor practice. *IEEE Transactions on Industrial Informatics* 18 (9): 6068–6078. https://doi.org/10.1109/TII.2021.3127204.

17 Chen, Z., Song, Z., and Ge, Z. (2023). Variational inference over graph: knowledge representation for deep process data analytics. *IEEE Transactions on Knowledge and Data Engineering* 1–16:https://doi.org/10.1109/TKDE.2023.3327415.

18 Raissi, M., Perdikaris, P., and Karniadakis, G.E. (2019). Physics-informed neural networks: a deep learning framework for solving forward and inverse problems involving nonlinear partial differential equations. *Journal of Computational Physics* 378: 686–707.

19 Zhiyong, W., Wang, H., He, C. et al. (2023). The application of physics-informed machine learning in multiphysics modeling in chemical engineering. *Industrial & Engineering Chemistry Research* 62 (44): 18178–18204. https://doi.org/10.1021/acs.iecr.3c02383.

20 Wang, R., Kashinath, K., Mustafa, M. et al. (2020). Towards physics-informed deep learning for turbulent flow prediction. *Proceedings of the 26th ACM SIGKDD International Conference on Knowledge Discovery & Data Mining*. New York, NY, USA: Association for Computing Machinery. pp. 1457–1466.

21 Zheng, Y. and Zhe, W. (2023). Physics-informed online machine learning and predictive control of nonlinear processes with parameter uncertainty. *Industrial & Engineering Chemistry Research* 62 (6): 2804–2818.

22 Alhajeri, M.S., Abdullah, F., Zhe, W., and Christofides, P.D. (2022). Physics-informed machine learning modeling for predictive control using noisy data. *Chemical Engineering Research and Design* 186: 34–49.

23 Zeng, L., Zheng, J., Yao, L., and Ge, Z. (2022). Dynamic Bayesian networks for feature learning and transfer applications in remaining useful life estimation. *IEEE Transactions on Instrumentation and Measurement* 72: 1–12.

24 Khosbayar, A., Valluru, J., and Huang, B. (2021). Multi-rate Gaussian Bayesian network soft sensor development with noisy input and missing data. *Journal of Process Control* 105: 48–61.

25 Wu, D. and Zhao, J. (2021). Process topology convolutional network model for chemical process fault diagnosis. *Process Safety and Environmental Protection* 150: 93–109.

26 Blei, D.M., Kucukelbir, A., and McAuliffe, J.D. (2017). Variational inference: a review for statisticians. *Journal of the American Statistical Association* 112 (518): 859–877.

27 Zhang, C., Bütepage, J., Kjellström, H., and Mandt, S. (2019). Advances in variational inference. *IEEE Transactions on Pattern Analysis and Machine Intelligence* 41 (8): 2008–2026. https://doi.org/10.1109/TPAMI.2018.2889774.

28 Tai, K.S., Socher, R., and Manning, C.D. (2015). Improved semantic representations from tree-structured long short-term memory networks. *Proceedings of the 53rd Annual Meeting of the Association for Computational Linguistics and the 7th International Joint Conference on Natural Language Processing (ACL-IJCNLP 2015)*. Beijing, China: Association for Computational Linguistics. pp. 1556–1566.

29 Elinas, P., Bonilla, E.V., and Tiao, L. (2020). Variational inference for graph convolutional networks in the absence of graph data and adversarial settings. *Advances in Neural Information Processing Systems* 33: 18648–18660.

30 Ying, Z., Bourgeois, D., You, J. et al. (2019). GNNExplainer: generating explanations for graph neural networks. *Advances in Neural Information Processing Systems* 32: 9244–9255.

31 Jang, E., Gu, S., and Poole, B. (2016). Categorical reparameterization with Gumbel-Softmax. *International Conference on Learning Representations*. San Juan, Puerto Rico. pp. 1–12.

32 Vaswani, A., Shazeer, N., Parmar, N. et al. (2017). Attention is all you need. *Neural Information Processing Systems*. Long Beach, California, USA: Curran Associates, Inc. pp. 6000–6010.
33 Gilmer, J., Schoenholz, S.S., Riley, P.F. (2017). Neural message passing for quantum chemistry. *Proceedings of the 34th International Conference on Machine Learning*, vol. 70. Sydney, Australia: PMLR. pp. 1263–1272.
34 Kipf, T.N. and Welling, M. (2017). Semi-supervised classification with graph convolutional networks. *International Conference on Learning Representations (ICLR)*. Toulon, France. pp. 1–14.
35 Fan, X., Zhang, S., Chen, B., and Zhou, M. (2020). Bayesian attention modules. *Advances in Neural Information Processing Systems* 33: 16362–16376.
36 Baldi, P. and Sadowski, P.J. (2013). Understanding dropout. *Advances in Neural Information Processing Systems* 26 (NeuralIPS2013): 2814–2822.
37 Yao, L. and Ge, Z. (2021). Cooperative deep dynamic feature extraction and variable time-delay estimation for industrial quality prediction. *IEEE Transactions on Industrial Informatics* 17 (6): 3782–3792. https://doi.org/10.1109/TII.2020.3021047.
38 Peter Atkins, P. and De Paula, J. (2014). *Atkins' Physical Chemistry*. Oxford University Press.
39 McCabe, W.L., Smith, J.C., and Harriott, P. (1993). *Unit Operations of Chemical Engineering*, vol. 5. New York: McGraw-Hill.
40 Ross, J.R.H. (2019). Chapter 8: Mass and heat transfer limitations and other aspects of the use of large-scale catalytic reactors. In: *Contemporary Catalysis* (ed. J.R.H. Ross), 187–213. Amsterdam: Elsevier.
41 Zhou, H., Zhang, S., Peng, J. et al. (2021). Informer: beyond efficient transformer for long sequence time-series forecasting. *The Thirty-Fifth AAAI Conference on Artificial Intelligence, AAAI 2021, Virtual Conference*, vol. 35. pp. 11106–11115.
42 Brody, S., Alon, U., and Yahav, E. (2021). How attentive are graph attention networks? *International Conference on Learning Representations*. pp. 1–26.
43 Tay, Y., Bahri, D., Metzler, D. et al. (2021). Synthesizer: rethinking self-attention for transformer models. *International Conference on Machine Learning*, PMLR. pp. 10183–10192.
44 Wu, Z., Pan, S., Long, G. et al. (2019). Graph wavenet for deep spatial-temporal graph modeling. *Proceedings of the 28th International Joint Conference on Artificial Intelligence, IJCAI'19*. Macao, China: Curran Associates, Inc. pp. 1907–1913.
45 Wu, Z., Pan, S., Long, G. et al. (2020). Connecting the dots: multivariate time series forecasting with graph neural networks. *Proceedings of the 26th ACM SIGKDD International Conference on Knowledge Discovery & Data Mining*. New York, NY, USA: Association for Computing Machinery. pp. 753–763.
46 Oreshkin, B.N., Amini, A., Coyle, L., and Coates, M. (2021). FC-GAGA: fully connected gated graph architecture for spatio-temporal traffic forecasting. *Proceedings of the AAAI Conference on Artificial Intelligence*, vol. 35. pp. 9233–9241.
47 Xie, R., Jan, N.M., Hao, K. et al. (2019). Supervised variational autoencoders for soft sensor modeling with missing data. *IEEE Transactions on Industrial Informatics* 16 (4): 2820–2828.

48 Shen, B., Yao, L., and Ge, Z. (2020). Nonlinear probabilistic latent variable regression models for soft sensor application: from shallow to deep structure. *Control Engineering Practice* 94: 104198.

49 Kim, H., Mnih, A., Schwarz, J. et al. (2019). Attentive neural processes. *International Conference on Learning Representations*. pp. 1–18.

50 Cremer, C., Li, X., and Duvenaud, D. (2018). Inference suboptimality in variational autoencoders. *International Conference on Machine Learning*. PMLR. pp. 1078–1086.

51 Kim, Y., Wiseman, S., Miller, A. et al. (2018). Semi-amortized variational autoencoders. *International Conference on Machine Learning*, vol. 80. PMLR. pp. 2678–2687.

52 Kingma, D.P. and Ba, J. (2015). Adam: a method for stochastic optimization. *International Conference on Learning Representations*. San Diega, CA, USA. pp. 1–8.

53 Paszke, A., Gross, S., Massa, F. et al. (2019). Pytorch: an imperative style, high-performance deep learning library. *Advances in Neural Information Processing Systems* 32: 1–12.

8

Data-Driven and Physics-Based Reduced-Order Modeling and Optimization of Cooling Tower Systems

Chang He[1,2] *and Zhiqiang Wu*[3]

[1] *Sun Yat-Sen University, School of Chemical Engineering and Technology, No. 135, Xingang Xi Road, Zhuhai, Guangdong, 519082, China*
[2] *The Key Laboratory of Low-carbon Chemistry & Energy Conservation of Guangdong Province, Guangdong Engineering Center for Petrochemical Energy Conservation, 132 Waihuan East Road, University City, Panyu District, Guangzhou, 510275, China*
[3] *Sun Yat-Sen University, School of Materials Science and Engineering, Guangzhou, No. 135, Xingang Xi Road, 510275, China*

8.1 Introduction

The process industry has long relied on cooling towers to enhance sustainability by removing waste heat from circulating water, particularly in the face of increasing water scarcity [1, 2]. However, in certain regions, traditional open cooling towers struggle to effectively cool circulating water or experience high evaporative losses due to significant seasonal fluctuations. In response to these challenges, more sophisticated closed-type cooling systems, such as closed-wet cooling towers [3, 4] and hybrid closed-circuit cooling towers [5, 6], have been recently developed to address these issues and are particularly well-suited for heat rejection and water conservation in diverse weather conditions. To ensure their flexible operation, it is crucial to thoroughly understand the interplay between operating conditions, weather conditions, and the thermodynamic performance of these cooling tower systems. For instance, in hot and humid weather, cooling towers may not operate at their optimal level, necessitating measures to enhance their efficiency. During cold and dry weather, conversely, they may perform excessively, requiring conservative measures to prevent over-cooling. This entails a need not only to develop accurate prediction models but also to utilize these models for optimal design and management.

Closed-type cooling tower models often involve complex interactions between fluid dynamics and heat-mass transfer processes as well as various contacting patterns of vapor–liquid outside the coil tubes [7, 8]. With the advancement of computational fluid dynamics (CFD), numerical studies have become the primary approach to gaining a deeper understanding of these processes through a detailed full-order model (FOM). The FOM can provide an accurate spatially distributed

Applied AI Techniques in the Process Industry: From Molecular Design to Process Design and Optimization, First Edition. Edited by Chang He and Jingzheng Ren.

parameter description of the coupled heat and mass transfer processes by considering the detailed multiphase interactions, transport effects, and phenomena at a broader length scale. For example, the profiles of temperature and humidity of the inlet air in the air–liquid contact units of the cooling towers are unevenly distributed and varied with uncertain and control parameters, which define the spatial heterogeneities of the humid and thermal characteristics of the air in the flow field. Deciphering such distributed parameters in detail is conducive to improving the overall thermodynamic performance of the cooling towers.

In theory, the solution of the FOMs of cooling tower systems involves tracking the multiphase interactions and the spatial movement of each liquid droplet, characterized by computationally expensive partial differential equations (PDEs). It is noted that the benefits of constructing FOMs must be balanced against the computational resources and time required to develop them, which can be a major obstacle to improving cooling systems [3]. For example, a single task of the 3D CFD model of a closed-wet cooling tower system normally takes more than 30 CPU hours by using the Reynolds averaged Navier–Stokes turbulence model [7, 9]. The computational burden would significantly limit the use of the principled CFD simulation for real-time prediction and multivariable analysis. It becomes more critical as the FOM is embedded in an equation-oriented optimization task because a large size of recourses to the CFD models have to be executed before converging to the optimum.

The reduced-order model (ROM) of a CFD-based model is a surrogate model for input-to-output mapping due to the computationally expensive nature of high-fidelity simulation, which entails iterative calculations. Surrogate-based model reduction is one of the practical solutions to construct a more tractable model as an alternative to computationally expensive FOM without loss of fidelity by only approximating the input–output relationship of a system [10]. Given input parameters (i.e. initial, boundary, and operational conditions), the quantities of interest for calculating distributed parameters at each mesh point, such as velocity, pressure, temperature, and moisture content, can be obtained rapidly without conducting principled CFD simulation. The model reduction approach can be roughly categorized into two categories: physics-based models and data-driven models [11, 12]. In the former, a reduced basis is extracted from the simulation data using an unsupervised learning technique, e.g. principal component analysis (PCA), in which the FOM operator is projected onto the subspace spanned by the reduced basis [13]. As a result, the degrees of freedom of the original system can be significantly reduced by retaining the underlying basic structure of the FOM. Another way to enable rapid simulations is to build a data-driven model, where a response surface of the system is learned from the simulation data in a supervised manner. The deterministic or probabilistic mapping models of input–output can be constructed using, e.g. polynomial basis functions [14], radial basis functions [15], Gaussian process [16], and stochastic polynomial chaos expansion [17].

The objective of this chapter is to create an integrated optimization approach using bi-level model reduction to efficiently determine the optimal thermodynamic performances and their corresponding spatially distributed parameters. This approach involves combining a data-driven ROM and a physics-based ROM, achieved through a series of statistical methods including optimal experiment design, multi-sample CFD simulation, bi-level model reduction, and model evaluation. The design of experiments (DoE) aims to maximize the process information obtained from a finite number of CFD simulations by carefully selecting experimental points. Subsequently, two types of surrogate models can be developed using model order reduction and data-driven techniques, enabling their substitution for the original FOM in design and optimization tasks using efficient nonlinear programming techniques.

This chapter is derived from the work of He and coworkers [4, 9, 18, 19].

8.2 Full-Scale Physical Model of Cooling Towers

The physical model of cooling towers investigated in this study is based on a bench-scale closed-wet cooling tower experimental set-up [8] with parallel counter-flow construction. As shown in Figure. 8.1a, the ambient air flows upward and traverses the outer wall of the tube bundle using the intake fan in countercurrent flow with the spray water. The tube bundle's cross section shown in Figure 8.1b is arranged in a triangular pattern, with adjacent tubes in each row having the same centerline spacing. The hot circulating water pumped from the main tube is evenly divided and distributed across the serpentine tubes inside the tower casing, where its sensible heat is indirectly absorbed by film evaporation and upward air flow outside the tube bundle. The heat and mass transfer processes of water film are mainly sustained by the joint effect of countercurrent spray water and ambient air. A small proportion of spray water diffuses together with the upward airflow due to evaporation, and the remaining portion leaving the water film is collected in a water tank and finally returns to the top nozzles through the spray water pump. Therefore, the circulating water relies on the film evaporation to achieve the required cooling targets, such as temperature drop and heat dissipation.

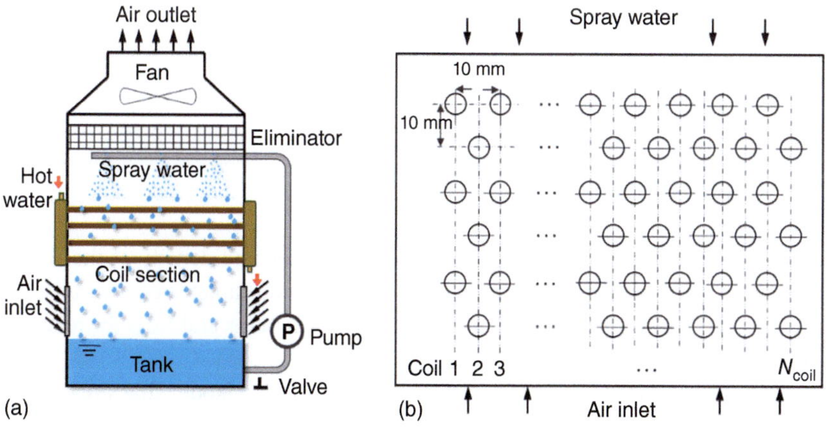

Figure 8.1 (a) Sketch of the cooling tower set-up and (b) cross section of the tube bundle [4, 9].

In the investigated cooling tower, the complex heat and mass transfer processes outside the tubes can be mathematically described by large-scale PDAEs that are constructed from mass and energy conservation and hydrodynamic equations. These equations can be written in general transport equation form, as presented in Eqs. (8.1)–(8.14). Among them, the Navier–Stokes and energy conservation equations are as follows:

$$\frac{\partial \rho}{\partial t} + \nabla \cdot (\rho \vec{v}) = S_m \tag{8.1}$$

$$\frac{\partial}{\partial t}(\rho \vec{v}) + \nabla \cdot (\rho \vec{v}\vec{v}) = -\nabla P_{\text{sp}} + \nabla \cdot (\tilde{\tau}) + \rho \vec{g} + \vec{F}b \tag{8.2}$$

$$\frac{\partial}{\partial t}(\rho E) + \nabla(\vec{v}(\rho E + P_{\text{sp}})) = \nabla \left(\text{kc}_{\text{eff}} \nabla T - \sum_{\text{js}} h_{\text{js}} D_{\text{js}} + (\tilde{\tau}_{\text{eff}} \cdot \vec{v}) \right) + S_h \tag{8.3}$$

where S_{m} is the mass added to the continuous phase from the dispersed second phase; P_{sp} and $\tilde{\tau}$ are the static pressure and the stress tensor; $\rho\vec{g}$ and $\vec{F}b$ are the gravitational body and external body forces; kc_{eff} is the effective conductivity; D_{js} is the diffusion flux of species js; S_h is a user-defined source term. In addition, the first three terms on the right-hand side of Eq. (8.3) represent energy transfer due to conduction, species diffusion, and viscous dissipation, respectively.

To numerically simulate the dynamics of multiphase flows, we use the Eulerian–Lagrangian approach, as described in previous studies [20–22]. In this approach, the air-fluid phase is treated as a continuous medium, and its behavior is modeled by solving the Navier–Stokes equations. On the other hand, the dispersed phase, represented by the spray water, is treated as a collection of particles or droplets. The motion of these particles or droplets is tracked by integrating the force balance equation, allowing for the exchange of momentum, mass, and energy with the fluid phase. The force balance equation is formulated to equate the particle's inertia with the forces acting upon it. Mathematically, this can be expressed as follows:

$$\frac{\text{d}v_{\text{p}}}{\text{d}t} = \frac{v_{\text{f}} - v_{\text{p}}}{\varsigma} + \frac{\vec{g}(\rho_{\text{p}} - \rho_{\text{f}})}{\rho_{\text{p}}} + \vec{F} \tag{8.4}$$

$$\varsigma = \frac{\rho_{\text{p}} d_{\text{p}}^2}{18\mu_{\text{f}}} \frac{24}{C_{\text{d}}\text{Re}} \tag{8.5}$$

$$\text{Re} = \frac{\rho_{\text{f}} d_{\text{p}} |v_{\text{p}} - v_{\text{f}}|}{\mu_{\text{f}}} \tag{8.6}$$

where v_{f}, μ_{f}, and ρ_{f} are the velocity, molecular viscosity, and density of the fluid, respectively; v_{p}, ρ_{p}, and d_{p} are the velocity, density, and diameter of the particle, respectively; $(v_{\text{f}} - v_{\text{p}})/\zeta$ is the drag force per unit particle mass; and Re is the relative Reynolds number.

According to previous studies [8, 23], the realizable k–ε model that presents a sufficient adaptation with the physics of flow turbulence is applied to model the turbulent flow outside the tubes in the tower. Based on the general form of the transport equation, the turbulence kinetic energy ke, and its rate of dissipation ε, are obtained from the realizable k–ε model equations, which are expressed as follows:

$$\frac{\partial}{\partial t}(\rho \mathrm{ke}) + \frac{\partial}{\partial x_{\mathrm{is}}}(\rho \mathrm{ke} u_{\mathrm{is}}) = \frac{\partial}{\partial x_{\mathrm{js}}}\left[\left(\mu + \frac{\mu_t}{C_{\mathrm{ke}}}\right)\frac{\partial \mathrm{ke}}{\partial x_{\mathrm{js}}}\right] + \mathrm{Gk} + \mathrm{Gb} - \rho\varepsilon - yr_M + S_{\mathrm{ke}} \tag{8.7}$$

$$\frac{\partial}{\partial t}(\rho \varepsilon) + \frac{\partial}{\partial x_{\mathrm{is}}}(\rho \varepsilon) = \frac{\partial}{\partial x_{\mathrm{js}}}\left[\left(\mu + \frac{\mu_t}{C_{\varepsilon}}\right)\frac{\partial \varepsilon}{\partial x_{\mathrm{js}}}\right] + \rho C_1 S\varepsilon - \rho C_2 \frac{\varepsilon^2}{\mathrm{ke} + \sqrt{\nu\varepsilon}} + C_{1\varepsilon}\frac{\varepsilon}{\mathrm{ke}} C_{3\varepsilon}\mathrm{Gb} + S_{\varepsilon} \tag{8.8}$$

where Gk and Gb represent the generation of turbulent kinetic energy due to the mean velocity gradients and buoyancy; yr_M is the contribution of the fluctuating dilatation in compressible turbulence to the overall dissipation rate; C_2 and $C_{1\varepsilon}$ are constants; and C_{ke} and C_{ε} are the turbulent Prandtl numbers for parameters ke and ε.

The mixture special transport model is employed to calculate the concentration of different species in the cooling tower, and the convection-diffusion equation is used to solve these given by Eq. (8.9).

$$\frac{\partial}{\partial t}(\rho \mathrm{fm}_{\mathrm{is}}) + \nabla \cdot (\rho \vec{v} y_{\mathrm{is}}) = -\nabla \cdot D_{\mathrm{is}} + R_{\mathrm{is}} + S_{\mathrm{is}} \tag{8.9}$$

$$D_{\mathrm{is}} = -\left(\rho D_{\mathrm{m,is}} + \frac{\mu_{\mathrm{t}}}{\mathrm{Sc}_{\mathrm{t}}}\right)\nabla y_i - D_{T,\mathrm{is}}\frac{\nabla T}{T} \tag{8.10}$$

where $\mathrm{fm}_{\mathrm{is}}$, R_{is}, $D_{\mathrm{m,is}}$, and $D_{T,\mathrm{is}}$ are the mass fraction, net rate, mass diffusion, and thermal diffusion coefficients for species *is*, respectively; Sc_{t} is the turbulent Schmidt number; and the default is 0.7.

The Eulerian wall film model, coupled with the mixture species transport model, is used to calculate phase changes between film liquid and gas vapor. The rate of phase change is governed by:

$$\dot{m}_{\mathrm{phase}} = \frac{(\rho_{\mathrm{mix}} D_{\mathrm{vap}}/\mathrm{WD})}{\rho_{\mathrm{mix}} D_{\mathrm{vap}}/\mathrm{WD} + C_{\mathrm{phase}}} C_{\mathrm{phase}}(\mathrm{fm}_{\mathrm{sat}} - \mathrm{fm}_{\mathrm{vap}}) \tag{8.11}$$

$$\mathrm{fm}_{\mathrm{sat}} = \frac{P_{\mathrm{sat}} T}{P_{\mathrm{mix}}} \frac{M_{\mathrm{vap}}}{M_{\mathrm{mix}}} \tag{8.12}$$

$$\log_{10} P_{\mathrm{sat}} = -2.1794 + 0.02953(T - 273.17) - 9.1837e^{-5}(T - 273.17)^2 + 1.4454e^{-7}(T - 273.17)^3 \tag{8.13}$$

where ρ_{mix} is the density of the gas mixture; D_{vap} is the mass diffusivity of the vapor species; WD is the cell-center-to-wall distance; C_{phase} is the phase change constant; $\mathrm{fm}_{\mathrm{vap}}$ is the mass fraction of the vapor species. In addition, $\mathrm{fm}_{\mathrm{sat}}$ and P_{sat} calculated via Eqs. (8.12) and (8.13) are the mass fractions and pressure of saturation species, M_{vap} and M_{mix} are the molecular weights of the vapor species and mixture, and P_{mix} is the absolute pressure of the gas mixture.

The droplet temperature is updated according to the energy balance shown in Eq. (8.14) that relates the sensible heat change between the droplet to the convective and latent heat transfer between the droplet and the continuous phase:

$$m_p c_{pp} \frac{dT_p}{dt} = htA_p(T_\infty - T_p) - \frac{dm_p}{dt} Q_{lh} - A_p \zeta_p C_{SB} T_p^4 \tag{8.14}$$

where c_{pp} is the droplet heat capacity, ht is the convective heat transfer coefficient, T_∞ and T_p are the temperatures of the continuous phase and droplet, dm_p/dt and Q_{lh} are the evaporation rate and latent heat, and C_{SB} is the Stefan–Boltzmann constant.

8.3 Bi-level Reduced-Order Models

In the context of model reduction for specific process equipment, three types of variable spaces are considered: the input space **X**, the output space **Y**, and the state space **Z** in the fluid field. Figure 8.2 illustrates the interconnection of matrices representing these three variable spaces, which is crucial for constructing the bi-level ROMs. In the cooling tower system depicted, there are n inputs and m outputs, while s state variables are monitored across a total of γ discretized elements. The steady-state nonlinear relations between the state and output variables, corresponding to the input variables, can be described by the following state-space model [13]:

$$\begin{bmatrix} \mathbf{y} \\ \mathbf{z} \end{bmatrix} = \begin{bmatrix} \mathbf{B}(\mathbf{x}) \\ \mathbf{D}(\mathbf{x}) \end{bmatrix} \tag{8.15}$$

where **B** and **D** are the mapping coefficient matrices and $\boldsymbol{x}, \boldsymbol{y}$, and $\boldsymbol{z}$ are the input, output, and state vectors, respectively, which contain the variables related to calculating the objective function and physical field.

The input, output, and state spaces for the CFD model of cooling towers should be defined before the experimental design. The input vector $\boldsymbol{x}$ in the input space is provided in Table 8.1. For a given inlet boundary condition, it contains the values of all variables in the input streams entering the cooling tower system, i.e. uncertain variable $\boldsymbol{x}^{(1)} = [T_a^{in}, \phi_a^{in}]^T$ and operating variable $\boldsymbol{x}^{(2)} = [\dot{m}_a, \dot{m}_{sw}]^T$. Similarly, the output vector $\boldsymbol{y}$ in the output space contains the values of all variables in the output streams, $\boldsymbol{y} = [T_{cw}^{out}, T_a^{out}, \phi_a^{out}]^T$. Note that the objective function calculation requires the results of both the input vector $\boldsymbol{x}$ and the corresponding output vector $\boldsymbol{y}$. The

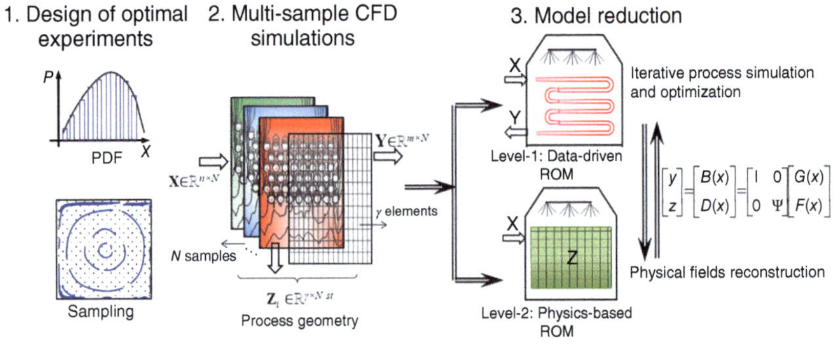

Figure 8.2 The construction of bi-level ROMs.

Table 8.1 Input space for the model parameters.

Input domain	Symbol	Unit	Specification
Control variable $\mathbf{X}^{(1)}$			
Volume flow rate of air	$\dot{m}_{\mathrm{a}}$	$\mathrm{m^3\,s^{-1}}$	U[1.0, 3.0]
Mass flow rate of spray water	$\dot{m}_{\mathrm{sw}}$	$\mathrm{kg\,s^{-1}}$	U[1.0, 2.0]
Uncertain variable $\mathbf{X}^{(2)}$			
Dry-bulb temperature of air	T_{a}	°C	Uncertain variable
Relative humidity of air	ϕ_{a}	—	Uncertain variable

state space is spatially distributed and bounded by the equipment geometry. The state vector $\boldsymbol{z}$ in the state space accommodates the values of distributed parameters at each discretized node of the mesh.

For constructing the bi-level ROMs of cooling towers with high fidelity, it is necessary to build the mappings between the aforementioned input and output vectors, as well as between the input and state vectors. For this purpose, much effort is required to collect and handle a huge amount of data obtained from a batch of principled CFD simulations of the equipment item. The main steps are summarized as follows:

Step 1: Implement a DoE in the specific ranges to obtain a finite set of samples over the input domain.

Step 2: Solve the CFD cases one by one with the obtained samples under the defined mesh system. The results of the state and output variables of interest are retrieved from the information stored in the CFD solutions.

- **Step 3.1:** Implement the mapping $\boldsymbol{x} \rightarrow \boldsymbol{y}$ and then formulate the data-driven ROM.
- **Step 3.2:** PCA is performed for the dimension reduction of obtained z to obtain the ranked principal components, loading matrix, and score matrix $\boldsymbol{\Phi}$, and then implement the mapping $\boldsymbol{x} \rightarrow \boldsymbol{\Phi}$ and formulate the physics-based ROM.

Evaluate the generalization performance of the developed bi-level ROMs.

More details about these four steps are provided in the following Sections 8.3.1–8.3.3.

8.3.1 Design of Optimal Experiments

Batch simulations of principled CFD models are often computationally intensive and time-consuming, necessitating the reduction of computational resources. DoE is a suitable approach for maximizing the amount of process information obtained from a fixed number of observations by selecting appropriate design points. Various methods can be used to implement DoE, such as Monte Carlo sampling, Taguchi experiments, probabilistic collocation, and stochastic collocation. Herein, a low-dimensional discrete approximation method called the stochastic reduced-order model (SROM) is utilized to generate a finite set of stochastic samples with varying probabilities. The SROM method can be considered a “smart”

Monte Carlo method for uncertainty quantification, as it can efficiently discretize the input space and reduce the computational complexity associated with propagated uncertainty, while maintaining the nonintrusive nature of the method [18]. A detailed description of the SROM method is provided below.

Let us consider the probabilistic behavior of $\mathbf{X} = \{\mathbf{x}_i\}_{i=1}^{n}$ to be specified with known expressions for its marginal distributions, moments of order q, and correlation matrix as follows:

$$\mathrm{F}_i(\theta) = \Pr(\mathbf{x}_i \leq \theta), \quad i = 1, \dots, n \tag{8.16}$$

$$\mu_i(q) = \mathrm{E}\left[\mathbf{x}_i^q\right], \quad i = 1, \dots, n \tag{8.17}$$

$$\mathbf{r} = \mathrm{E}[\mathbf{X}\mathbf{X}^{\mathrm{T}}] \tag{8.18}$$

An SROM $\widetilde{\mathbf{X}} \in \mathbb{R}^{n \times N}$ is an approximation of the $\mathbf{X}$ in the sense that $\widetilde{\mathbf{X}}$ and $\mathbf{X}$ have similar statistical properties. The SROM $\widetilde{\mathbf{X}}$ consists of a sample set $\{\widetilde{\mathbf{x}}^{(j)}\}_{j=1}^{N}$ with the corresponding probabilities $\{\mathrm{prob}^{(j)}\}_{j=1}^{N}$, thus $\sum_{j=1}^{N} \mathrm{prob}^{(j)} = 1$. Each sample $\widetilde{\mathbf{x}}^{(j)} = \left\{\widetilde{x}_i^{(j)}\right\}_{i=1}^{n}$ contains one or multiple values, depending on the dimension of $\mathbf{X}$.

Similar to $\mathbf{X}$, the statistics of the SROM $\widetilde{\mathbf{X}}$ are as follows:

$$\widetilde{F}_i(\theta) = \sum_{j=1}^{N} \mathrm{prob}^{(j)} I\left(\widetilde{x}_i^{(j)} \leq \theta\right) \quad i = 1, \dots, n \tag{8.19}$$

$$\widetilde{\mu}_i(q) = \sum_{j=1}^{N} \mathrm{prob}^{(j)} \left(\widetilde{x}_i^{(j)}\right)^q \quad i = 1, \dots, n \tag{8.20}$$

$$\widetilde{r}(i_1, i_2) = \sum_{j=1}^{N} \mathrm{prob}^{(j)} \widetilde{x}_{i_1}^{(j)} \widetilde{x}_{i_2}^{(j)} \quad i_1, i_2 = 1, \dots, n \tag{8.21}$$

where $I(A)$ is a binary function, $I(A) = 1$ if A is true, otherwise $I(A) = 0$. The discrepancy between $\widetilde{\mathbf{X}}$ and $\mathbf{X}$ in the statistical sense is represented by the distribution deviation e_1, the moment deviation e_2, and the correlation matrix deviation e_3:

$$e_1(\widetilde{\mathbf{X}}, \mathbf{prob}) = \frac{1}{2} \sum_{i=1}^{n} \sum_{j=1}^{N} \left[\widetilde{\mathrm{F}}_i\left(\widetilde{x}_i^{(j)}\right) - \mathrm{F}_i\left(\widetilde{x}_i^{(j)}\right)\right]^2 \tag{8.22}$$

$$e_2(\widetilde{\mathbf{X}}, \mathbf{prob}) = \frac{1}{2} \sum_{i=1}^{n} \sum_{q=1}^{\overline{q}} \left(\frac{\widetilde{\mu}_i(q) - \mu_i(q)}{\mu_i(q)}\right)^2 \tag{8.23}$$

$$e_3(\widetilde{\mathbf{X}}, \mathbf{prob}) = \frac{1}{2} \sum_{i_1, i_2 = 1; i_2 > i_1}^{n} \left(\frac{\widetilde{r}(i_1, i_2) - r(i_1, i_2)}{r(i_1, i_2)}\right)^2 \tag{8.24}$$

The optimal SROM $\widetilde{\mathbf{X}}$ for $\mathbf{X}$ with the minimum deviation can be obtained using the following optimization algorithm:

$$\begin{aligned} \widetilde{\mathbf{X}} &\equiv \underset{\widetilde{\mathbf{x}}, \mathbf{prob}}{\arg\min} \left(\sum_{i=1}^{3} \delta_i e_i(\widetilde{\mathbf{X}}, \mathbf{prob})\right) \\ \text{s.t.} &\quad \sum_{j=1}^{N} \mathrm{prob}^{(j)} = 1 \\ &\quad \mathrm{prob}^{(j)} \geq 0, \quad j = 1, \dots, N \end{aligned} \tag{8.25}$$

where the coefficients $\{\delta_i \geq 0\}_{i=1}^{3}$ are the weighting factors to control the contribution of each error term to the objective function. The determination of sample size N is dependent on computational considerations, and the optimal solution can be considered a good approximation of the original distribution.

8.3.2 Multi-sample CFD Simulations

To obtain the mapping data for constructing the ROMs, the uncertainty associated with the input variables is propagated by executing the rigorous CFD simulation of the cooling tower model for each of the SROM samples. The multi-sample CFD simulation is performed on the meshed geometry of the cooling tower system to calculate the coupled heat and mass transfer processes, as well as the mass and energy balances of the PDEs. Note that the circulating water inside the tubes relies on film evaporation to achieve the required cooling targets, such as temperature drop and heat dissipation. Thus, we only focus on the heat and mass transfer processes of the water film outside the tube bundle at the coil section, which is stimulated by a 3D CFD model of flow, temperature, and pressure fields. The following assumptions are made to reduce the complexity of the cooling tower model:

(1) The outer wall of the coil tubes is assumed to have no slips.
(2) Both the circulating water and inlet air are considered incompressible fluids.
(3) The surface of the coil tubes is uniformly covered by a water film, while the air flow and spray water are uniformly distributed throughout the tower.
(4) The temperature of the water film is equal to the average temperature of the circulating water.
(5) Thermal radiation between the tower and its surroundings is neglected.

The process geometry of the coil section shown in Figure 8.1b is constructed based on the standard diameter and length in the COMSOL modeling environment. It consists of 19 rows and 12 columns of bare-type copper tubes in the $0.61 \times 1.2 \times 0.68\ \text{m}^3$ domain. The preliminary sketch is meshed to adequately capture the change of fluids by using the meshing module. To enhance model accuracy and reduce computational load, it is necessary to fine-tune the mesh system near the tube walls before running the numerical program. As shown in Figure 8.3, a five-layer boundary layer grid scheme is implemented with a stretching factor of 1.2 for the inner and outer walls of the tube. This choice aims to improve the accuracy and stability of predictions. Tetrahedral meshes are employed for the remaining space. Moreover, the element size is gradually stretched to effectively capture high-gradient regions, ensuring a proper resolution of temperature and velocity fields. Thus, the computational domain is extended by 0.09, 1.2, and 0.68 m in x-, y-, and z-axes, respectively. The mesh structure has 1 114 893 nodes, and only 3 304 nodes are presented in the cross section along the center of the coil tube. The quality of this mesh structure, measured by the skewness value, should be satisfied at the required level.

In each CFD model, the RNG k–ε turbulence modeling approach is applied to simulate the mean flow characteristics of the air and water in turbulent flow conditions. The governing equations with boundaries are solved by the finite

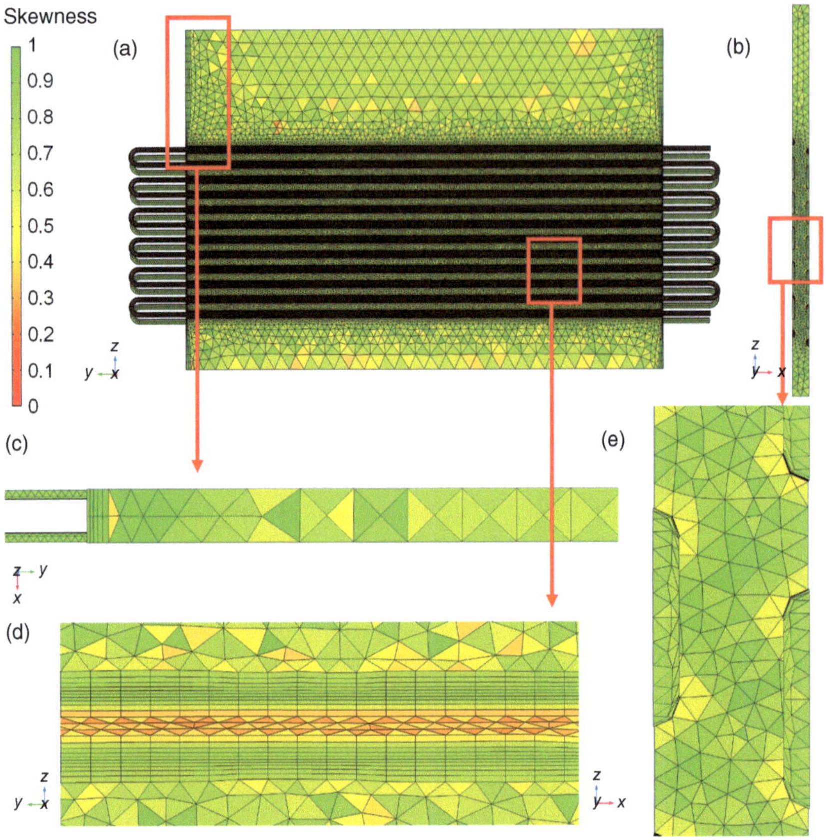

Figure 8.3 The mesh structure of the computational domain generated by COMSOL@ at (a) entirety, (b) lateral section, (c) mesh along the air outlet, (d) mesh along the tube walls, and (e) mesh near the tube walls [9, 18]. Color represents the value of skewness.

element method, and a segregated solver sequence was selected to handle the highly nonlinear models. Moreover, the source terms of moisture transport and heat transfer [4, 18] are added to the governing equations for accurately describing the heat and mass transfer processes.

8.3.3 Model Reduction

The data-driven ROM for cooling towers can be expressed as a set of equations that represent the mapping from the input vector to the output vector, $\boldsymbol{x} \rightarrow \boldsymbol{y} = \mathbf{G}(\boldsymbol{x})$. This mapping is crucial for integrating the equipment model into the simulator for simulation and optimization purposes. Various surrogate-based methods, such as artificial neural networks (ANNs), partial least squares, and Kriging, can be employed to realize this mapping. Here, a two-layer ANN with n inputs, l neurons in the hidden layer, and m targets in the output layer is applied to implement this mapping. Besides, a hyperbolic tangent sigmoid function $f(\cdot) = \text{tansig}(x)$ is used as the active function of the hidden layer due to its desirable nonlinearity and continuous first-order derivative. Therefore, the ANN model can be represented as follows:

$$x_i' = \frac{2(x_i - x_{\min,i})}{(x_{\max,i} - x_{\min,i})} - 1 \quad i = 1, \ldots, n \tag{8.26}$$

$$a_j' = \sum_{i=1}^{n} \mathrm{IW}_{j,i} x_i' + b_{1,j} \quad j = 1, \ldots, l \tag{8.27}$$

$$a_j = f\left(a_j'\right) = \frac{2}{\left(1 + e^{-2a_j'}\right)} - 1 \quad j = 1, \ldots, l \tag{8.28}$$

$$y_k' = \sum_{j=1}^{l} \mathrm{LW}_{j,k} x_j' + b_{2,k} \quad k = 1, \ldots, m \tag{8.29}$$

$$y_k = \frac{1}{2}\left(y_k' + 1\right)\left(y_{\max,k} - y_{\min,k}\right) + y_{\min,k} \quad k = 1, \ldots, m \tag{8.30}$$

where x', a', and y' are the scaled inputs, intermediate outputs from neurons, and scaled outputs, respectively.

This neural network mapping is trained with Bayesian regularization to obtain the following weighting parameters: $\mathbf{IW} \in \mathbb{R}^{l \times n}$, $\mathbf{LW} \in \mathbb{R}^{l \times m}$, $\mathbf{b}_1 \in \mathbb{R}^{l}$, $\mathbf{b}_2 \in \mathbb{R}^{m}$. In this way, the mapping is expressed by purely linear correlations in a matrix form:

$$y = G(x) = \mathrm{LW}^{\mathrm{T}}(f(\mathrm{IW} \cdot x + b_1)) + b_2 \tag{8.31}$$

The physics-based ROM of cooling towers is constructed by implementing the input-to-state mappings through a snapshot PCA–ANN approach. As shown in Figure 8.4, the PCA decomposition is first used to reduce the dimension of the state space. It can construct a reduced number of orthonormal basis vectors in a subspace that expresses a random vector of higher dimensionality without significant loss of data information. In particular, a specific state-space snapshot matrix $\mathbf{Z}_i \in \mathbb{R}^{\gamma \times N}$ can be decomposed after eliminating zeros as

$$\mathbf{Z}_i = \mathbf{M}\boldsymbol{\Sigma}\mathbf{V}^{\mathrm{T}} \quad i = 1, \ldots, s \tag{8.32}$$

where the diagonal matrix $\boldsymbol{\Sigma} = \mathrm{diag}(\sigma_1, \sigma_2, \ldots, \sigma_N)$ contains singular values of matrix $\mathbf{Z}_i$ in descending order, and square matrix $\mathbf{M} \in \mathbb{R}^{\gamma \times \gamma}$ and $\mathbf{V} \in \mathbb{R}^{N \times N}$ are the eigenvectors of $\mathbf{Z}_i\mathbf{Z}_i^{\mathrm{T}}$ and $\mathbf{Z}_i^{\mathrm{T}}\mathbf{Z}_i$ with the corresponding eigenvalues $\lambda_j = \sigma_j^2$, respectively.

The singular values of each variable in each eigenvector signify the importance of this variable to the new vector space. Note that σ declines rapidly with the rank order of $\sigma_1 \geq \sigma_2 \ldots \geq \sigma_N$; in most cases, the sum of the top 10% of σ can account for more than 95% of the total singular values. Reducing to rank α resulting from the cutoff criterion, the α-order reduced data set is expressed as

$$\mathbf{Z}_i \approx \mathbf{Z}_i^{(\alpha)} = \mathbf{M}^{(\alpha)}\boldsymbol{\Sigma}^{(\alpha)}(\mathbf{V}^{(\alpha)})^{\mathrm{T}} \tag{8.33}$$

where the superscript (α) that indicates the first α columns are taken from the original matrix to formulate a new matrix ($\alpha \ll \gamma$). Then, the matrix $\mathbf{Z}_i$ can be expressed as follows:

$$\mathbf{Z}_i = \boldsymbol{\Psi}\boldsymbol{\Phi} + \boldsymbol{\varepsilon} \tag{8.34}$$

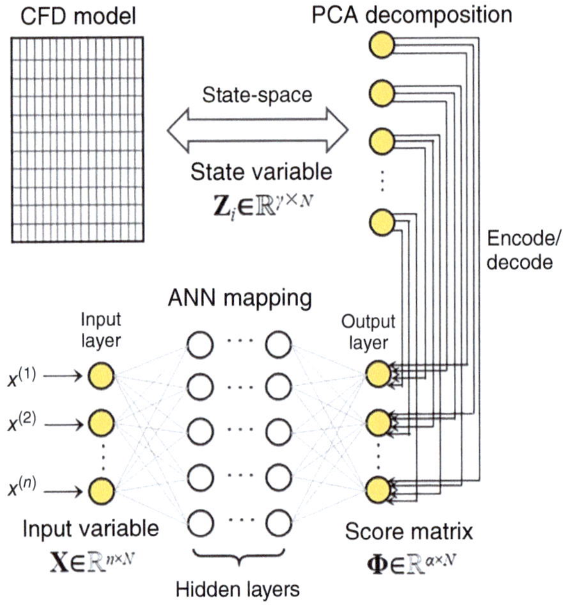

Figure 8.4 The PCA–ANN approach for the construction of physics-based ROM. Source: [4]/with permission of Elsevier.

where matrix $\mathbf{\Psi}^{\gamma \times \alpha} = \mathbf{M}^{(\alpha)} \in \mathbb{R}^{\gamma \times \alpha}$ is the principal component (PC) matrix; $\mathbf{\Phi}^{\alpha \times N} = \mathbf{\Sigma}^{(\alpha)}(\mathbf{V}^{(\alpha)})^{\mathrm{T}} \in \mathbb{R}^{\alpha \times N}$ is the score matrix of the transformed elements in the new vector space; $\boldsymbol{\varepsilon}$ is the remaining noise.

The dominant variance in the PC space captured by the α-order reduced data set should be as large as possible, it usually needs to be greater than 95% for better accuracy, as given by

$$R_\alpha = \sum_{j=1}^{\alpha} \lambda_j \Big/ \sum_{j=1}^{N} \lambda_j = \sum_{j=1}^{\alpha} \sigma_j^2 \Big/ \sum_{j=1}^{N} \sigma_j^2 \tag{8.35}$$

In this way, the snapshot matrix $\mathbf{Z}_i$ of the monitored state variable can be encoded/decoded into a lower-dimensional score matrix $\mathbf{\Phi}$. Once this dimension reduction has been completed, the input-to-ranked score mapping is built as $\boldsymbol{x} \rightarrow \mathbf{\Phi} = \mathrm{F}(\boldsymbol{x})$. Like data-driven ROM, a two-layer ANN with α outputs is used to implement the mapping, as expressed by

$$z = \mathbf{\Psi}\mathrm{F}(x) \tag{8.36}$$

The bi-level ROMs can be formulated in the following state-space model:

$$\begin{bmatrix} y \\ z \end{bmatrix} = \begin{bmatrix} B(x) \\ D(x) \end{bmatrix} = \begin{bmatrix} \mathbf{I} & 0 \\ 0 & \mathbf{\Psi} \end{bmatrix} \begin{bmatrix} G(x) \\ F(x) \end{bmatrix} \tag{8.37}$$

where $\mathbf{I}$ is a unit matrix.

It is well-known that the ROM performs well at training points with ANN mapping. The true performance of a developed ROM mainly depends on how well it can predict interpolated points. In this study, 85% of the SROM samples are randomly selected as the training set to train the neural network mappings. The remaining ones included in the test set are used as interpolated points to measure the generalization error between the outputs of the ROM and CFD models. For the data-driven and physics-based ROM, the generalization errors in terms of the average relative error (avgRE) are given by

$$\text{avgRE} = \begin{cases} \dfrac{1}{N_{\text{TE}}} \displaystyle\sum_{j}^{N_{\text{TE}}} \left| \dfrac{y_j - y_j^{\text{CFD}}}{y_j^{\text{CFD}}} \right|, & j = 1, \ldots, N_{\text{TE}} \\ \dfrac{1}{N_{\text{TE}} \cdot \gamma} \displaystyle\sum_{j}^{N_{\text{TE}}} \sum_{i=1}^{\gamma} \left| \dfrac{z_{i,j} - z_{i,j}^{\text{CFD}}}{z_{i,j}^{\text{CFD}}} \right|, & i = 1, \ldots, \gamma;\ j = 1, \ldots, N_{\text{TE}} \end{cases} \tag{8.38}$$

where $z_{i,j}$ and $z_{i,j}^{\text{CFD}}$ are the results of state variables at the ith node of the mesh corresponding to the jth sample in the test set obtained from physics-based ROM and CFD models, respectively, and N_{TE} is the sample size of the test set.

8.4 Thermodynamic Performance Indicators

In cooling tower systems, coupled heat and mass transfer processes between humid air and water are the most common and fundamental phenomena [19]. Exergy analysis is an effective theoretical tool for investigating the thermodynamic irreversibility of such processes. It exploits both the qualitative and quantitative natures of energy, and thus represents the true potential of a cooling tower to perform optimal work relative to a dead state. The past studies mainly applied it to investigate entropy generation [24], exergy destruction [25], and system optimization [26] of cooling towers. These studies based on traditional exergy analysis and system-level flowsheets only consider the lumped parameter description of input–output streams with many ideal assumptions. There are rarely studies on the description of the spatially distributed parameters inside the cooling tower systems, despite their importance.

The assessment of exergy allows for a comprehensive evaluation of the potential of a cooling tower system to achieve optimal work relative to a reference state. Figure 8.5a illustrates the division of total supply exergy (Ex_{sup}) introduced by a cooling tower into three components: exergy destruction (Ex_{des}) representing irreversibilities in heat and mass transfer, exergy loss (Ex_{loss}) dissipated to the environment, and productive exergy (Ex_{prod}). Figure 8.5b provides an overview of the input and output of exergy flows within the investigated cooling tower system. The supply exergy originates from the inlet air ($\text{Ex}_{\text{a,in}}$), circulating water ($\text{Ex}_{\text{cw,in}}$), and auxiliary equipment (Ex_{aux}), while the utilizable exergy solely comprises the flow exergy of the outlet circulating water ($\text{Ex}_{\text{cw,out}}$). In summary, the overall exergy balance of the cooling tower system can be expressed as follows:

$$\underbrace{\text{Ex}_{\text{cw,in}} + \text{Ex}_{\text{a,in}} + \text{Ex}_{\text{aux}}}_{\text{Ex}_{\text{sup}}} = \underbrace{\text{Ex}_{\text{cw,out}}}_{\text{Ex}_{\text{prod}}} + \underbrace{\text{Ex}_{\text{a,out}}}_{\text{Ex}_{\text{loss}}} + \text{Ex}_{\text{des}} \tag{8.39}$$

In the investigated cooling tower system, the exergy of humid air represents the maximum useful work done by inlet air as it is gradually saturated and approaches

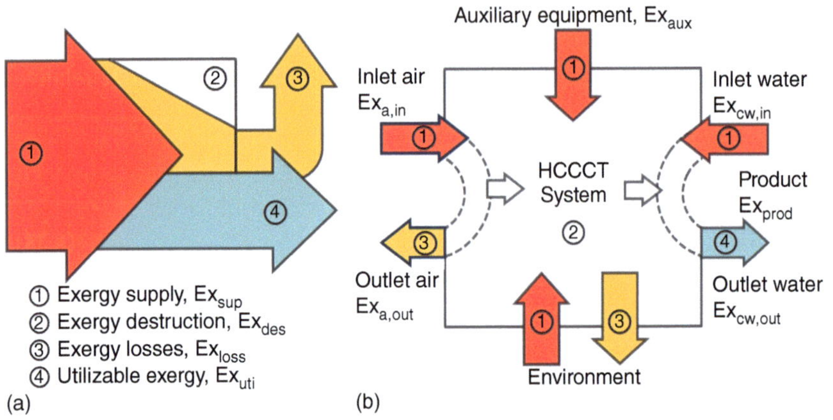

Figure 8.5 Exergy flow schematics of (a) the general cooling tower systems and (b) the investigated cooling tower system. Source: [18]/with permission of Elsevier.

an equilibrium state relative to the dead state. It can be broken down into thermal, mechanical, and chemical ones, as defined by Eqs. (8.40)–(8.42), respectively.

$$\mathrm{ex}_{\mathrm{a,them}}(\mathrm{T},\omega) = (\mathrm{C_a} + \omega_\mathrm{a} C_\mathrm{v}) \times \left(\mathrm{T_a} - \mathrm{T_0} - T_0 \ln \frac{T_\mathrm{a}}{T_0}\right) \tag{8.40}$$

$$\mathrm{ex}_{\mathrm{a,mech}}(T,\omega) = RT_0(1 + 1.608\omega_\mathrm{a}) \ln \frac{P_\mathrm{a}}{P_0} \tag{8.41}$$

$$\mathrm{ex}_{\mathrm{a,chem}}(\omega) = RT_0 \left\{ (1 + 1.608\omega_\mathrm{a}) \ln \frac{1 + 1.608\omega_0}{1 + 1.608\omega_\mathrm{a}} + 1.608\omega_\mathrm{a} \ln \frac{\omega_\mathrm{a}}{\omega_0} \right\} \tag{8.42}$$

where C and R are the specific heat capacity and gas constant, respectively; T, P, and ω are the dry-bulb temperature, pressure, and humidity, respectively; and the subscripts 0, a, and v are the dead state, air, and vapor, respectively. Note that the reference condition at ambient temperature with saturated humidity ratio is selected as the dead state [27].

Multiplying the exergy by the mass flow rate gives the flow exergy of humid air:

$$\dot{\mathrm{E}}\mathrm{x}^j_{\mathrm{a},i}(T,\omega,\dot{m}) = \mathrm{ex}^j_{\mathrm{a},i} \times \dot{m}_\mathrm{a}, \quad i = \{\text{them, mech, chem}\}\ j = \{\text{in, out}\} \tag{8.43}$$

The inlet circulating water is cooled down by the water film outside the tubes as it passes through the coil section in a steady state. Here, the chemical exergy is ignored, and thus the flow exergy is the sum of thermal and mechanical ones, as given in Eq. (8.44). Meanwhile, the circulation of air and spray water streams entering the cooling tower system is sustained by the intake fan and spray pump. The total exergy of electricity consumed by the auxiliary equipment can be calculated via Eq. (8.45).

$$\dot{\mathrm{E}}\mathrm{x}^j_{\mathrm{cw}}(\dot{m},T) \approx \dot{m}_\mathrm{cw} C_\mathrm{cw} \left(T^j_\mathrm{cw} - T_0 - T_0 \ln \frac{T^j_\mathrm{cw}}{T_0}\right) + \dot{m}_\mathrm{cw}\left(\frac{P^j_\mathrm{cw}}{P_0}\right),$$
$$j = \{\text{in, out}\} \tag{8.44}$$

$$\dot{\mathrm{E}}\mathrm{x}_\mathrm{aux}(\dot{m}) = Q_\mathrm{sp} + Q_\mathrm{fan} = \dot{m}_\mathrm{sw} gh/\eta_\mathrm{sp} + \dot{m}_\mathrm{a}\Delta P/(\rho_\mathrm{a}\eta_\mathrm{fan}\eta_\mathrm{belt}\eta_\mathrm{motor}) \tag{8.45}$$

where g, h, Q, and η are the gravitational acceleration, differential head, energy consumption, and mechanical efficiency, respectively; the subscripts sp, cw, sw, fan, belt, and motor are the spray pump, circulating water, spray water, fan blade, belt, and motor, respectively.

Figure 8.5b illustrates the input and output of exergy flows in the investigated cooling tower system. The fundamental goal of this system is to cool down the input circulating water by exchanging sensible heat with liquid film and air flow outside the tubes. To achieve this goal, the supplied flow exergy of all participating streams (spray water and air, etc.) and electrical energy (fans, and pumps, etc.) should be effectively transformed into the productive thermal exergy of circulating water while minimizing the destroyed thermal and humid exergies. Herein, the exergy efficiency ratio (EER) defined in Eq. (8.46) is employed to reveal the degree of effective use of available supplied exergies that are required to maintain the heat and mass transfer processes in a cooling tower system.

$$\mathrm{EER} = \frac{\text{productive exergy}}{\text{supplied exergy}} = \frac{\dot{\mathrm{E}}\mathrm{x}_{\mathrm{cw}}^{\mathrm{in}} - \dot{\mathrm{E}}\mathrm{x}_{\mathrm{cw}}^{\mathrm{out}}}{\dot{\mathrm{E}}\mathrm{x}_{\mathrm{a}}^{\mathrm{in}} + \dot{\mathrm{E}}\mathrm{x}_{\mathrm{cw}}^{\mathrm{in}} + \dot{\mathrm{E}}\mathrm{x}_{\mathrm{aux}}} \tag{8.46}$$

where Ex is the exergy flow rate; the superscripts in and out are the input and output streams, respectively.

Herein, a state variable is introduced, namely exergy flux ($\mathrm{W\,m^{-2}}$, similar to heat flux), for exploring the density of flow exergy of the humid air in the fluid field. The exergy flux includes the thermal, mechanical, and chemical components, which can be regarded as different types of distributed flow exergies per unit of area (dA). As defined in Eq. (8.47), it is a product of the exergy and the mass flux (j_{a}).

$$\vec{J}_{\mathrm{a},i} = \dot{\mathrm{E}}\mathrm{x}_{\mathrm{a},i}/\mathrm{d}A = \mathrm{ex}_{\mathrm{a},i}j_{\mathrm{a}}, \quad i = \{\text{therm, mech, chem}\} \tag{8.47}$$

8.5 Optimization Model

The developed data-driven ROM, which establishes a mapping between input and output variables, can serve as a surrogate module for the cooling tower system. This surrogate module can be integrated into a sampling-based stochastic optimization model. The objective of this optimization model is to obtain optimal solutions by maximizing the expected value of the model's objective distribution, which is represented by the exergy efficiency ratio of the cooling tower system. Specifically, the objective function formulation for maximizing the expected value of exergy efficiency ratios based on the SROM samples is expressed as follows:

$$\begin{aligned}
\max \mathrm{EER}^{\mathrm{Expected}} &= \sum_{j=1}^{N} \mathrm{prob}^{(j)} \times \left[\frac{\dot{\mathrm{E}}\mathrm{x}_{\mathrm{cw}}^{\mathrm{in}} - \dot{\mathrm{E}}\mathrm{x}_{\mathrm{cw}}^{\mathrm{out}}}{\dot{\mathrm{E}}\mathrm{x}_{\mathrm{a}}^{\mathrm{in}} + \dot{\mathrm{E}}\mathrm{x}_{\mathrm{cw}}^{\mathrm{in}} + \dot{\mathrm{E}}\mathrm{x}_{\mathrm{aux}}}\right]^{(j)} \\
&= \sum_{j=1}^{N} \mathrm{prob}^{(j)} \times f(x^{(1)}, x^{(2)})^{(j)} \\
\text{s.t.}\quad & \text{Eq. (8.1)} \sim \text{Eq. (8.7)} \\
& \text{Eq. (8.10)} \sim \text{Eq. (8.19)} \\
T_{\mathrm{cw}}^{\mathrm{out}} &\le T_{\mathrm{cw}}^{\mathrm{design}}
\end{aligned} \tag{8.48}$$

where $\text{prob}^{(j)}$ is the probability related to the occurrence of a specific SROM sample j, $T_{\text{cw}}^{\text{design}}$ is the design constraint that defines the upper outlet temperature of the circulating water for meeting the requirements of the cooling targets.

Therefore, the final solutions are obtained by repeatedly solving the optimization model N times according to the SROM samples. After that, the physics-based ROM can be determined according to the obtained optimal input and output variables, which enables the rapid reconstruction of the thermal-, mechanical-, and chemical-exergy flux fields in the cooling tower system. It is important to note that in the stochastic process optimization model, only the data-driven ROM is utilized to acquire the output stream information from the cooling tower system. This data-driven ROM serves the purpose of optimizing the stochastic process. However, after the optimization process is completed, the physics-based ROM is then employed for a more in-depth investigation of the exergy field.

8.6 Illustrative Example

A small-size cooling tower is presented to introduce the application of the bi-level ROMs. It is assumed the cooling tower system is located in a petrochemical park in Singapore. The original statistics of weather parameters during the years 2008–2018 are retrieved from the government database (http://www.noaa.gov). All seasons within the chosen years are taken into account. SROMPy is used to generate 100 samples from the input space ($N = 100$). The CFD modeling and stochastic optimization programming are implemented via COMSOL Multiphysics 5.4 and MATLAB R2018b modeling environment, respectively. Both models are solved on a workstation with Intel four processor Xeon E5 CPU@2.5 GHz and 32 GB RAM. In CFD modeling, the normalized residuals used for checking convergence are less than 10^{-4} for all governing equations.

In order to determine the most appropriate size of the PCs, the singular values and cumulative variance captured with varied PCs for the investigated exergy fields are presented. From Figure 8.6a, it is found that the singular values drop sharply with the increase in PC, and the first five singular values are several orders of magnitude larger than the remaining ones. Figure 8.6b further shows the proportion of the cumulative variance of the field data captured with the increase in the retained PCs. For this figure, the optimal number of PCs is selected as $\alpha = 5$, and the captured proportion of variance is close to 100%. Based on this selection, the dimensions of the field representation are all significantly reduced from 3304 to 5, while the amounts of snapshot data stored in the thermal-, mechanical-, and chemical-exergy fields are compressed by 99.85%. To conclude, snapshot PCA shows great potential for reducing the multidimensional data set of the exergy field on the premise of maintaining good accuracy.

To ensure the generalization ability of the reduced models, it is necessary to investigate the statistics of relative errors of the monitored variables for the selected samples in the test set between the outputs of bi-level ROMs and CFD model. As shown in Figure 8.7a, the relative errors of the output temperature of the circulating water,

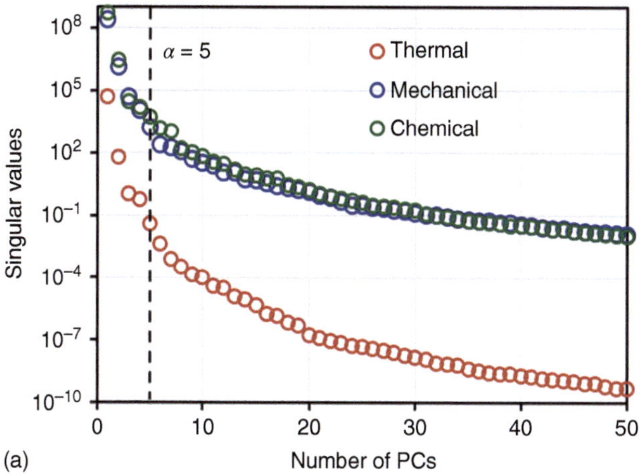

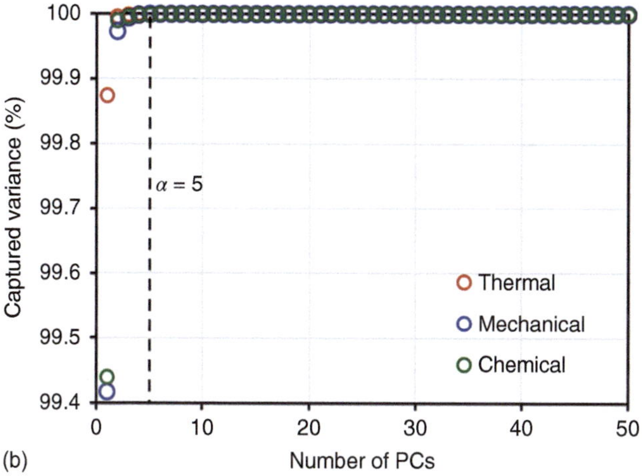

Figure 8.6 (a) The singular values of the PCs and (b) the cumulative variance captured with varied PCs. Source: [4]/with permission of Elsevier.

as well as the output relative humidity and temperature of the air, are all less than the magnitude of 1.00×10^{-3}. The corresponding avgRE for these three stream variables is 6.79×10^{-6}, 2.38×10^{-5}, and 7.29×10^{-5}. Figure 8.7b shows the relative errors of the monitored state variables between the physics-based ROM and the CFD model. As shown, the relative errors of the thermal-, mechanical-, and chemical-exergy fluxes are all less than the magnitude of 1.00×10^{-2}, and the corresponding avgRE are 2.41×10^{-4}, 2.86×10^{-6}, and 1.00×10^{-3}, respectively. From the statistics of these relative errors, it can be concluded that the constructed bi-level ROMs have a good generalization ability and provide sufficient confidence to perform stochastic optimization.

The optimal solutions of the exergy efficiency ratios, corresponding to the SROM samples under different weather conditions, are presented in Figure 8.7a. The green circles represent the optimal exergy efficiency ratios, which exhibit an increasing

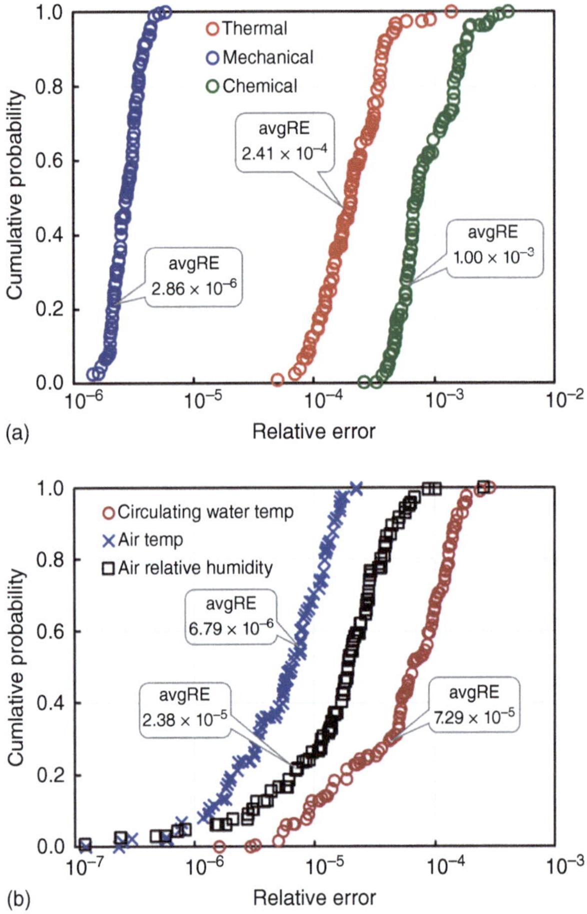

Figure 8.7 Relative errors between bi-level ROMs and CFD solutions. (a) Data-driven ROM, (b) physics-based ROM. Source: [4]/with permission of Elsevier.

trend from 0.37 to 0.74 with the rise in the dry-bulb temperature of the inlet air. The expected value of these optimal exergy efficiency ratios is 0.56. On the other hand, the corresponding probabilities of occurrence, indicated by black forks, demonstrate a scattered and wide distribution ranging from 0.001 to 0.032.

Figure 8.8b displays the optimal solutions of the air flow rate (represented by square dots) and spray water flow rate (represented by triangle dots) at various dry-bulb temperatures of the inlet air. Overall, the optimal values of the air flow rate tend to approach their upper limit, with six samples reaching this limit. Conversely, the optimal values of the spray water flow rate tend to be closer to their lower limit, with 12 samples reaching this limit. These trends suggest that while increasing the air flow rate and decreasing the spray water flow rate contribute to improving the exergy efficiency, blind adjustments cannot be made in practice due to significant constraints, such as the outlet temperature of the spray water, imposed by the developed bi-level ROMs-based optimization model.

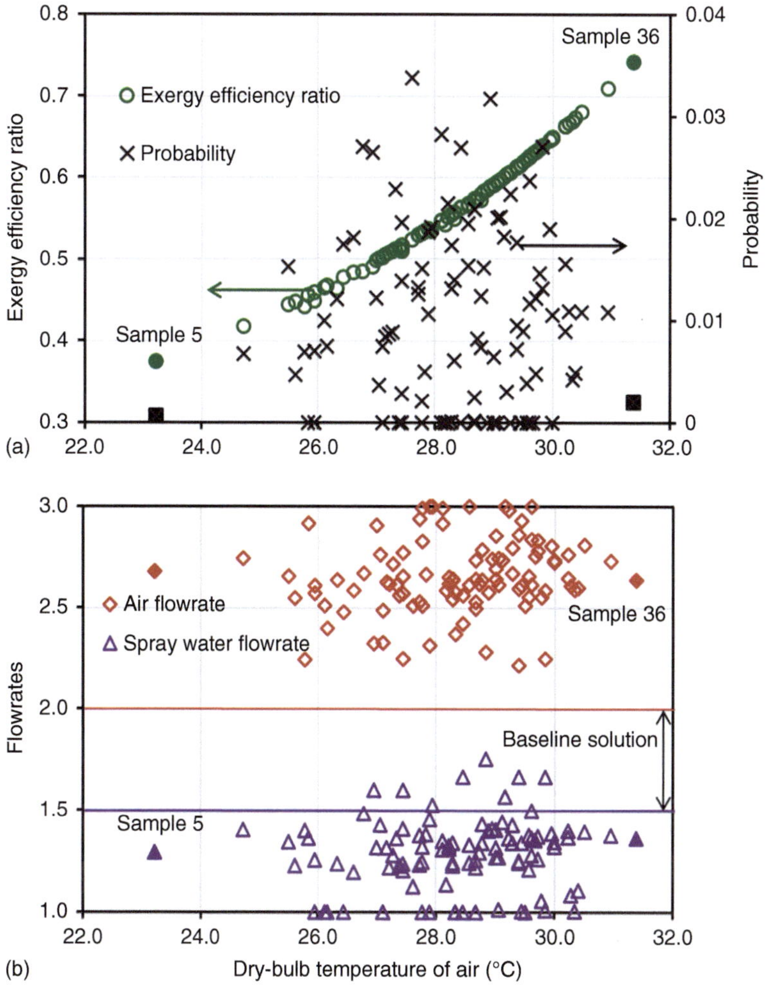

Figure 8.8 Optimization results based on trained data-driven ROM. (a) Exergy efficiency ratios and their probabilities; and (b) control variables. Source: [4]/with permission of Elsevier.

The proposed bi-level ROMs approach not only presents the optimal values of the model objective and the corresponding control variables for the cooling tower system but also reconstructs the high-fidelity fluid dynamics and reveals the complex thermal and flow fields via physics-based ROM. To better demonstrate the difference between the outputs of the physics-based ROM and CFD simulation, two representative samples with extreme and distinct weather conditions are investigated using a baseline solution. This baseline solution directly employs a deterministic control strategy that corresponds to the expected values of the control variables listed in Table 8.1 ($m_a = 2.0\,\mathrm{m^3\,s^{-1}}$, $m_{sw} = 1.5\,\mathrm{kg\,s^{-1}}$). The two representative samples correspond to the coldest and hottest weather conditions in the SROM samples, namely

Sample 5 (T_a^{in} = 23.2 °C, ϕ_a^{in} = 0.87, EER = 0.37) and Sample 36 (T_a^{in} = 31.4 °C, ϕ_a^{in} = 0.57, EER = 0.74).

From Figure 8.9, it is evident that the profiles of the three exergy fluxes obtained using the physics-based ROM are in good agreement with those obtained from the CFD simulation. There are minimal distortions observed in some localized areas. Comparing these profiles, it can be concluded that the outputs of the physics-based ROM exhibit high accuracy and show no noticeable differences at the interpolated points. Besides, note that the bi-level ROMs approach offers great advantages in terms of computational time and resources compared to using rigorous CFD models. For instance, the total time required to reconstruct the exergy profiles within the cooling tower system using CFD simulations takes approximately three days. In

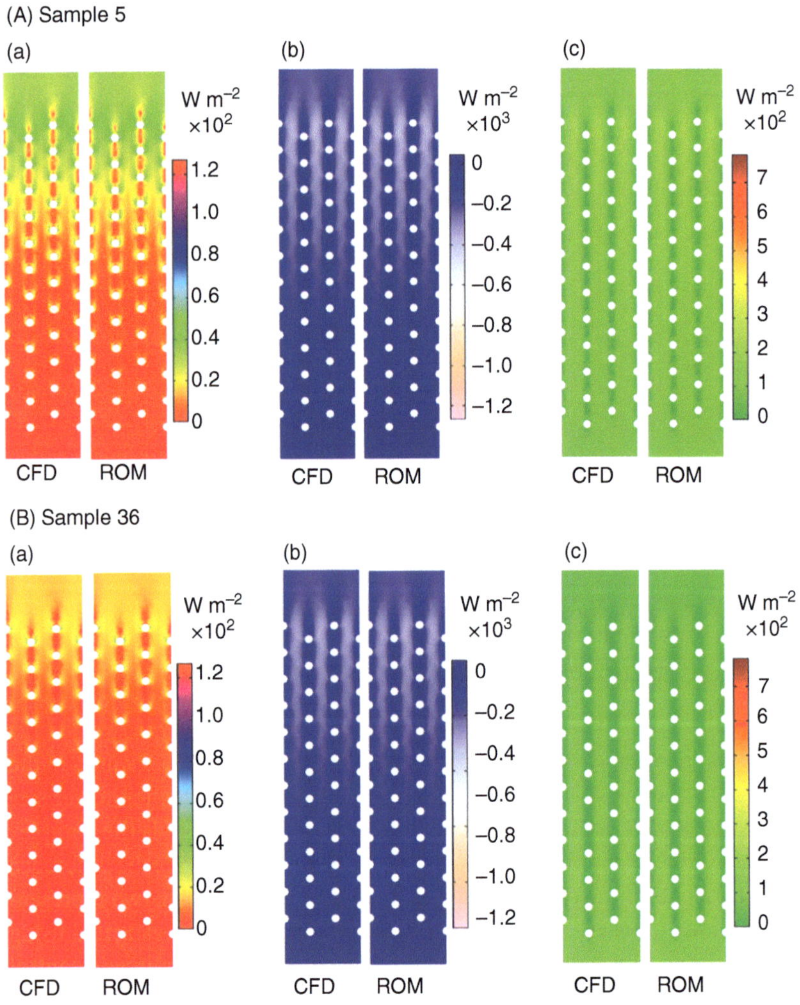

Figure 8.9 The profiles of (a) thermal-, (b) mechanical-, and (c) chemical-exergy fluxes obtained from CFD solution and physics-based ROM. Source: [4]/with permission of Elsevier.

contrast, the developed approach only takes 20–30 seconds for the same case. This demonstrates the efficiency and effectiveness of the proposed approach in reducing computational time and resources.

As the inlet air ascends and comes into contact with the descending spray water, it gradually becomes saturated with the water and heated by the circulating water inside the coil tubes. This leads to a decrease in the chemical-exergy flux and an increase in the thermal-exergy flux, resulting in a bottom-to-top directionality within the cooling tower system, as illustrated in Figure 8.9. Comparing the field profiles of Sample 5 and Sample 36, noticeable variations in the distributions of thermal- and chemical-exergy fluxes can be observed under different weather conditions. For instance, Sample 5 exhibits a larger temperature difference between the circulating water and inlet air compared to Sample 36, leading to a more pronounced change in the distribution of thermal exergy within the fluid field. Interestingly, the optimal values of mechanical-exergy flux for both Sample 5 and Sample 36 are negative, indicating that the applied pressure in the cooling tower system is lower than the reference pressure in the dead state. Moreover, the contours of the mechanical-exergy flux are nearly identical for both samples, ranging from -1.2×10^3 to $0\,\mathrm{W\,m^{-2}}$, suggesting that the weather conditions have a negligible influence on the distribution of mechanical-exergy flux.

To further reveal the influence of the optimal solution on the heat and mass transfer processes, the three exergy fields (see Figures 8.8 and 8.9) obtained from the baseline solution are compared with those obtained from the optimal solution (see Section 8.4). Figure 8.10 illustrates the exergy flux profiles in the cooling tower system obtained from the optimal and baseline solutions.

As shown in Figure. 8.10a1,b1, by and large, the two solutions yield similar thermal-exergy flux profiles for both samples. The amplified areas of the selected top tubes further show that the values of the contours of the thermal-exergy fluxes are in the ranges of 3.0–55.1 $\mathrm{W\,m^{-2}}$ (Sample 5) and 1.3–34.4 $\mathrm{W\,m^{-2}}$ (Sample 36) for the baseline solution, 8.7–94.7 $\mathrm{W\,m^{-2}}$ (Sample 5) and 1.8–45.1 $\mathrm{W\,m^{-2}}$ (Sample 36) for the optimal solution. As for the mechanical-exergy flux shown in Figure 8.10a2,b2, the longitudinal variations of the fluid field appear more evident between the two solutions for both samples. Taking Sample 5 as an example, the mechanical-exergy fluxes have about three times increased from −323.0 to $-36.5\,\mathrm{W\,m^{-2}}$ for the baseline solution to $-774.5\sim-117.6\,\mathrm{W\,m^{-2}}$ for the optimal solution. Similar to the mechanical-exergy flux, the chemical-exergy flux obtained from the optimal solution is much higher than that of the baseline solution. As shown in Figure 8.10a3,b3, the values of contours of the chemical-exergy fluxes of Sample 5 remarkably increase from 41.7 to 119.1 for the baseline solution to 113.8–651.4 $\mathrm{W\,m^{-2}}$ for the optimal solution. In the case of Sample 36, they have correspondingly increased by three times from 18.3–56.4 $\mathrm{W/m^2}$ to 55.0–163.2. These increases in mechanical- and chemical-exergy fluxes are mainly because the optimal solution can effectively strengthen the heat and mass transfer processes inside of the cooling tower system in extreme weather conditions by using a higher air flow rate ($m_a = 2.68\,\mathrm{m^3\,s^{-1}}$ for the Sample 5 and $2.63\,\mathrm{m^3\,s^{-1}}$ for the Sample 36) and a lower spray water flow rate ($m_{sw} = 1.29\,\mathrm{kg\,s^{-1}}$ for the Sample 5 and $1.36\,\mathrm{kg\,s^{-1}}$

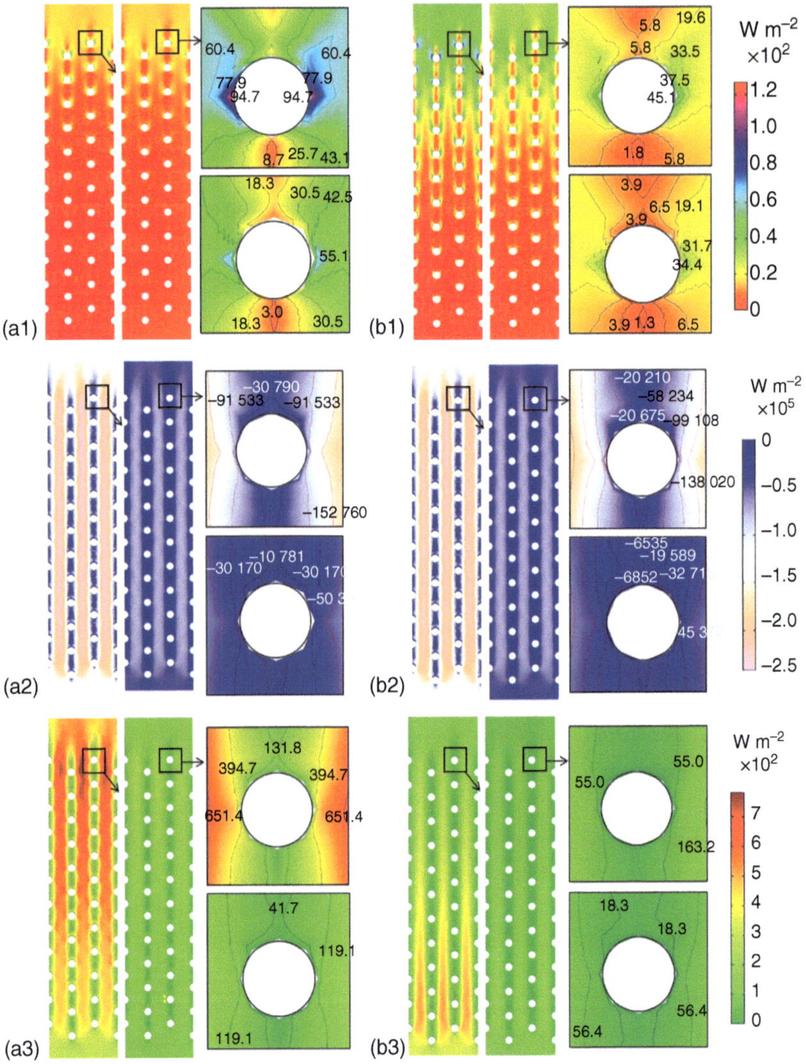

Figure 8.10 Comparison of the thermal exergy fields in the cooling tower system. (a, b) Correspond to Sample 5 and Sample 36; **1**, **2**, and **3** correspond to the thermal exergy field, mechanical exergy field, and chemical exergy field. Source: [4]/with permission of Elsevier.

for the Sample 36). This highlights the importance of considering the uncertainty of weather parameters in the optimal operation of the cooling water system.

The presented case study highlights the advantages of the bi-level ROMs approach in efficiently obtaining optimal exergy efficiency ratios and exergy flux profiles for cooling tower systems. The computational time is significantly reduced from several days to just a few seconds, making stochastic optimization tasks feasible. This enhances the robustness of the cooling tower system by considering environmental variations. Moreover, the analysis of the exergy field provides valuable insights into the distributed exergy parameters within the cooling tower.

8.7 Conclusion

This chapter presents a bi-level model reduction approach that includes a data-driven ROM and a physics-based ROM. The goal was to minimize the expected value of exergy efficiency ratios and achieve optimal exergy fields in the cooling tower system, taking into account variations in weather conditions. The development of the bi-level ROMs involved several key steps: DoE, multi-sample CFD simulations, and model reduction. The uncertainty associated with input variables was addressed through multi-sample CFD simulations of the cooling tower model for each sample. The resulting state and output variables from the CFD solutions were used to build the physics-based and data-driven models, which utilized a combination of PCA and ANN methods. These developed ROMs were then integrated into a sampling-based stochastic optimization model to minimize the expected value of exergy efficiency ratios and obtain optimal exergy flux fields within the cooling tower system. The proposed approach's applicability was demonstrated using a small-sized cooling tower. The case study results revealed that the optimal exergy efficiency ratios ranged from 0.37 to 0.74 as the dry-bulb temperature of the inlet air increased, with the expected value of the exergy efficiency ratio being 0.56. The investigation of exergy fields indicated that both mechanical- and chemical-exergy fluxes obtained from the optimal solution were approximately three times higher than those of the baseline solution, emphasizing the significance of considering environmental variations in the optimal operation of the cooling water system.

References

1 Guerras, L.S. and Martín, M. (2020). On the water footprint in power production: Sustainable design of wet cooling towers. *Applied Energy* 263: 114620.

2 Wang, W., Zhang, H., Liu, P. et al. (2017). The cooling performance of a natural draft dry cooling tower under crosswind and an enclosure approach to cooling efficiency enhancement. *Applied Energy* 186: 336–346.

3 Wu, Z., Lu, Z., Zhang, B. et al. (2022). Stochastic bi-objective optimization for closed wet cooling tower systems based on a simplified analytical model. *Energy* 250: 123703.

4 Qu, J., Li, M., He, C. et al. (2022). Deciphering the optimal exergy field in closed-wet cooling towers using Bi-level reduced-order models. *Energy* 238: 121766.

5 Sarker, M.M.A., Shim, G.J., Lee, H.S. et al. (2009). Enhancement of cooling capacity in a hybrid closed circuit cooling tower. *Applied Thermal Engineering* 29 (16): 3328–3333.

6 Zuazua-Ros, A., Ramos, J.C., Martín-Gómez, C. et al. (2020). Performance and feasibility assessment of a hybrid cooling system for office buildings based on heat dissipation panels. *Energy* 205: 117975.

7 Xie, X., Liu, H., He, C. et al. (2019). Deciphering the heat and mass transfer behaviors of staggered tube bundles in a closed wet cooling tower using a 3-D VOF model. *Applied Thermal Engineering* 161: 114202.

8 Xie, X., He, C., Xu, T. et al. (2017). Deciphering the thermal and hydraulic performances of closed wet cooling towers with plain, oval and longitudinal fin tubes. *Applied Thermal Engineering* 120: 203–218.

9 Zhu, Q., Zhang, B., Chen, Q. et al. (2020). Model reductions for multiscale stochastic optimization of cooling water system equipped with closed wet cooling towers. *Chemical Engineering Science* 224: 115773.

10 Barrasso, D., Tamrakar, A., and Ramachandran, R. (2014). A reduced order PBM–ANN model of a multi-scale PBM–DEM description of a wet granulation process. *Chemical Engineering Science* 119: 319–329.

11 Lang, Y., Zitney, S.E., and Biegler, L.T. (2011). Optimization of IGCC processes with reduced order CFD models. *Computers and Chemical Engineering* 35 (9): 1705–1717.

12 Bellemans, A., Aversano, G., Coussement, A., and Parente, A. (2018). Feature extraction and reduced-order modelling of nitrogen plasma models using principal component analysis. *Computers and Chemical Engineering* 115: 504–514.

13 Lang, Y.-d., Malacina, A., Biegler, L.T. et al. (2009). Reduced order model based on principal component analysis for process simulation and optimization. *Energy & Fuels* 23 (3): 1695–1706.

14 Ren, W.-X. and Chen, H.-B. (2010). Finite element model updating in structural dynamics by using the response surface method. *Engineering Structures* 32 (8): 2455–2465.

15 Regis, R.G. and Shoemaker, C.A. (2007). A stochastic radial basis function method for the global optimization of expensive functions. *INFORMS Journal on Computing* 19 (4): 497–509.

16 Atkinson, S. and Zabaras, N. (2019). Structured Bayesian Gaussian process latent variable model: applications to data-driven dimensionality reduction and high-dimensional inversion. *Journal of Computational Physics* 383: 166–195.

17 Najm, H.N. (2009). Uncertainty quantification and polynomial chaos techniques in computational fluid dynamics. *Annual Review of Fluid Mechanics* 41 (1): 35–52.

18 Liu, H., Wu, Z., Zhang, B. et al. (2023). A large-scale stochastic simulation-based thermodynamic optimization for the hybrid closed circuit cooling tower system with parallel computing. *Energy* 283: 128434.

19 Zhang, L., Liu, X., and Jiang, Y. (2015). Exergy analysis of parameter unmatched characteristic in coupled heat and mass transfer between humid air and water. *International Journal of Heat and Mass Transfer* 84: 327–338.

20 Apte, S.V., Gorokhovski, M., and Moin, P. (2003). LES of atomizing spray with stochastic modeling of secondary breakup. *International Journal of Multiphase Flow* 29 (9): 1503–1522.

21 Pawar, S.K., Abrahams, R.H.M., Deen, N.G. et al. (2014). An experimental study of dynamic jet behaviour in a scaled cold flow spray dryer model using PIV. *Canadian Journal of Chemical Engineering* 92 (12): 2013–2020.

22 Pawar, S., Padding, J., Deen, N. et al. (2015). Numerical and experimental investigation of induced flow and droplet-droplet interactions in a liquid spray. *Chemical Engineering Science* 22 (138): 17–30.

23 Bazdidi-Tehrani, F. and Zeinivand, H. (2010). Presumed PDF modeling of reactive two-phase flow in a three dimensional jet-stabilized model combustor. *Energy Conversion and Management* 51 (1): 225–234.

24 Akbarpour Ghazani, M., Hashem-ol-Hosseini, A., and Emami, M.D. (2017). A comprehensive analysis of a laboratory scale counter flow wet cooling tower using the first and the second laws of thermodynamics. *Applied Thermal Engineering* 125: 1389–1401.

25 Zhang, L., Song, X., and Zhang, X. (2019). Theoretical analysis of exergy destruction and exergy flow in direct contact process between humid air and water/liquid desiccant solution. *Energy* 187: 115976.

26 Singh, K. and Das, R. (2017). Exergy optimization of cooling tower for HGSHP and HVAC applications. *Energy Conversion and Management* 136: 418–430.

27 Chengqin, R., Nianping, L., and Guangfa, T. (2002). Principles of exergy analysis in HVAC and evaluation of evaporative cooling schemes. *Building and Environment* 37 (11): 1045–1055.

9

AI-Aided High-Throughput Screening and Optimistic Design of MOF Materials for Adsorptive Gas Separation

Li Zhou, Min Cheng, Shihui Wang, and Xu Ji

Sichuan University, School of Chemical Engineering, No.24 South Section 1, Yihuan Road, Chengdu, Sichuan Province, 610065, China

9.1 Introduction

Gas separation/purification is of paramount significance in the petrochemical industry. Distillation is one of the most practiced techniques in this regard, such as the separation of alkenes from alkanes [1, 2], O_2/N_2 [3], which is energy- and capital-intensive. Selective adsorption separation based on solid porous materials can be a promising alternative owing to its low energy consumption and cost [4]. To date, multiple types of porous materials (e.g. resins, zeolites, and carbon-based materials) have been examined and applied for gas purification applications. Metal–organic frameworks (MOFs), as a special class of porous materials, have attracted widespread attention in gas separation due to their tunable pore sizes and chemistry characterizations and are considered promising replacements and/or enhancements to traditional adsorbents. In recent decades, considerable investigations have been carried out to accelerate their applications in industry, covering expert knowledge-guided experimental synthesis and evaluation [5, 6], molecular simulation-assisted screening of high-performing candidates [7–9], as well as techno-economic analysis for their large-scale synthesis [10]. At the early stage, experiment-driven investigations for better MOFs were conducted and reported for various applications [11, 12]. Although traditional expertise-aided trial-and-error experimental approaches are effective in discovering better materials for certain applications within a restrained chemical space, their effectiveness and efficiency pale when facing enormous potential chemical spaces for MOFs for which the theoretical number of realizable structures is infinite. This brought the rise of computer-aided material discovery. Computer-aided material discovery heavily relies on the accurate and rapid evaluation of properties of the candidate materials, such that materials with high potential can be rapidly screened out for further experimental validation. To better support the discovery of desirable MOFs, several open-source authoritative MOF databases were contributed by the community, including the CSDSS-MOFs (Cambridge Structural Database non-disordered

Applied AI Techniques in the Process Industry: From Molecular Design to Process Design and Optimization, First Edition. Edited by Chang He and Jingzheng Ren.

metal–organic frameworks subset) [13] and the hMOFs (Northwestern Hypothetical MOF Database) [14]. The CSDSS contains experimentally synthesized MOFs of one-dimensional (1D), two-dimensional (2D), and three-dimensional (3D). In 2014, Chung et al. [15] initiated the CoREMOF 2014 (Computation-Ready, Experimental MOF Database) with 4764 3D MOFs derived from the CSDSS. In 2019, the database was updated to include more new 3D MOFs contributed by CoRE-MOF users, updated from the CSD, and the Web of Science search [16].

9.2 Methodology

Generally, the procedure of computer-aided high-throughput screening and optimistic design of MOF materials for adsorptive gas separation consists of several steps, which are illustrated in Figure 9.1. As shown, three coupled steps are required, covering molecular level simulation and screening driven by rigorous molecular simulation as well as machine learning, unit operation level modeling and validation, and data mining assisted new molecular design to enrich the targeted chemical space and aim for better performances in the particular application.

9.2.1 Molecular Level Simulation and Screening Driven by Rigorous Molecular Simulation and Machine Learning

9.2.1.1 Molecular Characterization

The premise for machine learning modeling is to translate abstract MOF structures into machine-readable numerical features (also known as descriptors). To comprehensively characterize a MOF, a set of indicators that can qualitatively and/or quantitatively characterize the structural and chemical distinctions of a MOF is adopted to assist the overall screening and optimistic design procedure. Regarding structural features of a MOF, several computationally simple and physically intuitive properties can be employed to describe its pore geometry and size, such as the largest cavity diameter (LCD), pore limiting diameter (PLD), volumetric/accessible/gravimetric surface area (*VSA*/*ASA*/*GSA*), pore volume (PV), density (ρ), and porosity (ϕ). In some studies, the pore size distribution (PSD) and its derived indicators are also introduced to depict the inhomogeneity of the inner pore structures of a MOF [17]. The optional chemical indicators cover the total degree of molecular unsaturation (TDU), the metallic percentage (MP), the oxygen–metal ratio (*OMR*), the nitrogen–oxygen ratio (NOR), the Henry's coefficient, the electronegative atoms–total atoms ratio (ETR), the weighted electronegativity per atom (WEPA), the number of different constituent atoms (x_H, x_C, x_O, etc.), and so on. These descriptors are widely used in the identification of suitable MOFs for the adsorptive separation/purification of ethane–ethylene gas mixture [18], O_2/N_2 gas mixture [19], xenon recovery from exhaled anesthetic gas mixture [17] and industrial hydrogen purification [20]. According to the latest technology known to the authors up to now, the calculation of ϕ as well as Henry's coefficients can be realized by RASPA package [21], while the rest of the aforementioned

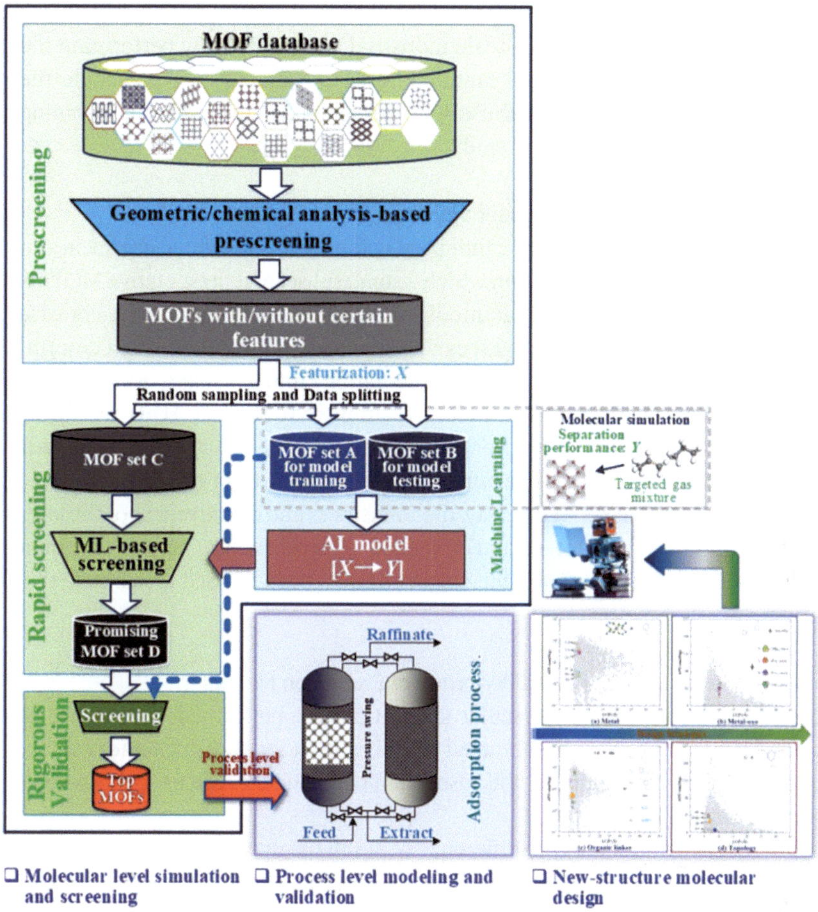

Figure 9.1 The outline of AI-aided high-throughput screening of MOF Materials for Gas Separation. Source: [20]/with permission of Elsevier.

structural properties can be computed by Zeo++ software [22]. For the calculation of the remaining chemical descriptors, a MOF chemical feature calculation tool is available (https://github.com/barisoo/MOF-Features).

9.2.1.2 Structural/Chemical Analysis-Based Prescreening

The key for MOFs to be high-potential alternatives in various selective gas adsorption and separation processes is their highly tunable structural and chemical features, which endow them flexibility to adapt to different scenarios. Hence, a prescreening procedure based on structural/chemical analysis according to the respective application is considered a preliminary step. Based on the molecular characterization results, structures with *PLD* less than the kinetic diameter of the targeted molecule and/or $VSA = 0\,m^2\ cm^{-3}$ are excluded, as they are considered uncompetitive in the targeted adsorptive separation. Subsequently, chemical analysis is performed, and MOFs with the undesired chemical characteristics (e.g. MOFs with rare and/or toxic metals) are identified and removed, since their performance may be low or unpractical in large-scale industrial applications. By performing the prescreening procedure, the MOF candidate pool can be curated to exclude the undesirable structures, and only those with the potential to be high-performing adsorbents are reserved for further study.

9.2.1.3 Diversity Analysis and Dataset Splitting

The distribution of the prescreened candidates in their self-constituted chemical and structural space is analyzed, based on which a small subset of representative MOFs is randomly sampled for later artificial intelligence (AI) model development. For ease of visualization, t-SNE technology [23] can be applied to project the chosen descriptors (structural, chemical, and/or their combination) into 2D maps. As shown in Figure 9.2, the distribution of the randomly sampled MOFs (set A∪B) should be evenly overlapped with that of the overall MOF candidates (set A∪B∪C) so that their representativity can be assured. The same rule applies to the training MOF set (set A) and the test MOF set (set B). In this way, the robustness and effectiveness of the later stage AI model can be guaranteed and enhanced to some extent. There are optional algorithms, such as the Kennard–Stone algorithm, that can be used to facilitate this procedure.

9.2.1.4 Molecular Simulation and Performance Evaluation Metrics

For the sampled candidates, molecular simulations are performed to mimic their adsorption behaviors toward the targeted gas mixture. A set of performance evaluation metrics is defined to quantify the adsorption performance. Table 9.1 lists several of the commonly used indicators.

Working capacity measures the difference between the uptake amounts of the targeted molecule under the adsorption and the desorption conditions. The greater the working capacity, the more the targeted component can be captured through the process. Adsorption selectivity indicates the degree of affinity of a MOF toward the targeted gas molecule. It is calculated by the ratio of adsorbed gas uptakes of

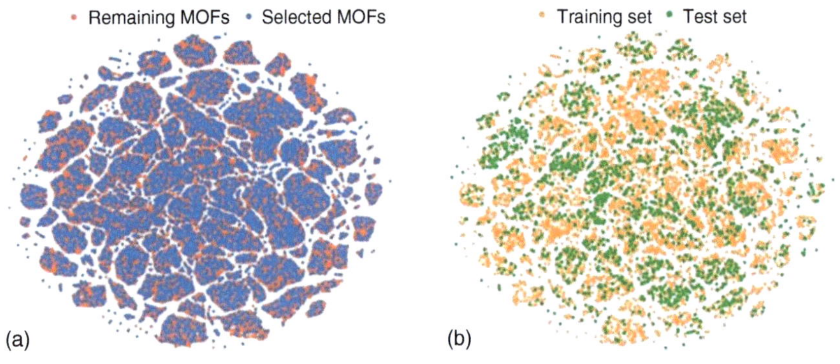

Figure 9.2 An example of t-SNE maps based on the combination of the structural and chemical descriptors for (a) the randomly selected MOFs and the remaining MOFs and (b) the training and test sets. Source: [2]/with permission of Elsevier.

Table 9.1 Performance evaluation metrics.

Metrics	Formula[a)]
Selectivity	$S_{i/J}^{ads} = \frac{N_i^{ads}/\sum_{j=1,j\in J}^{n} N_j^{ads}}{f_i/\sum_{j=1,j\in J}^{n} f_j}$
Working capacity	$\Delta N_i = N_i^{ads} - N_i^{des}$
Trade-off between working capacity and selectivity	$APS = S_{i/J}^{ads} \times \Delta N_i$
Regenerability	$R\% = \frac{\Delta N_i}{N_i^{ads}} \times 100\%$

a) N_i^{ads}, N_i^{des}: gravimetric uptakes for the targeted component i at adsorption/desorption conditions; j, the nontarget component; f: fugacity in gas mixture.

the targeted component and the remaining, normalized by their partial fugacities. It has been proven that a trade-off exists for the working capacity and selectivity of MOF absorbents. Hence, the trade-off between the working capacity and adsorption selectivity is usually adopted to evaluate the performance of a MOF candidate more thoroughly. Regenerability ($R\%$) is another performance indicator that is considered crucial for the MOFs as adsorbents in a gas adsorption process. It is a measurement of the percentage regeneration of the adsorption sites during the desorption of the adsorbed gases, and it is defined as the ratio of working capacity to the gas uptakes of the targeted gas molecule at the adsorption pressure.

Grand canonical Monte Carlo (GCMC) simulations are performed to estimate the gas uptake amounts of the components in an adsorption and desorption process by setting the bulk pressure, temperature, and gas mixture compositions. For the modeling of adsorbates and adsorbents as well as the interactions between them, different settings should be conducted according to the specific applications in order to obtain reliable results. Also, a trade-off should be made while carrying out the GCMC simulation cycles to balance the computational cost and result accuracy. It should be noted that validation of the parameter settings for the GCMC simulations should also be performed by comparing the results from simulations with experiments.

9.2.1.5 AI-Aided Quantitative Structure–Property Relationship Development and Rapid Screening

By estimating the separation performance (Y, such as APS/TSN, *and* $R\%$) of all MOFs in datasets A and B, along with the chemical and/or structural descriptors (X), the quantitative structure–property relationships (QSPR) model can be established based on AI technology. This model would effectively correlate the molecular features to their separation performances ($X \rightarrow Y$). Various well-established AI algorithms, including random forest regression and artificial neural networks, can be used for the development of the QSPR model. In order to increase the robustness and predictive accuracy of the AI model, feature engineering techniques are sometimes necessary to eliminate insignificant features. Commonly used methods for this

Table 9.2 Evaluation criteria of models.

Evaluation criterion	Formula
Determination coefficient	$R^2 = 1 - \frac{\sum_{i=1}^{m}(y_i - \hat{y}_l)^2}{\sum_{i=1}^{m}(y_i - \bar{y}_l)^2}$
Root-mean-square error	$\text{RMSE} = \sqrt{\frac{1}{m}\sum_{i=1}^{m}(y_i - \bar{y}_l)^2}$
Mean absolute error	$\text{MAE} = \frac{1}{m}\sum_{i=1}^{m} \mid y_i - \bar{y}_l \mid$

y_i and $\hat{y}_l$ represent the simulated and predicted performance values of the ith structure in the MOF database, respectively. $\bar{y}_l$ represents the mean of the simulated performance values of all the MOFs, and m is the number of MOFs.

purpose include Pearson correlation analysis and the recursive feature elimination algorithm. Besides, hyperparameter tuning is decisive for the predictive accuracy of the developed AI model in certain cases. After the training process, model performance can be properly measured by scoring criteria such as mean absolute error (*MAE*), root-mean-square error (*RMSE*), and R^2 (Table 9.2).

Finally, the best-performing machine learning model developed based on MOF samples (set $A \cup B$) is used to predict the separation performance of the MOFs in the remaining dataset (set C) to perform high-throughput screening for high-performance MOFs. The top MOFs with high predictive separation performances for the targeted gas separation task are identified in this step, for which GCMC simulations are again performed to calibrate their adsorption performance. The final high-performing MOFs with desirable separation capability are determined from the resultant dataset.

9.2.1.6 Process Level Simulation

To gain a deeper understanding of the high-performing MOF adsorbents, it is beneficial to conduct process level studies by simulating the adsorption and desorption processes specific to the targeted separation task. This allows for an examination of the gas recovery rate under various industrial scenarios, providing insights into the practical application potential of the identified MOFs. In practical applications, the presence of impurities in material streams is inevitable, with water being the most common impurity. Therefore, it is essential to investigate the influence of impurities on the separation performance of the adsorbents. Furthermore, practical factors such as temperature and pressure can be manipulated to affect the adsorbent's separation performance. Considering the fluctuating nature of inlet gas composition during production, it is also important to assess the flexibility of the adsorbent in industrial practice by investigating its separation performance under different inlet gas concentrations.

9.2.1.7 Reverse Molecular Design

The merit of computer-aided material discovery is that a rich molecular space can be explored at an affordable cost. However, in the majority of the cases,

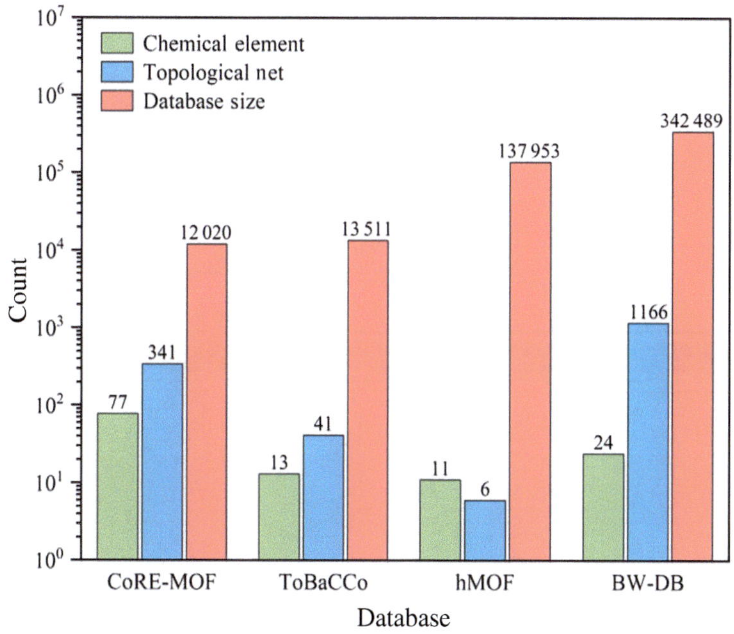

Figure 9.3 Diversity analysis of the MOF databases. Source: [2]/with permission of Elsevier.

the existing molecular space composed of the known molecules synthesized by experiments and/or assembled by algorithms targeting other applications is still rather restricted. For example, Figure 9.3 presents the statistical analysis for the diversity of the MOFs in four of the representative databases, covering CoRE, ToBaCCo, hMOF, and BW-DB. As can be seen, the chemical elements as well as the topological net covered by these databases are limited, especially in the case of ToBaCCo and hMOF. Moreover, as mentioned at the beginning of this chapter, the most attractive aspect of MOFs is that their tunable pore structure and the chemical microenvironments allow them to be tailored for different applications. Hence, it can be beneficial if directed structural design can be achieved based on a statistical analysis of the MOFs' characteristics with high separation potential.

As commonly understood, a MOF consists of three essential components: metal nodes, organic linkers, and a matching topology net. Based on this understanding, three design strategies can be applied accordingly: substitution of the metal node, regulation of the topology net, and adjustment of the organic linker.

Figure 9.4a and b show the adsorption performance of a series of MOF analogs with the same topology and organic linker but different metal nodes. Despite having a similar *LCD*, these MOFs exhibit varying adsorption behaviors, indicating that substituting the metal node can enhance the separation performance of a MOF to some extent. Our previous study [17] also revealed that the substitution of the metal node slightly tunes the largest pore size, resulting in improvements in working capacity and selectivity.

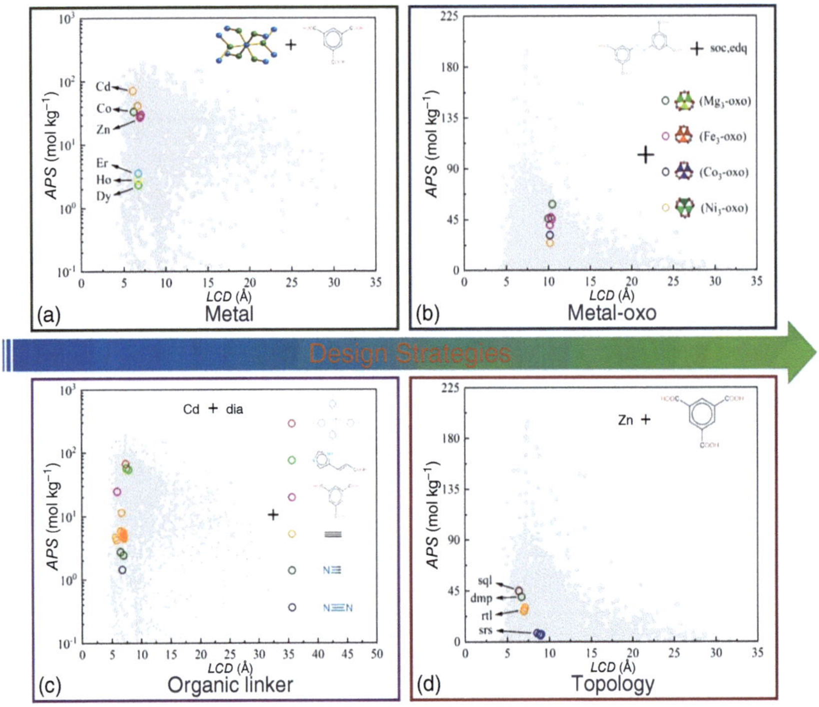

Figure 9.4 Three design strategies for new MOF structure. Source: [17]/with permission of Elsevier.

Pore size and shape are important factors that influence the adsorption performance of MOFs, which can be tailored by altering the organic linker. Figure 9.4c shows that the adsorption performance can be improved by adjusting the organic linker. Clues about the pros and cons of the organic linkers toward different guest molecules can be derived via expert knowledge or analysis of the behavior of relative candidates. For example, in a study by Cheng et al. [17], it was observed that substituting the aliphatic linker with an aromatic linker significantly enhanced the adsorption separation of Xe from an exhaled gas mixture. This finding aligns with expert knowledge, as MOFs with aromatic rings offer more adsorption sites and thus exhibit a stronger affinity for the targeted Xe molecule.

Topology is another factor affecting the pore geometry of MOFs. Regulating the topology of a MOF was proven to be an effective strategy for the enhancement of separation performance in several applications [19, 24]. Figure 9.4d illustrates four different topology nets assembled by Zn and an aromatic linker. As can be seen, despite the identical building blocks, the pore sizes of these four MOFs can be adjusted by regulating their topology nets, thereby improving separation performance.

9.3 Case Studies

This section exemplifies the application of the AI-aided high-throughput screening of MOF materials for adsorptive gas separation with a case study.

9.3.1 High-Throughput Screening of Metal–Organic Frameworks for Hydrogen Purification

In industry, the separation/purification of hydrogen from methane impurities is of significant importance, especially in light of carbon dioxide mitigation toward carbon neutrality. Adsorption separation is an effective route for this application regarding energy cost and separation efficiency, for which the selection of adsorbents is the key.

9.3.1.1 Prescreening

The CoRE-MOF 2019 database is employed as the starting point for this case. Firstly, the structural/chemical analysis-based prescreening is performed. *LCD*, *PLD*, *VSA*, ρ, and ϕ are chosen as the structural descriptors and calculated to feature each MOF. Based on the result, structures with *PLD* less than the kinetic diameter of CH_4 (3.8 Å) and with rare and/or toxic metals are excluded. This procedure shortlisted 5096 candidates out of the original 12 020 MOFs. Subsequently, Pearson correlation analysis is performed on the five structural descriptors to reveal the correlations. The result in Figure 9.5 suggests that the correlation between *LCD* and *VSA* is the weakest, as the calculated coefficient between these two gives the smallest. Hence, these two are chosen as the structural descriptors for the structure–property relationship analysis.

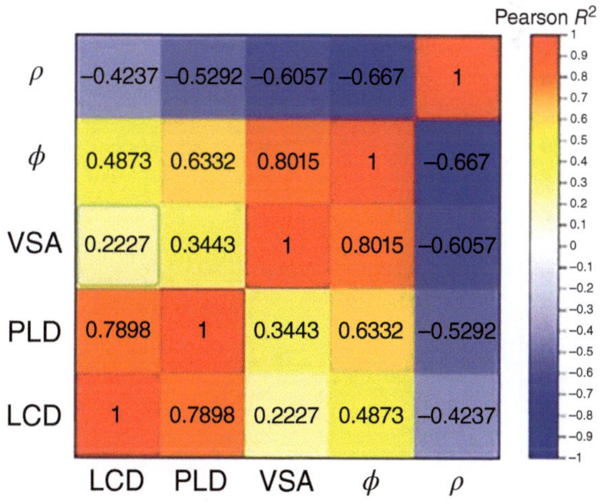

Figure 9.5 Correlations of various structural features for 5096 CoRE-MOFs. Source: [20]/with permission of Elsevier.

9.3.1.2 Rapid Screening

Two hundred MOFs are randomly sampled from the 5096 MOF candidates, for which GCMC simulations are conducted to study their adsorption behaviors toward CH_4/H_2 mixture. On this basis, the structure-performance relationship is analyzed to study how the adsorbent performance of the MOFs varies with their structural features. Figure 9.6 gives the relationship plot between *LCD* and *VSA* at different *APS* of the 200 sampled MOFs. As shown, MOFs with high *APS* values ($APS \geq 150\,\text{mol kg}^{-1}$) exist in the region where *VSA* value falls into the range of 600–1800 $\text{m}^2\,\text{cm}^{-3}$ and *LCD* value ranges from 4 to 11 Å. Therefore, two screening criteria for high-performing MOFs are derived, namely, $4 \leq LCD \leq 11$ Å and $600 \leq VSA \leq 1800\,\text{m}^2\,\text{cm}^{-3}$. With the new screening criteria, the 5096 resultant candidates from the prescreening are further filtered, resulting in a subset of 2408 MOFs with the desired structural properties.

9.3.1.3 Rigorous Validation

The adsorption performance of the 2408 CoRE-MOFs toward CH_4/H_2 gas mixture is then evaluated based on GCMC simulation, and the corresponding performance metrics ($S^{ads}_{CH_4/H_2}$, ΔN_{CH_4}, *APS*, ΔQ, and *R*%) are obtained to sort the separation performance of MOFs. The adsorption performance of the 2408 CoRE-MOFs is illustrated in Figure 9.7a. The horizontal and vertical axes give the simulated working capacity and CH_4-selectivity respectively, while the *APS* is color-coded. The blue dots denote MOFs with *APS* values lower than $150\,\text{mol kg}^{-1}$, while the purple ones distributed on the top-right side of the purple line are the ones with *APS* values equal to or larger than $150\,\text{mol kg}^{-1}$. A wide distribution of selectivity is observed among MOFs, with values ranging from 2 to 26 724. The majority of MOFs fall within the range of 20–100. Similarly, the working capacity of MOFs varies significantly, from 0.08 to 6.80 mol kg^{-1}, and the majority fall within the range of 1–3 mol kg^{-1}.

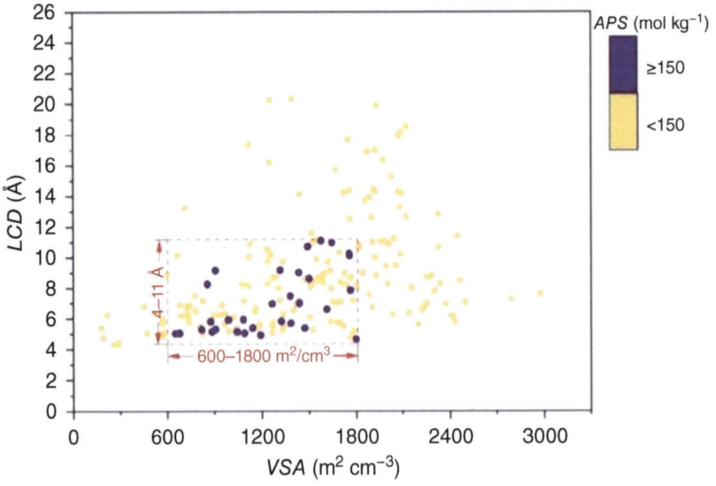

Figure 9.6 Relationship plots of *LCD* versus *VSA* of the 200 sampled CoRE-MOFs, color-coded with adsorbent performance score. Source: [20]/with permission of Elsevier.

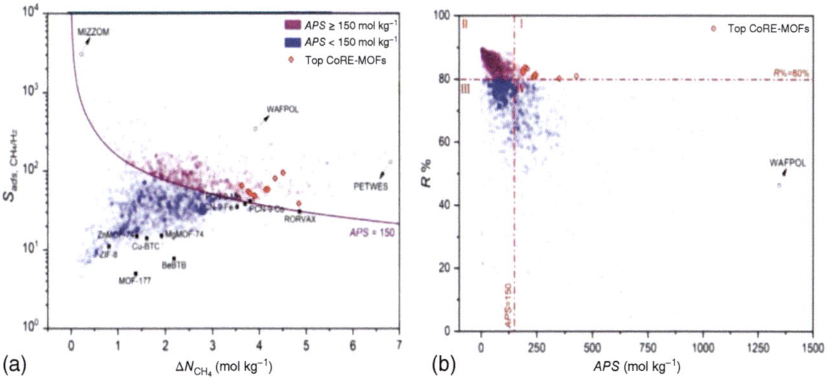

Figure 9.7 (a) Selectivity of the screened MOFs as a function of their working capacity, color-coded with *APS*; (b) Regenerability of the screened MOFs as a function of *APS*. Source: [20]/with permission of Elsevier.

Within this diverse distribution, it is possible to identify MOFs that exhibit both high selectivity and working capacity. For example, a MOF with refcode MIZZOM possesses a high theoretical selectivity of 3028, but its working capacity is quite low, 0.22 mol kg^{-1}. On the other hand, a MOF with refcode PETWES exhibits a high working capacity of 6.80 mol kg^{-1}, while the corresponding selectivity is relatively low, which is 131. This reveals the widely proved trade-off between MOFs' selectivity and working capacity. Therefore, one may infer that a desirable MOF candidate with both high selectivity and working selectivity should be located in the middle of the purple area. Following this clue, a MOF with refcode WAFPOL comes into the spotlight for its well-balanced working capacity and selectivity, yet its suitability needs to be further examined by more practical factors affecting its performance in industrial applications. In practice, regenerability is another important factor that needs to be considered. The *R*% of the 2408 CoRE-MOFs is presented in Figure 9.7b. Generally speaking, a preferable adsorbent material should have regenerability not lower than 80%. As shown, two auxiliary lines with regenerability equal to 80% and *APS* equal to 150 mol kg^{-1} are drawn, dividing the space into four quadrants. In the fourth quadrant, where MOFs with high *APS* but low regenerability are gathered, the MOF with refcode WAFPOL is found. It turns out that its regenerability is quite low ($R\% = 46\%$), which makes it unsuitable for practical applications. MOFs with desirable separation performance are distributed in the first quadrant, where both high *APS* and regenerability can be provided. The candidates with $R\% \geq 80\%$ are ranked by their *APS* values, and the practicability of the top-ranked MOFs is investigated by tracing the corresponding experimental data in the literature, and the MOFs that have been tested for gas adsorption are selected; then, nine promising CoRE-MOFs toward CH_4/H_2 separation are identified, as visualized by red diamonds in Figure 9.7. Table 9.3 lists the detailed information about them. The stability of the top performers is further examined in the literature. The MOF with refcode DAFJOM has been tested to be chemically stable under 75% humidity at ambient

Table 9.3 The detailed information on the nine promising CoRE-MOFs.

MOF refcodes	$S^{ads}_{CH_4/H_2}$	ΔN_{CH_4} (mol kg^{-1})	*APS* (mol kg^{-1})	*R*%	Metal	Has OMS	R_k
TAKTAD	94.75	4.51	427.39	80.97	Ga	No	6.00×10^3
XILLEL	80.61	4.34	349.74	80.20	Zn	No	0.10
CUYWUP	58.47	4.19	244.87	81.54	Cu	Yes	24.68
DAFJOM	58.36	4.14	241.66	80.90	Mg	No	1.48
SETDUS	64.80	3.63	235.40	80.71	Cu	Yes	101.90
KUKPUD	54.94	3.76	206.85	83.34	Cu	Yes	108.44
DIDBEZ	52.04	3.81	198.09	83.71	Co	No	0.15
SOZHOG	48.59	3.90	189.48	81.94	Cu	No	0.18
JALCAD	38.81	4.84	188.03	82.66	Al	Yes	3.98×10^5

OMS: Open metal site. R_k: Ratio of Henry's coefficients of H_2O and CH_4.

temperature [25], while the MOF with refcode JALCAD exhibited outstanding thermal stability and high tolerance to water vapor [26]. The remaining seven proved to be thermally stable at a high temperature of up to 400 K [27–32].

9.3.1.4 Structure–Property Relationship Analysis

Relationships between the calculated separation performance of the 2408 CoRE-MOFs and the corresponding geometric characteristics are visualized in Figure 9.8. Figure 9.8a gives the relationship $S^{ads}_{CH_4/H_2} \sim PLD$ at different *LCD*. It can be seen that, generally, with increasing *PLD* or *LCD*, a lower selectivity is observed; meanwhile, a MOF with a larger *PLD* tends to have a larger *LCD*. This implies that a MOF with a more compact pore size is more preferential for selective adsorption of CH_4 over H_2; nevertheless, the corresponding regenerability gets hindered. It can be seen that a MOF with *PLD* around 5 Å and *LCD* near 7 Å is preferable. This finding is consistent with the observations from Altintas et al. [33]. Figure 9.8b plots the relationship $\Delta N_{CH_4} \sim \rho$ at different ϕ. With increasing density, the working capacity seems to decrease; and a high density usually corresponds to low porosity. Apparently, it is more likely for MOFs with higher porosity to possess higher working capacity, as they contain more space available for adsorption. Figure 9.8c presents the relationship $S^{ads}_{CH_4/H_2} \sim K^0_{CH_4}$ at different ΔN_{CH_4}. It is shown that Henry's coefficient of CH_4 is another important factor that affects the selectivity of a MOF. A higher Henry's coefficient means a stronger affinity of the MOF toward CH_4, resulting in a higher selectivity. Yet too strong an affinity between the adsorbent and adsorbate may pose difficulties for its regeneration, as shown in Figure 9.8d.

Further, the influence of the types of metal sites is also investigated. Figure 9.9a gives the proportion of metals in all 2408 CoRE-MOFs, in which 13 types of metals appear more than 20 times. Among them, Zn is the most observed metal site, followed by Cd, Cu, Co, Mn, and Ni, accounting for more than 5%. It can also be seen in Figure 9.9b that Zn, Cu, and Co appear more frequently in the desired MOF

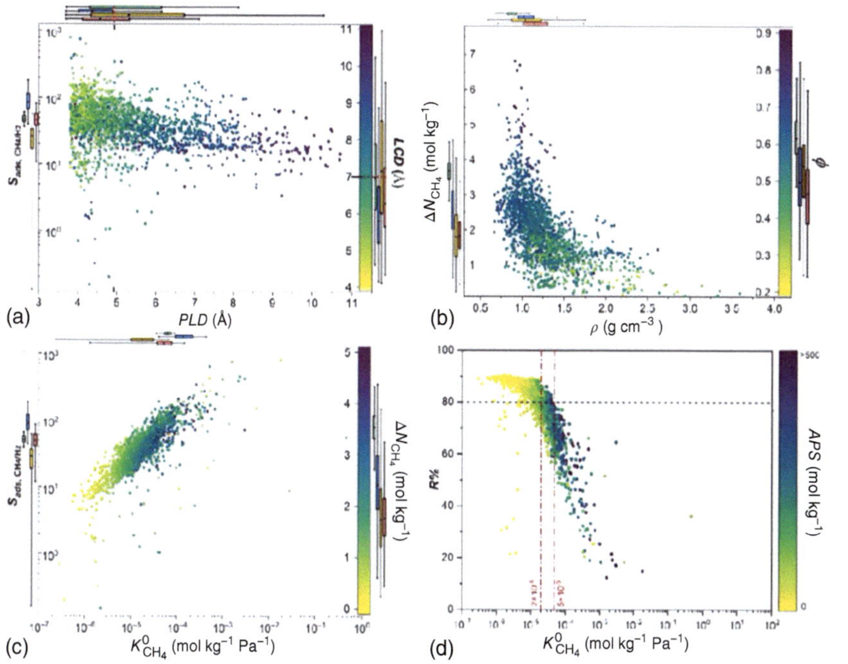

Figure 9.8 Relationship plots of (a) selectivity versus *PLD*, color-coded with *LCD*, (b) working capacity versus ρ, color-coded with ϕ, (c) selectivity versus Henry's coefficient of CH_4, color-coded with working capacity, and (d) regenerability versus Henry's coefficient of CH_4, color-coded with adsorbent performance score, for 2408 CoRE-MOFs. Source: [20]/with permission of Elsevier.

category than in all CoRE-MOFs. Particularly, structures with Cu sites have the highest probability of possessing both high *APS* and *R%*. Furthermore, the structural properties of the CoRE-MOFs containing Zn, Cd, Cu, Co, Mn, and Ni were analyzed, as visualized in Figure 9.9c. It can be seen that, from a statistical point of view, the pore sizes (*LCD* and *PLD*) of the MOFs are positively correlated with the atomic radii of the metal sites (Zn > Cu > Co). It is more likely for Cu to form "preferable" pores with ligands for CH_4/H_2 separation (approximately 5 and 7 Å for *PLD* and *LCD*, respectively). This also agrees well with the findings from He et al. [34]. Notice that the pore sizes of most of the MOFs with Cd (the largest atom among the six) are relatively low. It is conjectured that this may be owing to the deformation of large pores. Statistical analysis for Mn and Ni shows slight deviations, probably because of the insufficient sample.

9.3.1.5 Investigation on Practical Factors

Humidity In industrial scenarios, the presence of impurity in the material streams is inevitable; hence, it is of practical importance to study its influence on the adsorbents' separation performance. Among others, water is the most common impurity in practice. Figure 9.10 plots the calculated APS values of nine promising CoRE-MOFs facing various water content in the gas mixture stream. Meanwhile,

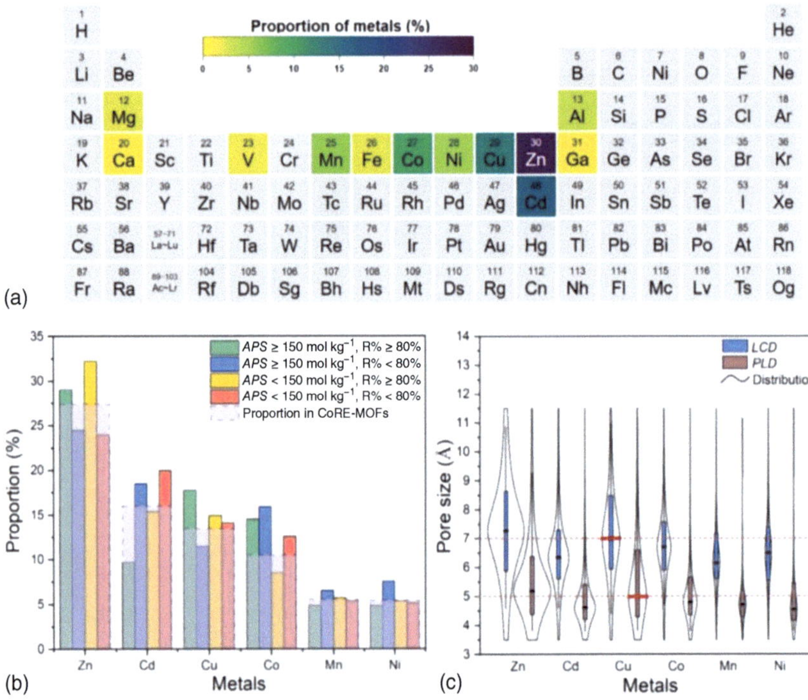

Figure 9.9 (a) Proportion of metals in 2408 CoRE-MOFs. Metals that appear more than 20 times are displayed. (b) Proportion of Zn, Cd, Cu, Co, Mn, and Ni in the categories of MOFs with different adsorption behaviors and in all CoRE-MOFs. (c) Box plots and distribution for *LCD* and *PLD* of MOFs with Zn, Cd, Cu, Co, Mn, and Ni. Source: [20]/with permission of Elsevier.

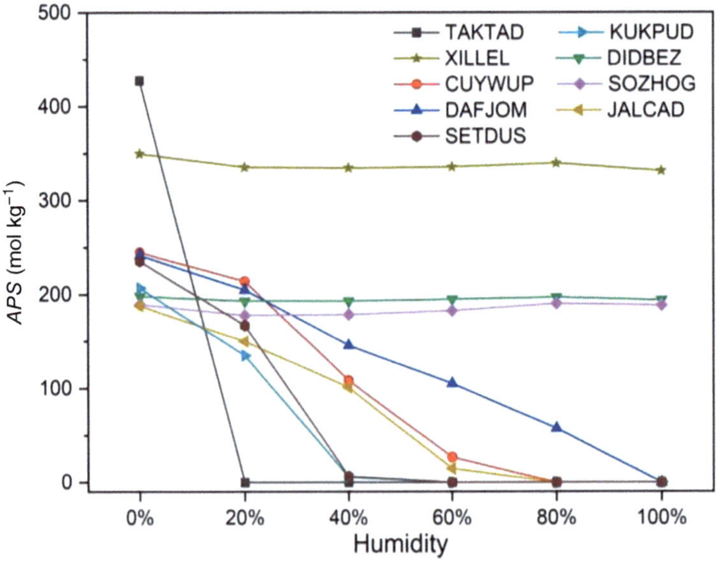

Figure 9.10 Adsorbent performance score of the nine promising CoRE-MOFs under different humidity conditions. The saturation pressure of water is 4.24 kPa. Source: [20]/with permission of Elsevier.

Table 9.4 Separation performance comparison of the top CoRE-MOFs under different operation condition settings [20].

MOFs	Original operation condition						Original operation condition					
	$CH_4/H_2 = 0.3/0.7$		$CH_4/H_2 = 0.5/0.5$		$CH_4/H_2 = 0.6/0.4$		$CH_4/H_2 = 0.3/0.7$		$CH_4/H_2 = 0.5/0.5$		$CH_4/H_2 = 0.6/0.4$	
	APS (mol kg^{-1})	*R*%	*APS* (mol kg^{-1})	*R*%	*APS* (mol kg^{-1})	*R*%	*APS* (mol kg^{-1})	*R*%	*APS* (mol kg^{-1})	*R*%	*APS* (mol kg^{-1})	*R*%
TAKTAD	427.39	80.97	468.35	73.63	463.79	70.36	620.38	86.4	666.44	80.96	680.98	78.2
XILLEL	349.74	80.2	352.34	73.5	342.19	70.56	499.84	86.13	491.62	80.34	469.52	78.55
CUYWUP	244.87	81.54	272.87	75.86	275.28	73	356.05	86.86	387.09	82.8	370.09	80.99
DAFJOM	241.66	80.9	278.96	76.25	278.4	74.07	344.05	86.64	393.83	82.25	406.51	81.38
SETDUS	235.4	80.71	246.91	75.39	245.16	72.6	331.73	87.11	324.75	82.33	321.43	80.96
KUKPUD	206.85	83.34	239.61	78.83	249.76	75.93	304.32	88.72	334.54	84.84	322.7	82.89
DIDBEZ	198.09	83.71	238.61	79.22	242.51	76.8	293.22	88.8	330.52	85.09	349.32	83.16
SOZHOG	189.48	81.94	218.45	77.17	214.73	74.36	264.64	87.61	288.49	82.89	289.24	81.19
JALCAD	188.03	82.66	215.48	77.72	216.85	75.85	259.13	87.99	306.38	84.08	295.59	82.31

Original operation conditions: adsorption and desorption conditions are (303 K, 10 bar) and (303 K, 1 bar), respectively. Adjusted operation conditions: adsorption and desorption conditions are (293 K, 10 bar) and (313 K, 1 bar), respectively.

the ratio of Henry's coefficients, R_k ($R_k = K^0_{H_2O}/K^0_{CH_4}$), is also given in Table 9.4, indicating the relative magnitude of affinity of the MOFs toward H_2O and CH_4. When the value of R_k is more than 1, the MOF has a stronger affinity toward H_2O; on the contrary, it is more inclined to adsorb CH_4. For MOFs with refcodes XILLEL, DIDBEZ, and SOZHOG with $R_k < 1$, no significant deterioration in separation performance is observed. At the same time, these MOFs maintain a high regeneration capacity (*R%* are 80%, 83%, and 82%, respectively) when the humidity reaches 100%. On the other hand, it is observed that some MOFs offer $R_k > 1$, and they possess exposed OMSs and/or hydrophilic sites that may form strong interactions with water [35, 36]. Yet these adsorbents have different sensitivities to water. For MOFs with refcodes TAKTAD, SETDUS, and KUKPUD, their tolerance to water is quite low, as the corresponding *APS* quickly deteriorates to zero. While the MOFs with refcodes CUYWUP, DAFJOM, and JALCAD maintain a relatively gradual decrease with increasing gas humidity, eventually dropping to almost zero when the gas mixture's humidity is over 80%.

Figure 9.11 shows the distributions of adsorbed CH_4 and H_2O in the MOFs with refcodes TAKTAD, CUYWUP, and DIDBEZ at 0%, 20%, and 100% humidity, and the subgraphs represent radial distribution functions of corresponding atom pairs of framework atoms with adsorbate. As can be seen in Figure 9.11a, in the dry structure, CH_4 molecules are confined within the pores by the framework. At 20% humidity, H_2O molecules introduced into the structure are preferentially adsorbed near the hydrophilic sites on the framework, such as hydroxyl groups, as the first atomic pair peak belongs to the atom pairs of O with H_2O, and quickly occupies the adsorbing sites of CH_4. Interestingly, in Figure 9.11b, the MOF with refcode CUYWUP can contain a considerable number of CH_4 molecules when the humidity is low, even though its R_k is much larger than one. It can be attributed to the different adsorption sites of CH_4 and H_2O. The water molecules first adsorb near the open Cu sites of the structure (the first atomic pair peak belongs to the atom pairs of Cu with H_2O), while the CH_4 molecules are preferentially adsorbed near the framework atoms. Yet as the humidity increases, H_2O takes up more space, and thus CH_4 molecules are excluded from the structure. As for the MOF with refcode DIDBEZ in Figure 9.11c, there is almost no H_2O molecule adsorbed in the structure even at 100% humidity.

Operation Temperature and Inlet Gas Composition Temperature is a practical factor that can be manipulated to affect the separation performance of an adsorbent. Adsorption behaviors of the nine promising CoRE-MOFs were simulated under various temperatures (293, 298, 303, 308, and 313 K), and the results are illustrated in Figure 9.12. It is observed that low temperatures are beneficial for adsorption but unfavorable for the regeneration of MOFs. The flexibility of the adsorbent in industrial practice is another concern, as the inlet gas composition may fluctuate during production. To this end, the separation performance of top CoRE-MOFs under different inlet CH_4 concentrations (30%, 40%, 50%, and 60%) is simulated, as shown in Figure 9.13. It can be seen that the *APS* values of most MOFs increase when CH_4 concentration rises from 30% to 40%; as the concentration continues to increase, the *APS* values tend to be stable. On the other hand, the regenerability drops below 80%

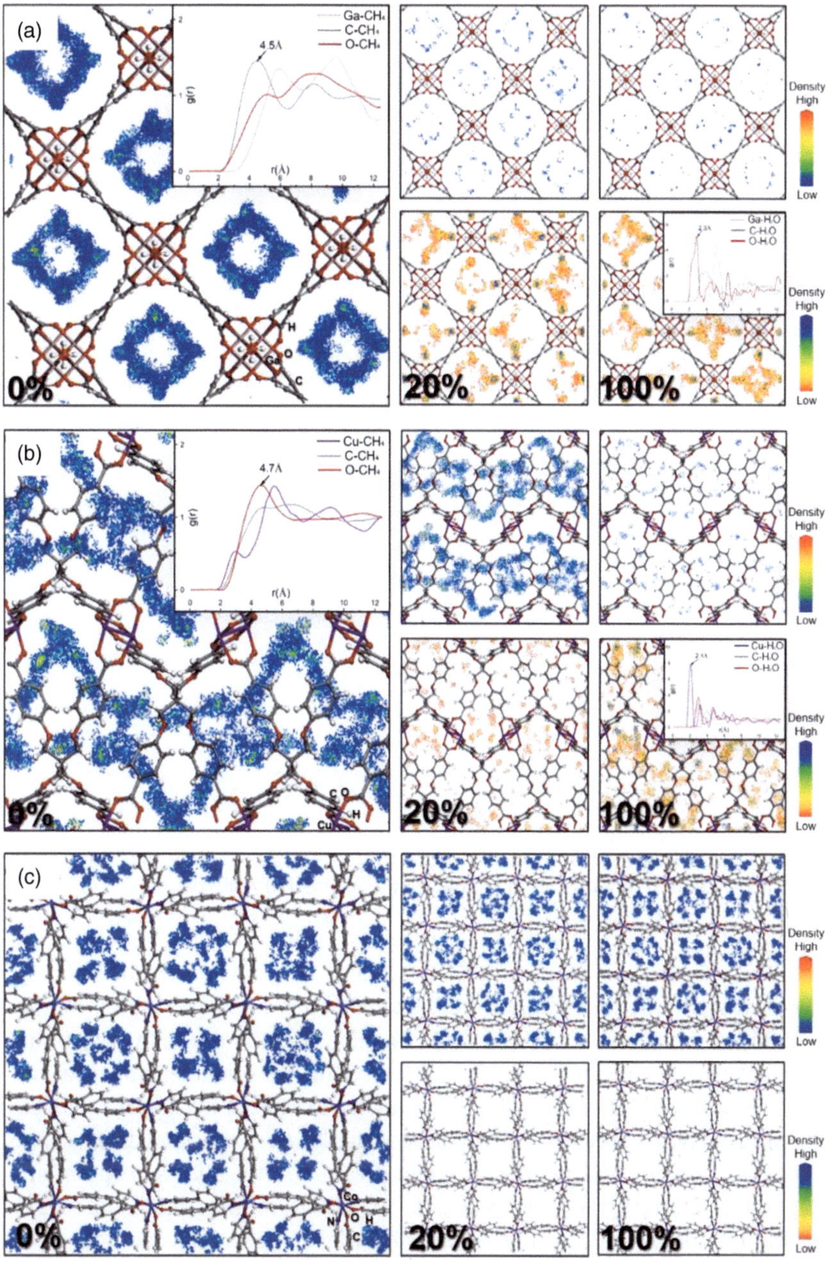

Figure 9.11 (Left panel) CH_4 distribution at 0% humidity; (middle panel) CH_4 (top) and H_2O (bottom) distributions at 20% of humidity; (right panel) CH_4 (top) and H_2O (bottom) distributions at 100% humidity in the MOFs with refcodes (a) TAKTAD, (b) CUYWUP and (c) DIDBEZ. The subgraph represents the radial distribution functions of different atom pairs between the framework atoms and CH_4 (H_2O). Source: [20]/with permission of Elsevier.

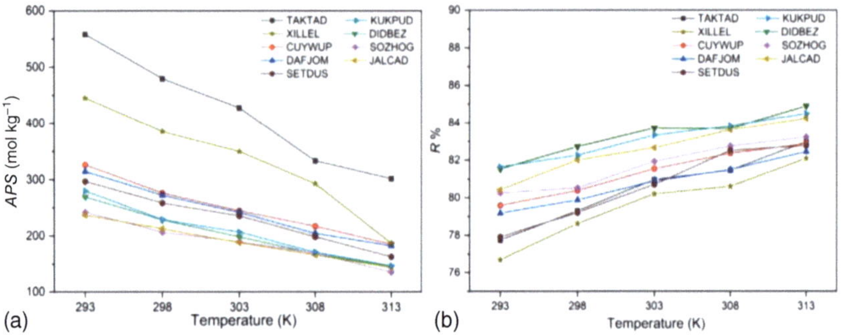

Figure 9.12 (a) Adsorbent performance score and (b) regenerability of the nine promising CoRE-MOFs under different operation temperatures. Source: [20]/with permission of Elsevier.

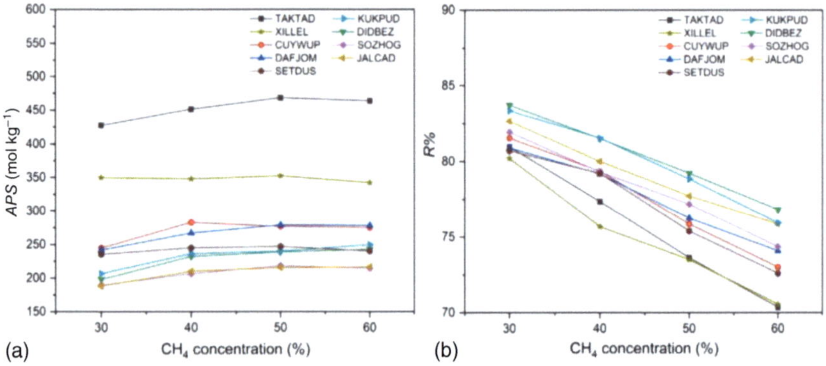

Figure 9.13 (a) Adsorbent performance score and (b) regenerability of the nine promising CoRE-MOFs under different operation temperatures. Source: [20]/with permission of Elsevier.

when CH_4 concentration increases to 50%, which is unfavorable from an economic perspective. Based on the findings from the temperature investigation, this may be remediable by temperature manipulation. To guide practical usage, the separation performance of the nine promising CoRE-MOFs is further examined under various operation conditions based on GCMC simulation. Apart from the original inlet gas molar ratio of 0.3/0.7 for CH_4/H_2, two more scenarios were considered, i.e. 0.5/0.5 and 0.6/0.4. Two different temperature and pressure settings for adsorption and desorption were investigated. The first is (303 K, 10 bar) for adsorption and (303 K, 1 bar) for desorption, which is the same as the original operation condition, while the other is (293 K, 10 bar) for adsorption and (313 K, 1 bar) for desorption, regarded as an adjusted operation condition. Simulation results are listed in Table 9.3. As can be seen, for the molar ratio of 0.3/0.7, the adjusted *APS* and regenerability are increased by more than 37% and 6%, respectively. It is worth noting that when the inlet CH_4 concentration is 50% or 60%, not only the *APS* values are enhanced, but also the regeneration of most MOFs recovers to satisfactory values by regulating the adsorption and desorption temperatures.

9.4 Conclusions

In this chapter, we introduce an AI-aided high-throughput screening and optimistic design of MOF materials for adsorptive gas separation. MOFs free of rare and/or toxic metals and MOFs uncompetitive in the targeted separation task are first excluded by a prescreening procedure. This is followed by a molecular simulation and machine learning coupled analysis for small sampled MOFs to acquire QSPR. Based on the obtained QSPR model, MOFs with the potential to perform well in the targeted application can be identified, and hence the candidates are further narrowed down. Then, rigorous molecular simulation is employed to assess the separation performances of these MOFs. Based on the result, top-performing MOFs are identified. Additionally, the impact of practical factors, such as humidity, operation temperature, and inlet gas composition, on the performance of the selected adsorbents in practice can be explored.

References

1 Xue, X., Cheng, M., Wang, S. et al. (2023). High-throughput screening of metal-organic frameworks assisted by machine learning: propane/propylene separation. *Industrial & Engineering Chemistry Research* 62 (2): 1073–1084.

2 Cheng, M., Feng, M., Zhou, L. et al. A machine learning boosted high-throughput screening of metal-organic frameworks for ethane/ethylene separation: from molecular simulation to process modeling. *Applied Energy* Submitted to Journal.

3 He, S., Cheng, M., Liu, C. et al. High-throughput virtual screening of bio-metal-organic frameworks for O_2/N_2 separation. *Industrial & Engineering Chemistry Research* 63: 2347–2360.

4 Wang, Y., Peh, S.B., and Zhao, D. (2019). Alternatives to cryogenic distillation: advanced porous materials in adsorptive light olefin/paraffin separations. *Small* 15: 190005825.

5 Mao, D., Griffin, J.M., Dawson, R. et al. (2021). Metal organic frameworks for hydrogen purification. *International Journal of Hydrogen Energy* 46 (45): 23380–23405.

6 Li, H., Li, L., Lin, R. et al. (2019). Porous metal-organic frameworks for gas storage and separation: status and challenges. *EnergyChem* 1: 1000061.

7 Daglar, H. and Keskin, S. (2020). Recent advances, opportunities, and challenges in high-throughput computational screening of MOFs for gas separations. *Coordination Chemistry Reviews* 422: 213470.

8 Clayson, I.G., Hewitt, D., Hutereau, M. et al. (2020). High throughput methods in the synthesis, characterization, and optimization of porous materials. *Advanced Materials* 32: 200278044SI.

9 Qiao, Z., Li, L., Li, S. et al. (2021). Molecular fingerprint and machine learning to accelerate design of high-performance homochiral metal-organic frameworks. *AIChE Journal* 67: e1735210.

10 Li, J., Sculley, J., and Zhou, H. (2012). Metal-organic frameworks for separations. *Chemical Reviews* 112 (2SI): 869–932.

11 Kwon, H.T. and Jeong, H. (2013). *In situ* synthesis of thin zeolitic-imidazolate framework ZIF-8 membranes exhibiting exceptionally high propylene/propane separation. *Journal of the American Chemical Society* 135 (29): 10763–10768.

12 Barea, E., Montoro, C., and Navarro, J.A.R. (2014). Toxic gas removal – metal-organic frameworks for the capture and degradation of toxic gases and vapours. *Chemical Society Reviews* 43 (16): 5419–5430.

13 Moghadam, P.Z., Li, A., Wiggin, S.B. et al. (2017). Development of a Cambridge Structural Database Subset: a collection of metal-organic frameworks for past, present, and future. *Chemistry of Materials* 29 (7): 2618–2625.

14 Wilmer, C.E., Leaf, M., Lee, C.Y. et al. (2012). Large-scale screening of hypothetical metal-organic frameworks. *Nature Chemistry* 4 (2): 83–89.

15 Chung, Y.G., Camp, J., Haranczyk, M. et al. (2014). Computation-ready, experimental metal-organic frameworks: a tool to enable high-throughput screening of nanoporous crystals. *Chemistry of Materials* 26 (21): 6185–6192.

16 Chung, Y.G., Haldoupis, E., Bucior, B.J. et al. (2019). Advances, updates, and analytics for the computation-ready, experimental metal-organic framework database: CoRE MOF 2019. *Journal of Chemical & Engineering Data* 64 (12): 5985–5998.

17 Cheng, M., Wang, S., Zhang, Z. et al. (2023). High-throughput virtual screening of metal-organic frameworks for xenon recovery from exhaled anesthetic gas mixture. *Chemical Engineering Journal* 451: 1382181.

18 Halder, P. and Singh, J.K. (2020). High-throughput screening of metal-organic frameworks for ethane-ethylene separation using the machine learning technique. *Energy & Fuels* 34 (11): 14591–14597.

19 Yan, Y., Shi, Z., Li, H. et al. (2022). Machine learning and *in-silico* screening of metal-organic frameworks for O_2/N_2 dynamic adsorption and separation. *Chemical Engineering Journal* 427: 131604.

20 Wang, S., Cheng, M., Luo, L. et al. (2023). High-throughput screening of metal-organic frameworks for hydrogen purification. *Chemical Engineering Journal* 451: 1384362.

21 Dubbeldam, D., Calero, S., Ellis, D.E. et al. (2016). RASPA: molecular simulation software for adsorption and diffusion in flexible nanoporous materials. *Molecular Simulation* 42 (2): 81–101.

22 Willems, T.F., Rycroft, C., Kazi, M. et al. (2012). Algorithms and tools for high-throughput geometry-based analysis of crystalline porous materials. *Microporous and Mesoporous Materials* 149 (1): 134–141.

23 van der Maaten, L. and Hinton, G. (2008). Visualizing data using t-SNE. *Journal of Machine Learning Research* 9: 2579–2605.

24 Tang, H. and Jiang, J. (2021). In silico screening and design strategies of ethane-selective metal-organic frameworks for ethane/ethylene separation. *AIChE Journal* 67: e170253.

25 Zhai, Q., Bu, X., Mao, C. et al. (2016). An ultra-tunable platform for molecular engineering of high-performance crystalline porous materials. *Nature Communications* 7: 13645.

26 Belmabkhout, Y., Pillai, R.S., Alezi, D. et al. (2017). Metal-organic frameworks to satisfy gas upgrading demands: fine-tuning the soc-MOF platform for the operative removal of H_2S. *Journal of Materials Chemistry A* 5 (7): 3293–3303.

27 Xue, Z., Sheng, T., Zhu, Q. et al. (2013). Two cationic metal-organic frameworks featuring different cage-to-cage connections: syntheses, crystal structures, photoluminescence and gas sorption properties. *CrystEngComm* 15 (40): 8139–8145.

28 Liang, L., Zhang, J., Ren, S. et al. (2010). Rational synthesis of a microporous metal-organic framework with PtS topology using a semi-rigid tetrahedral linker. *CrystEngComm* 12 (7): 2008–2010.

29 Murdock, C.R., Lu, Z., and Jenkins, D.M. (2013). Effects of solvation on the framework of a breathing copper MOF employing a semirigid linker. *Inorganic Chemistry* 52 (4): 2182–2187.

30 Chen, M., Liu, D., Deng, Y. et al. (2015). A 3D pillared Cu-based metal-organic framework: synthesis, structure, adsorption and catalytic properties. *Inorganic Chemistry Communications* 55: 96–98.

31 Zeng, M., Tan, Y., He, Y. et al. (2013). A porous 4-fold-interpenetrated chiral framework exhibiting vapochromism, single-crystal-to-single-crystal solvent exchange, gas sorption, and a poisoning effect. *Inorganic Chemistry* 52 (5): 2353–2360.

32 Chen, D., Ma, J., and Cheng, P. (2015). Solvent-induced secondary building unit (SBU) variations in a series of Cu(II) metal-organic frameworks derived from a bifunctional ligand. *Dalton Transactions* 44 (19): 8926–8931.

33 Altintas, C., Avci, G., Daglar, H. et al. (2018). Database for CO_2 separation performances of MOFs based on computational materials screening. *ACS Applied Materials & Interfaces* 10 (20): 17257–17268.

34 He, X., Chen, D., and Wang, W. (2020). Bimetallic metal-organic frameworks (MOFs) synthesized using the spray method for tunable CO_2 adsorption. *Chemical Engineering Journal* 382: 122825.

35 Burtch, N.C., Jasuja, H., and Walton, K.S. (2014). Water stability and adsorption in metal-organic frameworks. *Chemical Reviews* 114 (20): 10575–10612.

36 Yu, J., Xie, L., Li, J. et al. (2017). CO_2 capture and separations using MOFs: computational and experimental studies. *Chemical Reviews* 117 (14SI): 9674–9754.

10

Surrogate Modeling for Accelerating Optimization of Complex Systems in Chemical Engineering

Jianzhao Zhou[1] *and Jingzheng Ren*[1,2,3]

[1] *The Hong Kong Polytechnic University, Department of Industrial and Systems Engineering, Yuk Choi Road, Hong Kong SAR, China*
[2] *The Hong Kong Polytechnic University, Research Center for Resources Engineering Towards Carbon Neutrality, Yuk Choi Road, Kowloon, Hong Kong SAR, China*
[3] *The Hong Kong Polytechnic University, Research Institute for Advanced Manufacturing, Department of Industrial and Systems Engineering, Yuk Choi Road, Hung Hom, Kowloon, Hong Kong SAR, China*

10.1 Introduction

Optimization in the field of chemical engineering plays a crucial role in improving process efficiency, reducing costs, ensuring product quality, enhancing safety, minimizing environmental impact, and supporting effective decision-making. It is an essential way to achieve environmentally friendly and economically viable chemical processes. The general form of optimization can be formulated as Eq. (10.1) [1]. The optimization problems are classified into different categories according to the type of variables (continuous and discrete optimization), the number of objective functions (single-objective and multi-objective optimization), the structure of equations (linear and nonlinear optimization), the availability of constraints (constrained optimization and unconstrained optimization), etc.

$$\min/\max\ \text{obj} = f(x)$$

$$\text{s.t.} \begin{cases} g_i(x) \le 0 \\ h_j(x) = 0 \\ x_{\min} \le x \le x_{\max} \end{cases} \tag{10.1}$$

where f is the evaluation function used to describe the relationship between the objective (obj) and the decision variables (x). g and f denote the inequality and equality constraints, respectively. The decision variables that appear in optimization can be continuous (temperature, pressure concentration, etc.) or discrete (number of plates in the distillation column, number of batches, etc.). The values of these variables are constrained to a range within the lower bound of $x_{\min}$ and the upper bound of $x_{\max}$.

Many contemporary optimization problems heavily rely on computer-based simulations, which offer accurate and useful paradigms for describing complex

Applied AI Techniques in the Process Industry: From Molecular Design to Process Design and Optimization, First Edition. Edited by Chang He and Jingzheng Ren.

physical and chemical systems. While first-principle-based models, grounded in fundamental physical and chemical laws like the conservation of mass, momentum, and energy, provide crucial insights into system behavior, they present significant challenges in optimization, which can be seen as an inverse problem of simulation. In the domain of chemical engineering, these models, distinguished by their high-fidelity attributes, often assume a complex form, including non-algebraic equations, nonlinearities, and rigorous convergence [2]. For example, closed-circuit reverse osmosis (CCRO) systems typically entail the coupling of partial differential equations (PDEs) and ordinary differential equations (ODEs) [3]. Furthermore, flowsheet optimization problems are commonly displayed in the form of nonlinear programming characterized by nonconvexities and multiple local optima [4]. Even seemingly straightforward tasks, such as flash calculations, necessitate numerous iterations when the mechanism-based model is implemented [5]. In contrast, when dealing with a chemical reactor system involving multiple simultaneous reactions, each influenced by complex kinetics and thermodynamics, the computational requirements for modeling using first principles are exacerbated. This presents a challenge in optimization, resulting in prolonged simulation times and impeding the exploration of optimal operating conditions. Consequently, optimization-based on such mechanistic models becomes time-consuming and even infeasible in some cases, posing a persistent challenge despite substantial advancements in process systems engineering. Compounding the issue, a deterministic relationship between decision variables and their responses within an actual system remains unclear to researchers or engineers in certain scenarios [6].

To address the challenge of mechanism-based optimization, the surrogate modeling methodology that involves replacing costly or unclear model calls with computationally more affordable surrogate models can be employed. Essentially, the goal is to determine the alternative model to represent the mapping functions including, f, g, and h presented in Eq. (10.1). This approach has demonstrated its advantages and capabilities to provide accurate approximations of complex original models through surrogating (the surrogate models are also known as meta-models, regression surfaces, or emulators). Based on the available data from actual system operation or computationally intensive simulation, surrogate models can be constructed. Figure 10.1 presents a two-input surrogate model created by fitting 30 randomly sampled data points. The utilization of surrogate models can provide the following benefits: (i) a more efficient evaluation and compact description of the underlying input–output relationships, (ii) a more outstanding ability to integrate complex models from diverse sources, and (iii) increasing the speed of analyzing the design space and optimization process. As the volume of available data continues to grow, surrogate modeling becomes increasingly appealing for optimization implementation [7]. In chemical engineering, the utilization of surrogate models can be traced back to the mid-twentieth century. In 1951, George E. P. Box and K. B. Wilson employed a second-degree polynomial model to obtain an optimal response based on data from a sequence of designed experiments [8]. Though it was acknowledged that this model was only an approximation, such a model was easy to estimate and

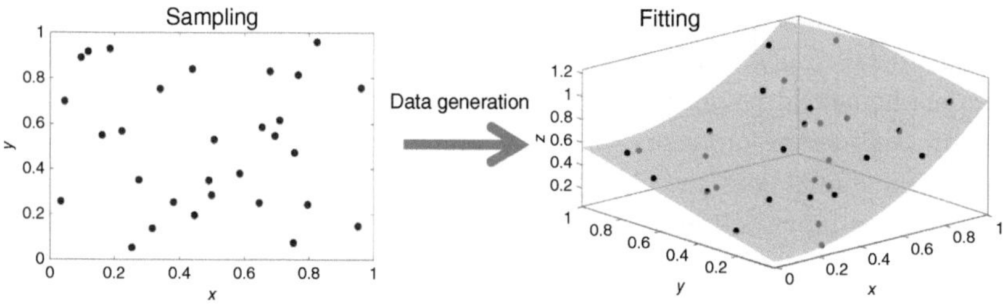

Figure 10.1 An example of surrogate model construction.

apply even with limited knowledge of the process or system. Subsequently, a series of surrogate models based on strategies such as order reduction [9] and machine learning [10] were proposed and applied in process optimization within the field of chemical engineering.

The basic workflow in surrogate modeling for optimization encompasses the following steps: (i) collect data from complex simulations or actual system operation of a chemical system, (ii) employ surrogate modeling techniques to construct surrogate models based on the available data, and (iii) assess the established surrogate model comprehensively and employ it in optimization. Among these steps, particular emphasis in this chapter will be placed on the construction of the surrogate model, considering its central role. Several characteristics of a given problem in the process influence its appropriateness for surrogacy, including linearity/nonlinearity, required accuracy, problem size (input dimensions), necessary information, computation speed, sample size, and the availability of convenient software or tools [11]. Section 10.2 introduces common surrogate models to guide selection, and Section 10.3 delves into the application specifics of surrogate modeling. Finally, Section 10.4 provides a concluding summary.

10.2 Surrogate Modeling Techniques

Focusing on an exploration of surrogate modeling techniques opens the door to a diverse array of mathematical tools and algorithms designed to approximate the behavior of complex systems. Basically, surrogate modeling approaches can be categorized as interpolation (when the surrogate model aligns with the true function value at each point in the training dataset) or regression (when it does not) [12]. Initially, researchers primarily employed mathematical interpolation methods to establish surrogate models. Specifically, a series of system input and output data are used to establish the algebraic relationship between input and output variables through mathematical interpolation, gradually creating a mathematical model. Mature interpolation methods developed over time include Newton interpolation [13], spline interpolation [14], Lagrange interpolation [15], and Hermite interpolation [16]. The aforementioned interpolation method is relatively simple in mathematical structure and thus more computationally efficient, but the effect is generally moderate. Following this, researchers proposed the polynomial response surface model method to address the shortcomings of interpolation. Myers et al. [17] provided a detailed description of the polynomial response surface model, which utilizes mathematical polynomials of varying orders to characterize approximate models in engineering problems. The polynomial response surface model boasts a low computational cost, the ability to derive explicit mathematical expressions between system input and output variables, good continuity and differentiability, fast convergence speed, and ease of optimization. With the rapid development of artificial intelligence (AI)/machine learning, a large number of novel modeling techniques have emerged and been used in the construction of regression-based surrogate models, including but not limited to artificial neural networks (ANN), support vector machines (SVM), and decision trees (DT). While lacking explicit mathematical structure expressions, these models showcase remarkable fitting capabilities and have garnered significant attention. Each technique brings its unique strengths, tailored to capture specific aspects of system responses. This section introduces the most commonly used surrogate models in the literature related to optimization involved in chemical process engineering, including polynomial regression (PR), polynomial chaos expansion (PCE), Kriging, ANN, radial basis function (RBF), high-dimensional model representation (HDMR), SVM, and DT.

10.2.1 Polynomial Regression (PR)

Polynomial functions stand out as the predominant surrogate models in engineering applications [18]. Polynomials play a pivotal role in conveniently revealing the underlying relationship within chemical systems conveniently by leveraging the magnitudes of their coefficients to provide general insights into the design and optimization problem. They are recognized as one of the computationally simplest models for regression purposes and are recommended for less complex underlying models. Considering higher-order interactions often lack significance and demand more data to accommodate the additional parameters, quadratic polynomial functions, as surrogate models, are usually constrained to main effects and first-order interactions to approximate the relationship between decision variables and system response as outlined in Eq. (10.2) [6]. This simplified cutoff helps avoid overfitting, particularly when dealing with small datasets.

$$\hat{y}(x) = a_0 + \sum_i a_i x_i + \sum_i b_i x_i^2 + \sum_i \sum_{j>i} c_{i,j} x_i x_j \tag{10.2}$$

where $\hat{y}$ is the predicted response of the system, a, b, and c are the coefficients to be determined, while x is the decision variables and the subscripts i and j indicate the ith and jth decision variables, respectively.

This class of regression models has a rich history in classical experimental design, particularly in scenarios whcrc system information was frequently unknown. In such cases, it became imperative to devise methods capable of exploring the significance of the main process variables and their interactions as indicated by the determined coefficients [19]. Polynomial surrogates prove highly advantageous when coping with systems that exhibit a smooth response and thus are useful for continuous optimization. Higher efficiency is notable for low-dimensional problems. However, engineering practices often encounter high-dimensional and highly nonlinear systems. In such instances, polynomial surrogates may fall short of reliably representing the response surface [20, 21] and are usually restricted to local regions.

10.2.2 Polynomial Chaos Expansion

PCE, also called Wiener chaos expansion [22], is a method for representing a random variable in terms of a polynomial function of other random variables. The selection of polynomials is based on their orthogonality with respect to the joint probability distribution of these random variables. This expansion offers a means to represent the response variable Y (seen as the random variable), with finite variance, as a function of an M-dimensional random vector X (decision variables). It employs a polynomial basis that is orthogonal to the distribution of this random vector. The standard form of PCE is given by the following equation [22]:

$$Y = \sum_{i \in \mathbb{N}} c_i \Psi_i(X) \tag{10.3}$$

where c_i denotes coefficient to be determined and Ψ_i denotes a polynomial basis function. In the one-dimensional scenario, considering only the Gaussian distribution, the orthogonal polynomial basis functions are the set of ith degree Hermite polynomials H_i. The PCE of Y in terms of the standard random normal decision variables θ_i is expressed as follows [23]:

$$Y = \sum_{i \in \mathbb{N}} c_i \mathrm{H}_i(\theta_i) \tag{10.4}$$

The ith degree Hermite polynomials H_i can be defined as follows [23]:

$$\mathrm{H}_i(\theta_1 \ldots \theta_i) = (-1)^i \mathrm{e}^{(1/2)\theta^T\theta} (\partial^i \mathrm{e}^{(-1/2)\theta^T\theta} / \partial\theta_1 \ldots \partial\theta_i) \tag{10.5}$$

Since the polynomial chaos terms are functions of random variables, they become random variables, and terms of different orders are orthogonal to each other [24]. This orthogonality is defined in Gaussian measures as the expected value of the product of the two random variables. Due to the mean-square convergence of PCE, it is advantageous to calculate the coefficients using least-squares minimization (LSM), considering sample input/output pairs from the model. Optimization is carried out until the best fit is achieved between the surrogate PCE and the nonlinear model or simulated/experimental data. PCE has found applications in various domains such as electrical measurement, electric circuit models, chemical processes, biotechnological processes, reaction engineering, transport phenomena, batteries, robot manipulators, helicopters, and mechanical systems [23].

10.2.3 Kriging

Kriging, with a mathematical basis as a Gaussian process regression (GPR) model [25], has emerged as a widely used method in surrogate modeling by capturing not only the mean response (expected value) but also the associated prediction uncertainty (variance). Originally developed to describe spatial distributions in geostatistics [26], Kriging gained popularity for deterministic computer experiments involving computationally demanding simulations [7, 27, 28]. Its flexibility in modeling various functions and interpolating data, along with the requirement for only a few fitted parameters, makes it a key technique among the surrogate modeling techniques. To achieve high performance in surrogating, as presented in Eq. (10.6), the model comprises a deterministic polynomial term representing the global trend of the data ($p^T(x)\beta$) and a stochastic process accounting for the lack of fit in the polynomial term ($z(x)$) [29]. The stochastic part involves selecting a correlation function (Eq. (10.7)), either a priori or by fitting a semi-variogram to trends in the data. The Gaussian kernel function, also known as the squared-exponential or RBF kernel, is famous in the realm of chemical engineering due to its smoothness as presented in Eq. (10.8). Predictions for unsampled points are linear functions of observed data (Eq. (10.9)), and prediction errors are calculated using Eq. (10.10). Regions with high uncertainty in the sampling space can be identified, prompting the addition of new samples to enhance model performance. Kriging-specific adaptive sampling techniques have been developed to address this goal explicitly, as highlighted by various studies [20, 30, 31]. It is crucial to note that predictions assume the correct fitting of model parameters from observed data, and the predicted variance is itself a prediction of the expected model uncertainty. Kriging is recommended for problems with dimensions below 20, continuous variables, and smooth underlying functions [32]. In regression, discontinuities may lead to poor results due to the stationary covariance assumption of the correlation. Fitting the Kriging model (Eq. (10.9)) involves matrix inversion, becoming computationally demanding with large observed sets. There exist various types of Kriging beyond those presented here, and a comprehensive list of variants can be found in Yondo et al. [33].

$$\hat{y}(x) = p^T(x)\beta + z(x) \tag{10.6}$$

$$cov(z(x_i), z(x_j)) = \sigma^2 R(x_i, x_j) \tag{10.7}$$

$$R(x_i, x_j) = \sigma^2 \exp\left(- \sum_{k=1}^{n} \theta_k x_i^k - x_j^{k2} \right) \tag{10.8}$$

$$\hat{y}(x) = p^T(x)\hat{\beta} + r(x)R^{-1}(y_D - F\hat{\beta}) \tag{10.9}$$

$$s^2(x) = \hat{\sigma}^2(1 - r(x)^T R^{-1} r(x) + (FR^{-1}r(x) - p(x))^T (FR^{-1r(x)} - p(x))/F^T R^{-1} F) \tag{10.10}$$

10.2.4 Radial Basis Functions (RBF)

RBFs is a powerful and versatile tool for approximating complex relationships between input and output variables. Belonging to the family of kernel-based methods, RBF is particularly well suited for applications in optimization, uncertainty quantification, and simulation of computationally expensive models. At its core, RBF leverages the concept of radial symmetry. It constructs a representation by combining local univariate functions, each centered around specific points, through a weighted linear combination [34]. This unique characteristic allows RBFs to efficiently capture complex and nonlinear patterns in data, enabling them especially effective in scenarios where the underlying relationships may be nonlinear or exhibit complex interactions. The RBF approximation takes the form of Eq. (10.10). This form of the RBF is identical to an ANN with a single hidden layer with RBFs [35]. Generally, RBF is applicable to situations where Kriging surrogates may be used, but are not as often used in chemical engineering modeling and optimization because the parameterized basis function of Kriging (which may be considered a special form of RBF) is preferred due to its higher accuracy, flexibility, and ability to make predictions of model variance [32]. Wang and Ierapetritou [36] recently addressed this issue by developing an adaptive sampling technique for cubic RBFs. In several cases, they showed that cubic RBFs improved flexibility in exploring the design space with higher accuracy while using fewer samples than Kriging.

$$\hat{y}(x) = \sum\nolimits_i \lambda_i \phi(\|x - x_i\|_2) \tag{10.11}$$

where x_i denotes the ith center of n basis functions ϕ, which can take several forms; $\|x - x_i\|_2$ evaluates the Euclidean distances between the prediction sites and the basis function centers; and λ_i are scalar weights during regression.

10.2.5 High-Dimensional Model Representation (HDMR)

HDMR stands out as an influential and efficient methodology employed in the generation of surrogate models. It has gained notable attention due to its exceptional precision in diverse industrial applications. HDMR's distinctive capability lies in its feature of decomposing a comprehensive function into a summation of constituent functions, each represented by a specific subset of input variables. This intricate decomposition, as illustrated in Eq. (10.12), allows for an exact representation of the function under consideration. In practice, the application of HDMR proves pragmatic, as terms encompassing functions of more than two input parameters often contribute negligibly compared to lower-order terms [37, 38]. This truncated approximation, as presented in Eq. (10.13), remains not only efficient but also remarkably effective for a broad spectrum of models and datasets encountered in practical industrial settings. Though it is possible to evaluate each of these terms using direct numerical integration, a more efficient method is to approximate the functions f_i and f_{ij} with analytic functions, as shown in Eq. (10.14) [39]. The truncated approximation provided by HDMR allows for flexibility in adjusting the model complexity based on the specific requirements of the problem and makes it possible to achieve the trade-off between accuracy and efficiency in surrogating [40]. As the dimensionality of the input space increases, the number of terms in the HDMR expansion grows exponentially. This can lead to challenges in terms of computational complexity and storage requirements, especially for large-scale systems. As PR, HDMR is more effective when dealing with smooth functions.

$$y = f_0 + \sum_{i=1}^{N} f_i(x_i) + \sum_{i=1}^{N} \sum_{j=i+1}^{N} f_{ij}(x_i x_j) + \cdots + f_{12\ldots N}(x_1 x_2 \cdots x_N) \tag{10.12}$$

$$y = f_0 + \sum_{i=1}^{N} f_i(x_i) + \sum_{i=1}^{N} \sum_{j=i+1}^{N} f_{ij}(x_i x_j) \tag{10.13}$$

$$y = C + \sum_{i=1}^{N} \sum_{k=1}^{K} A_{i,k} \times x_i^k + \sum_{i=1}^{N} \sum_{j=i+1}^{N} \sum_{k=1}^{K} \sum_{n=1}^{K} B_{i,j,k,n} \times x_i^k \times x_j^n \tag{10.14}$$

where y is the calculated function value, N is the number of input variables, and f denotes the effect between input variables. C is a constant term, $A_{i,k}$ and $B_{i,j,k,n}$ are the first and second-order coefficients, K is the highest degree of input variables, and the subscripts i and j denote the ith and jth input parameters, respectively, while the subscripts k and n denote the orders of ith and jth input parameters.

10.2.6 Decision Tree (DT)

The decision tree is a powerful tool for capturing linear and nonlinear relationships in data and thus has great potential in surrogate modeling. In this context, decision tree surrogates undertake a process of recursive partitioning, systematically dividing the input space into distinct regions while assigning a constant value to each delineated region. In the pursuit of leveraging decision trees for surrogate modeling, the classification and regression tree (CART) algorithm emerges as a pivotal choice [41]. CART methodically constructs a binary tree structure, where internal nodes serve as splitting conditions predicated on input features, and leaf nodes stand as repositories for predicted class labels or continuous output values. The underpinning of CART lies in the application of linear regression to both parent and child nodes within the tree. The linear regression process, elucidated through Eqs. (10.15)–(10.19) [42], stands as a cornerstone in the CART methodology. These equations delineate the progression of linear regression across nodes, laying the groundwork for assessing the total variance ($\boldsymbol{V}$), as depicted by a corresponding equation. This calculated total variance becomes instrumental in identifying the optimal division of the parent node, steering the decision tree toward a refined and data-informed structure. As a result, decision trees, with the CART algorithm at its core, offers a robust framework for constructing surrogate models that adeptly navigate nonlinearity, ensuring a precise representation of complex relationships within the underlying data. DT can be extended into ensemble methods like random forests, where multiple trees are combined to enhance predictive performance and reduce overfitting. One of the biggest challenges in using DT for surrogate modeling is the tendency of decision trees to overfit the training data, especially when the tree captures noise or specific patterns in the training data that do not generalize well to new, unseen data. This challenge arises due to the inherent flexibility of decision trees, which allows them to create complex, highly detailed structures to fit the training data perfectly [43].

$$\hat{y} = \mathbf{X} \cdot \mathbf{K} \tag{10.15}$$

$$\mathbf{K} = [\mathbf{X}^{\mathrm{T}} \cdot \mathbf{X}]^{-1} (\mathbf{X}^{\mathrm{T}} \cdot \mathbf{Y}) \tag{10.16}$$

$$\mathbf{X} = \begin{pmatrix} x_{11} & \cdots & x_{1i} \\ \vdots & \ddots & \vdots \\ x_{j1} & \cdots & x_{ji} \end{pmatrix} \tag{10.17}$$

$$\mathbf{Y} = \begin{pmatrix} y_{11} & \cdots & y_{1k} \\ \vdots & \ddots & \vdots \\ y_{j1} & \cdots & y_{jk} \end{pmatrix} \tag{10.18}$$

$$\mathbf{V} = \sum_{j=1}^{J} \sum_{k=1}^{K} (\hat{y}_{jk} - y_{jk})^2 \tag{10.19}$$

where **X** is the matrix of model inputs, **Y** is the matrix of model outputs, **K** is the regression matrix, and $\hat{y}$ is the prediction matrix.

10.2.7 Support Vector Machine (SVM)

SVM regression is a widely adopted and potent machine learning technique in surrogate modeling for optimization of chemical engineering systems, as evidenced by its successful applications in various industrial processes [44, 45]. SVM regression endeavors to determine a hyperplane within the feature space that maximizes the margin between data points and the hyperplane while minimizing regression errors. In the context of a training dataset comprising input vectors $X = \{x_1, x_2, \ldots, x_n\}$ and corresponding target values y, SVM regression seeks to discover a regression function $f(x)$ capable of accurately predicting the target variable for new input instances. In SVM regression, the polynomial kernel as defined by Eq. (10.20) [46] is frequently employed due to its advantageous combination of fast training and preservation of nonlinear capabilities [47]. It enables SVM regression to implicitly map features to a higher-dimensional space. Besides, Gaussian RBF kernel-based SVM, as shown in Eq. (10.21) [48] was recognized as the best choice for practical applications [49]. The objective of SVM regression is to find the optimal hyperplane that minimizes the regression error while satisfying the margin constraints. And the output (y_i) can be predicted by Eq. (10.22). Although SVM regression is a robust and effective tool for surrogate modeling, offering high predictive accuracy and adaptability to complex relationships, its computational demands and sensitivity to certain parameters should be considered when applying it in practice.

$$K_{\text{Polyomial}}(x_i, x_j) = (\gamma x_i \cdot x_j)^d \tag{10.20}$$

$$K_{\text{RBF}}(x_i, x_j) = \mathrm{e}^{-\|x - x_i\|_2 / 2\sigma^2} \tag{10.21}$$

where γ is a scale factor, $x_i \cdot x_j$ denotes the dot product of the input vectors x_i and x_j. d is the polynomial degree. $\|x - x_i\|_2$ indicates the Euclidean distances and σ is the width of the kernel in SVM.

$$y_i = \sum\nolimits_j \alpha_j K(x_i, x_j) + \beta \tag{10.22}$$

where α_j and β are SVM parameters to be determined during training.

10.2.8 Artificial Neural Network (ANN)

ANN has emerged as a versatile and powerful machine learning approach, aiming to explore the complex relationships between input and output parameters. The essence of ANN lies in the composition of artificial neurons arranged in connected layers, where information is processed and transmitted [50]. These networks consist of units called artificial neurons, which receive numerical information from each neuron in the last layer and produce an analogous response forwarded to neurons in the subsequent layer. The transformative process involves weights assigned to each transmission, optimized during training using algorithms like backpropagation. This iterative optimization enables the model to learn and map input data to process responses, making ANN adept at representing diverse systems and yielding profound results across various tasks [51]. The capability of ANN to capture the global nature of design spaces for high-dimensional nonlinear systems is a distinct advantage. However, a key challenge lies in designing the appropriate network architecture, encompassing considerations such as layout, hyperparameters (e.g. learning rate, transfer functions, and regularization methods), and other intricate details. This process may require additional training and validation costs, demanding significant amounts of data to accommodate the large number of weights without overfitting. Figure 10.2 presents the topology structure of a typical three-layer ANN [42, 52]. Based on a rectified linear unit (ReLU)-activated hidden layer, which is widely used in modeling chemical systems [53], Eq. (10.23) shows the values of neurons in the hidden layer (H_l) are computed based on the values of neurons in the input layer (x_i), which are subsequently employed in Eq. (10.24) to determine the output values (y_o). The model's parameters ω and b are determined by model training via the gradient descent optimization algorithm [54].

$$H_l = \max\left(\sum_{i=1}^{n} \omega_{i,l} x_i + b_l, 0\right) \tag{10.23}$$

$$y_o = \sum_{l=1}^{m} \omega_{l,o} H_l + b_o \tag{10.24}$$

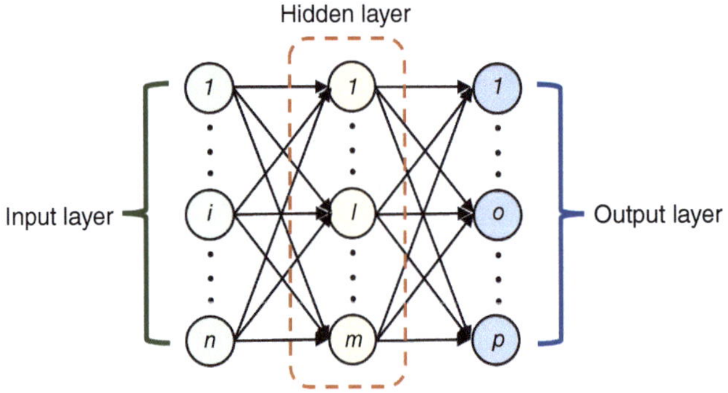

Figure 10.2 Schematic diagram of three-layer ANN. Source: [52]/with permission of Elsevier.

where $\omega_{i,l}$ indicates the connecting weight of ith node in the input layer and lth node in the hidden layer, b_l represents the bias of lth node in the hidden layer, and n is the number of nodes in the input layer. Similarly, $\omega_{l,o}$ denotes the connecting weight of lth node in the hidden layer and oth node in the output layer, while b_o is the bias of oth node in the output layer, and m is the number of nodes in the hidden layer.

10.3 Application of Surrogate Model in Optimization of Chemical Processes

In general, surrogate models of lower complexity that are inexpensive to evaluate and approximate accurately a large-scale model can greatly facilitate computationally intensive analysis tasks at hand. Such analysis specifically refers to optimizations, which are typically performed iteratively, or by relying on population-based algorithms. The prediction of surrogate models aids designers in performing optimization with an affordable number of computer simulations. Surrogate modeling has been widely used in the optimization and synthesis of chemical processes, which usually involves different units like reaction systems, separation systems, and heat exchange systems.

10.3.1 Reaction Engineering

Reaction engineering is a branch of chemical engineering that focuses on the design and optimization of chemical reactors and processes where chemical reactions occur. It encompasses the study of reaction kinetics, thermodynamics, and transport phenomena to understand and control chemical reactions within a chemical treatment step of a process. The primary objective is to design and operate reactors efficiently, ensuring optimal conversion of a variety of starting materials to desired products while considering performance indicators such as yield, selectivity, and safety. The application of surrogate models in accelerating the optimization of reactors can be classified based on the type of chemical reactors mainly including batch reactor, continuous stirred-tank reactor (CSTR), plug flow reactor, packed (fixed) bed reactor, fluidized bed reactor, membrane reactor, and microreactor.

Batch processes, accounting for 40–60% of chemical process industries [55], such as those in food products, electronic chemicals, biotechnology, polymers, and pharmaceuticals [56], encounter challenges such as the absence of steady-state operating points, nonlinear behavior, constrained operation with limited measurements, and the presence of disturbances [57]. This complexity poses challenges for engineers in understanding and optimizing batch chemical reactors. To address these issues, Tasnim et al. [23] employed PCE to develop a nonlinear surrogate model of a batch chemical reaction process based on an assumed sequence reaction scheme. This surrogate model was subsequently used to identify the optimal temperature profile needed to maximize the concentration of an intermediate product at the end of the batch. Validation and optimization results demonstrated that the PCE-based surrogate model can be used as a robust approach for the rapid design, control, and optimization of batch reactor systems. Similarly, Kumar and Budman [58] utilized PCE in modeling a fed-batch bioreactor for robust optimization, showcasing its superior performance compared to its nominal counterparts. Besides, ANN was also widely used in modeling and trajectory optimization of a batch or a semibatch process [59]. Recently, Zhan et al. [60] compared the performances of PR and ANN in modeling co-digestion of poultry litter with wheat straw in an anaerobic sequencing batch reactor for biomethane production and then optimized the operational conditions for maximum methane yield based on the PR, which had better performance. ANN, as one of the most widely used surrogate models for CSTR [61, 62], was applied by Sivapathasekaran and Sen [63] to develop a nonlinear model for maximizing lipopeptide biosurfactant concentration using a genetic algorithm (GA).

The plug flow reactor model is used to describe chemical reactions in continuous, flowing systems of cylindrical geometry based on certain assumptions. Significant work has been dedicated to surrogate modeling for such reactors. With data from a pilot-scale gasifier, reduced order model developed previously, Wang et al. [64] developed an ANN-based surrogate model, demonstrating its efficiency by being at least 4 orders of magnitude faster than reduced order models in optimizing operating conditions to achieve maximum carbon conversion and production of hydrogen gas. Simplifying the screening of input parameters, Jiang et al. [65] used HDMR in surrogate modeling for a rigorous kinetic-based plug flow reactor where hydrogen oxidation occurred to approximate the large system of ODE with simple algebraic equations to describe and solve the chemical kinetics. And then a deterministic global optimization method is used to optimize a hydrogen oxidation model using response surfaces obtained from the HDMR model. In addition, Shokry and Espuña [66] also used Kriging surrogate model to obtain simpler, accurate, robust, and computationally inexpensive predictive dynamic models for accelerating sequential dynamic optimization.

Fixed bed reactors and fluidized bed reactors, commonly used in chemical engineering, are essential components whose modeling is of significant importance. Ayub and Zhou et al. [67, 68] have employed RBF-based surrogate optimization to achieve the best performance of a fluidized bed gasifier based on data from rigorous simulation and revealed the high efficiency of such an approach (with fast convergence during optimization). Regarding the optimization of gasification, Kim et al. [69] employed DT-based RF and ANN algorithms to construct a surrogate model to predict the outcomes of biomass gasification in a fluidized bed with high prediction accuracy. Additionally, Fang et al. [42] also developed decision tree and SVM as surrogate models for predicting the yield and profit of an industrial propane dehydrogenation process. This enables efficient optimization-based on particle swarm optimization, with the optimal solution typically found within the fifth to tenth generation of the optimization process. Notably, the datasets used for training these surrogate models are based on rigorous simulation. Regarding surrogate modeling for a member reactor, Waqas et al. [70] implemented the ANN and SVM to depict the statistical modeling approach using experimental datasets. Besides, SVM was also used to build the data-driven surrogate model of a microreactor based on computational fluid dynamic (CFD) simulation data and was further used to optimize the shape of the reactor [71]. To address the high computational complexity derived from CFD simulation, deep neural network-based surrogate modeling has also become a promising solution [72]. With abundant work focusing on this domain, surrogate modeling can be applied to optimize various aspects of these reactors, contributing to improved efficiency, selectivity, and overall process performance. Overall, ANN is one of the most widely used methods due to its high accuracy in modeling, while in comparison PR still exhibits a more apparent advantage in optimization efficiency due to its simple structure.

10.3.2 Separation Engineering

Separation engineering, a specialized field within chemical engineering, focuses on the processes and methodologies involved in extracting components from mixtures to obtain pure substances. These separation processes are vital in various industries, including chemical manufacturing, pharmaceuticals, petrochemicals, and environmental engineering [73]. The goal is to efficiently isolate and purify specific substances from streams with complex mixtures, often involving physical or chemical methods. Typically, distillation is one of the most common techniques in separation engineering. Although rigorous distillation models are versatile and relatively accurate, ensuring convergence may necessitate good initial estimates. Additionally, their inherent computational complexity, like their discreteness in determining the plate number of columns, presents significant challenges when integrated into optimization frameworks. Recognizing its importance in chemical engineering, various surrogate modeling techniques have been explored for the optimization of distillation columns. Ibrahim et al. [74] have employed surrogate modeling to accelerate the optimization of crude oil distillation units. Their approach integrated surrogate models based on ANN with SVM to optimize column configuration and operating conditions. The SVM aided in filtering infeasible design options, reducing computational efforts, and enhancing the final solution's quality. Rigorous process simulations and pinch analysis were employed to build the surrogate model and determine maximum heat recovery and minimum utility costs.

To mitigate the nonlinearity and complexity of the integrated distillation optimization process, an RBF-based neural network was utilized as a surrogate model for function evaluation. This surrogate model helped to significantly reduce the computational expenses associated with the optimization process. Subsequently, efficient multi-objective optimization was achieved [75]. Besides, Quirante et al. used interpolation-based surrogate models like Kriging to obtain accurate models of distillation columns [76]. The surrogate-based optimization strategy ensures that convergence to a local optimum is guaranteed for numerical noise-free models. Kim et al. [77] used a computational-cost-efficient surrogate model based on machine learning, using data from a first-principle mathematical model implemented in the gPROMS® modeling environment [78]. The surrogate model considered adsorption pressure, desorption, feed rate, and rinse rate as input variables, successfully obtaining a Pareto front between productivity and purity in a vacuum pressure swing adsorption for CO separation process optimization.

The input variables were determined as the adsorption pressure, desorption, feed rate, and rinse rate. Similar surrogate-based multi-objective optimization was also implemented by Beck et al. [79] to address the computational requirements of high-fidelity simulations needed to evaluate alternative designs. The study specifically employed the Kriging model due to its ability to generate confidence bands for predictions effortlessly. This outstanding advantage of the Kriging model facilitated effective navigation of the design space. Crystallization, another vital separation technique, separates components based on solubility differences. Surrogate models contribute to optimizing crystallization conditions, including temperature profiles, cooling rates, and seeding strategies, for improved yield and purity [80, 81]. It could be seen that in the realm of separation engineering, the surrogate model plays an important role in accelerating optimization, and a large number of techniques have shown satisfactory capabilities.

10.3.3 Heat Exchange and Integration

Heat exchange, a fundamental process in thermodynamics and chemical engineering, involves transferring heat from one medium to another. Components like reboilers, condensers, and vaporizers, as the main heat exchangers, play key roles [82]. It is crucial for various industrial applications, such as power generation, refrigeration, and heating. Heat integration, on the other hand, refers to the systematic design and optimization of heat exchanger networks within a process to enhance energy efficiency. The goal of optimizing heat exchange and integration in chemical engineering is to minimize energy consumption, reduce utility costs, and improve overall process sustainability. Surrogate modeling has been employed in heat exchange and integration to optimize the design and operation of heat exchanger networks by a lot of researchers and engineers [83–85]. Typically, surrogate models have been utilized to optimize the selection and arrangement of heat exchangers within a network, involving determining the optimal number of stages, heat exchanger sizes, and their locations to maximize energy efficiency [86, 87]. In addition, numerous researches focus on the operating conditions of systems related to heat exchange and utilization. For example, Maakala et al. [88] employed the PR as a surrogate model to optimize the heat transfer performance of the recovery boiler superheaters with training data from CFD simulations. Furthermore, Zhou et al. [53] used linear regression and ANN in surrogate modeling organic Rankine cycle (ORC)-based combined system. The operations feasibility identification and energy prediction were achieved by surrogate models, and then the total efficiency optimization can be expressed as a mixed-integer linear programming (MILP) problem. Only ~0.1 seconds is required, compared to more than 10 hours required for optimization based on the mechanistic model. Besides, Vilasboas et al. [89] used RF and PR for surrogate modeling of the ORC and used these established surrogate models to optimize the specific cost and energy efficiency of the ORC system. The results showed that the computational time was reduced by more than 99.9% for all indicated surrogates. These studies demonstrated that surrogate modeling poses great potential for accelerating the optimization of heat recovery and utilization systems by simplifying the mechanism-based model.

10.3.4 Process Design and Synthesis

The approach of employing surrogate modeling for various units involved in chemical processes, also known as surrogate-based superstructure optimization, enables the achievement of process design and synthesis with high efficiency [90]. As shown in Figure 10.3, the optimization-based "superstructure" methods for process synthesis are to determine the optimal pathways for conversion of feed to achieve the best performances in efficiency, economy, environment, etc. It is more powerful than traditional sequential-conceptual methods as they account for all complex interactions between design decisions.

Palmer and Realff [91, 92] pioneered the optimization of chemical flow sheets using surrogates, employing PR and Kriging surrogates to handle limited datasets wisely. During the optimization, model reduction was conducted if insignificant input variables were identified. The results showed that this methodology yielded process configurations comparable to those reported in the literature. Davis and Ierapetritou [93] introduced one of the earliest surrogate-based optimization approaches, namely the Kriging-response surface methodology. This approach incorporated a sequential design method to model noisy black box functions within deterministic, feasible regions. Subsequently, they integrated a branch and bound method into the algorithm to handle integer variables, enabling the consideration of process synthesis and design problems [94]. In a follow-up article [95], discrete variables were incorporated directly into the black box models. Successful optimization of processes, including superstructures, such as purifying alcohol dehydrogenase and producing *tert*-butyl methacrylate, was achieved using Kriging models for the nonlinear programming (NLP) sub-problems. However, as the dimensionality of complete flow sheets increases, a modular or distributed approach becomes more practical. Henao and Maravelias [96] proposed a novel framework that achieved efficient superstructure optimization by simplifying formulation. The complex first-principle unit models were first replaced by compact

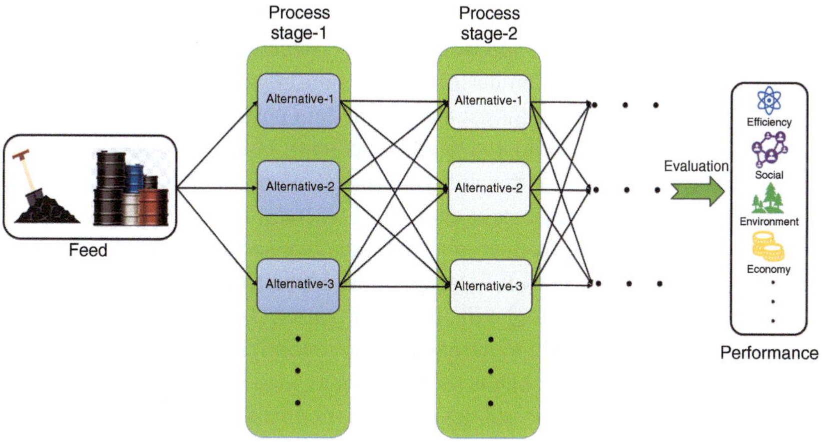

Figure 10.3 Superstructure in chemical process engineering.

and yet accurate ANN-based surrogate models and then binary variables that allow activation/deactivation of particular units were incorporated within the superstructure. Caballero and Grossmann [97] also applied surrogate models to replace individual units in a process flow sheet model, effectively lowering the dimensionality of each surrogate. In their study, they recommended a maximum dimensionality of 9 or 10 for each surrogate to prevent sampling from becoming the computationally limited step and to maintain surrogate accuracy. The choice of Kriging surrogate was motivated by its ability, at each iteration, to maintain the predicted variance of the model, facilitating convenient determination of stopping criteria and constraint feasibility. This allowed noise generated by simulation models to be considered as a stopping criterion for optimization. While this introduced higher computational costs for fitting and updating Kriging models compared to polynomials, the algorithm was explicitly designed not to update Kriging parameters at each iteration, aiming to generate satisfactory models initially with minimal adaptive improvement.

The applicability of surrogate modeling has been demonstrated in practice. Recently, Quirante et al. [98] provided an example of using Kriging modeling in the successful superstructure optimization of a vinyl chloride monomer production process. Fahmi and Cremeschi [99] conducted a study on the superstructure optimization of a biodiesel plant, with ANN-based surrogates replacing each unit operation, thermodynamics, and mixing models. Notably, ANNs in several works were limited to simple network architectures with a single hidden layer and only a few neurons. Martino et al. [100] explored a feed-forward ANN-based superstructure optimization approach and showed the advantage of simple ANN (one layer and fewer nodes). While networks with simple structures are easy to fit and do not require large amounts of data, they lack the predictive power of larger networks. Larger networks require more data to prevent overfitting, which can become counterproductive when limited computational expense is desirable. With the development of deep learning represented by deep neural networks, more attention should be applied to building high-quality ANN when selected as a surrogate model, drawing insights from the machine learning community [51].

A notable case of surrogate modeling in superstructure optimization is the ALAMO framework [101], which designs mathematically simple surrogates from a set of basic functions using the least amount of data possible. Cozad et al. [102] added constrained regression to the method, placing bounds on the surrogate output to enhance extrapolation reliability, which is an important feature for modeling physical or safety limitations in chemical processes. ALAMO also holds the potential to achieve similar accuracy through adaptive sampling, which requires fewer data points compared to full sampling and thus reduces the number of complex simulations needed [103]. With a similar pattern, Boukouvala and Floudas [104] developed ARGONAUT, a framework for optimizing constrained global derivative-free optimization problems using surrogate models. Surrogates are automatically chosen from a list of possibilities (i.e. polynomials, RBF, Kriging) based on the need to limit complexity and maintain accuracy of the objective and constraints of the underlying gray-box models. Since process superstructure typically results in large-scale non-convex mixed-integer nonlinear programs (MINLP), which are very hard to solve effectively, surrogate modeling for the first-principle-based model has become more and more important. In most of the examples so far, the Gaussian process-based Kriging method has been implemented. However, with an increasing number of available data, ANN has become more popular in chemical engineering for process design and synthesis.

10.4 Conclusion

Surrogate modeling has a huge advantage in simplifying the underlying relation involved in the study object and further reducing the computational burden for optimization. This chapter has provided an overview of the most commonly employed surrogate modeling techniques. The widespread adoption of these techniques, as highlighted in this discussion, has played a pivotal role in addressing challenging problems in chemical and process engineering. Due to the unremitting research and exploration of researchers, a large number of surrogate models have been successfully applied to accelerate the optimization of complex chemical systems. All of these efforts combine to establish a set of guidelines for surrogate model use and development. This, in turn, will lead to a more systematic and structured approach to surrogate modeling, enhancing the efficiency of modeling, optimizing, and studying complex processes. In the foreseeable future, with the development of AI, surrogate modeling coupled with cutting-edge AI methodologies holds promise for further accelerating the optimization of chemical processes.

Acknowledgment

The authors express their sincere thanks to the Research Committee of The Hong Kong Polytechnic University for the financial support of the project through a PhD studentship (project account code: RKQ1). The work described in this paper was also supported by a grant from Research Grants Council of the Hong Kong Special Administrative Region, China-General Research Fund (Project ID: P0042030, Funding Body Ref. No: 15304222, Project No. B-Q97U) and a grant from Research Grants Council of the Hong Kong Special Administrative Region, China-General Research Fund (Project ID: P0046940, Funding Body Ref. No: 15305823, Project No. B-QC83).

References

1 Dutta, S. (2016). Optimization in chemical engineering. In: *A Brief Discussion on Optimization*, 1–11. Cambridge University Press https://doi.org/10.1017/CBO9781316134504.

2 Zhou, J., Shi, T., Ren, J., and He, C. (2024). Accelerating operation optimization of complex chemical processes: a novel framework integrating artificial neural

network and mixed-integer linear programming. *Chemical Engineering Journal* 481: 148421. https://doi.org/10.1016/J.CEJ.2023.148421.
3 Li, M. (2023). An improved closed-circuit RO (CCRO) system: design and cyclic simulation. *Desalination* 554: 116519. https://doi.org/10.1016/J.DESAL.2023.116519.
4 Schweidtmann, A.M., Bongartz, D., Huster, W.R., and Mitsos, A. (2019). Deterministic global process optimization: flash calculations via artificial neural networks. *Computer Aided Chemical Engineering* 46: 937–942. https://doi.org/10.1016/B978-0-12-818634-3.50157-0.
5 Li, Y., Zhang, T., Sun, S., and Gao, X. (2019). Accelerating flash calculation through deep learning methods. *Journal of Computational Physics* 394: 153–165. https://doi.org/10.1016/J.JCP.2019.05.028.
6 Simpson, T.W., Mauery, T.M., Korte, J.J., and Mistree, F. (2012). Kriging models for global approximation in simulation-based multidisciplinary design optimization. 39: 2233–2241. https://doi.org/10.2514/2.1234.
7 Kleijnen, J.P.C. (2018). Design and analysis of simulation experiments. In: *Simulation Optimization*, 241–300. Springer https://doi.org/10.1007/978-3-319-18087-8.
8 Box, G.E.P. and Wilson, K.B. (1992). On the experimental attainment of optimum conditions. In: *Breakthroughs in Statistics. Springer Series in Statistics* (ed. S. Kotz and N.L. Johnson), 270–310. New York, NY: Springer https://doi.org/10.1007/978-1-4612-4380-9_23.
9 Udy, J., Blackburn, L., Hedengren, J.D., and Darby, M. (2017). Reduced order modeling for reservoir injection optimization and forecasting. *Proceedings of the FOCAPO/CPC Conference*, Tuscon, AZ, USA, pp. 8–12.
10 Yin, X., Wen, K., Wu, Y. et al. (2022). A machine learning-based surrogate model for the rapid control of piping flow: application to a natural gas flowmeter calibration system. *Journal of Natural Gas Science and Engineering* 98: 104384. https://doi.org/10.1016/J.JNGSE.2021.104384.
11 Alizadeh, R., Allen, J.K., and Mistree, F. (2020). Managing computational complexity using surrogate models: a critical review. *Research in Engineering Design* 31: 275–298. https://doi.org/10.1007/S00163-020-00336-7.
12 Hwang, J.T. and Martins, J.R.R.A. (2018). A fast-prediction surrogate model for large datasets. *Aerospace Science and Technology* 75: 74–87. https://doi.org/10.1016/J.AST.2017.12.030.
13 Reichel, L. (1990). Newton interpolation at Leja points. *BIT* 30: 332–346. https://doi.org/10.1007/BF02017352/METRICS.
14 Kaya, E. (2013). Spline interpolation techniques. *Journal of Technical Science and Technologies* 47–52. https://doi.org/10.31578/.V2I1.56.
15 Nevai, P. (1984). Mean convergence of Lagrange interpolation. III. *Transactions of the American Mathematical Society* 282: 669–698. https://doi.org/10.2307/1999259.
16 de Boor, C., Höllig, K., and Sabin, M. (1987). High accuracy geometric Hermite interpolation. *Computer Aided Geometric Design* 4: 269–278. https://doi.org/10.1016/0167-8396(87)90002-1.

17 Myers, R.H., Montgomery, D.C., and Anderson-Cook, C.M. (2016). *Response surface methodology: process and product optimization using designed experiments*. Wiley.

18 Forrester, A.I.J. and Keane, A.J. (2009). Recent advances in surrogate-based optimization. *Progress in Aerospace Sciences* 45: 50–79. https://doi.org/10.1016/J.PAEROSCI.2008.11.001.

19 Freeny, A., Box, G.E.P., and Draper, N.R. (1988). Empirical model building and response surfaces. *Technometrics* 30: 229. https://doi.org/10.2307/1270169.

20 Jones, D.R. (2001). A taxonomy of global optimization methods based on response surfaces. *Journal of Global Optimization* 21: 345–383. https://doi.org/10.1023/A:1012771025575/METRICS.

21 Forrester, A.I.J., Keane, A.J., and Bressloff, N.W. (2012). Design and analysis of "noisy" computer experiments. *AIAA Journal* 44: 2331–2339. https://doi.org/10.2514/1.20068.

22 Wiener, N. (1938). The homogeneous chaos. *American Journal of Mathematics* 60: 897. https://doi.org/10.2307/2371268.

23 Tasnim, N., Sanzida, N., and Momtaz, M. (2020). Surrogate modeling & optimization of a nonlinear batch reactor by polynomial chaos expansion. *Chemical Engineering Research Bulletin* 22: 121–126. https://doi.org/10.3329/cerb.v22i1.54310.

24 Field, R.V. and Grigoriu, M. (2004). On the accuracy of the polynomial chaos approximation. *Probabilistic Engineering Mechanics* 19: 65–80. https://doi.org/10.1016/J.PROBENGMECH.2003.11.017.

25 Rasmussen, C.E. and Williams, C.K.I. (2006). *Gaussian Processes for Machine Learning*. MIT Press.

26 Krige, D. (1951). A statistical approach to some basic mine valuation problems on the Witwatersrand. *Journal of the South African Institute of Mining and Metallurgy* 52: 119–139.

27 Santner, T.J., Williams, B.J., and Notz, W.I. (2003). *The Design and Analysis of Computer Experiments (Springer Series in Statistics)*, 2e. Springer.

28 Jones, D.R., Schonlau, M., and Welch, W.J. (1998). Efficient global optimization of expensive black-box functions. *Journal of Global Optimization* 13: 455–492. https://doi.org/10.1023/A:1008306431147/METRICS.

29 Cressie, N. (2015). *Statistics for Spatial Data*. Wiley.

30 Boukouvala, F. and Ierapetritou, M.G. (2014). Derivative-free optimization for expensive constrained problems using a novel expected improvement objective function. *AIChE Journal* 60: 2462–2474. https://doi.org/10.1002/AIC.14442.

31 Wang, Z. and Ierapetritou, M. (2018). Constrained optimization of black-box stochastic systems using a novel feasibility enhanced Kriging-based method. *Computers and Chemical Engineering* 118: 210–223. https://doi.org/10.1016/J.COMPCHEMENG.2018.07.016.

32 McBride, K. and Sundmacher, K. (2019). Overview of surrogate modeling in chemical process engineering. *Chemie Ingenieur Technik* 91: 228–239. https://doi.org/10.1002/CITE.201800091.

33 Yondo, R., Andrés, E., and Valero, E. (2018). A review on design of experiments and surrogate models in aircraft real-time and many-query aerodynamic analyses. *Progress in Aerospace Sciences* 96: 23–61. https://doi.org/10.1016/J.PAEROSCI.2017.11.003.

34 Fang, H. and Horstemeyer, M.F. (2006). Global response approximation with radial basis functions. *Engineering Optimization* 38: 407–424. https://doi.org/10.1080/03052150500422294.

35 Queipo, N.V., Haftka, R.T., Shyy, W. et al. (2005). Surrogate-based analysis and optimization. *Progress in Aerospace Sciences* 41: 1–28. https://doi.org/10.1016/J.PAEROSCI.2005.02.001.

36 Wang, Z. and Ierapetritou, M. (2017). A novel feasibility analysis method for black-box processes using a radial basis function adaptive sampling approach. *AIChE Journal* 63: 532–550. https://doi.org/10.1002/AIC.15362.

37 Rabitz, H. and Aliş, Ö.F. (1999). General foundations of high-dimensional model representations. *Journal of Mathematical Chemistry* 25: 197–233. https://doi.org/10.1023/A:1019188517934/METRICS.

38 Li, G., Wang, S.W., and Rabitz, H. (2002). Practical approaches to construct RS-HDMR component functions. *Journal of Physical Chemistry A* 106: 8721–8733. https://doi.org/10.1021/JP014567T.

39 Ayub, Y., Zhou, J., Ren, J. et al. (2023). High-dimensional model representation-based surrogate model for optimization and prediction of biomass gasification process. *International Journal of Energy Research*. https://doi.org/10.1155/2023/7787947.

40 Huang, Z., Qiu, H., Zhao, M. et al. (2015). An adaptive SVR-HDMR model for approximating high dimensional problems. *Engineering Computations (Swansea, Wales)* 32: 643–667. https://doi.org/10.1108/EC-08-2013-0208/FULL/XML.

41 Loh, W.Y. (2011). Classification and regression trees. *Wiley Interdisciplinary Reviews: Data Mining and Knowledge Discovery* 1: 14–23. https://doi.org/10.1002/WIDM.8.

42 Fang, H., Zhou, J., Wang, Z. et al. (2022). Hybrid method integrating machine learning and particle swarm optimization for smart chemical process operations. *Frontiers of Chemical Science and Engineering* 16: 274–287. https://doi.org/10.1007/S11705-021-2043-0/METRICS.

43 Kotsiantis, S.B. (2013). Decision trees: a recent overview. *Artificial Intelligence Review* 39: 261–283. https://doi.org/10.1007/S10462-011-9272-4/METRICS.

44 Herceg, S., Ujević Andrijić, Ž., and Bolf, N. (2019). Development of soft sensors for isomerization process based on support vector machine regression and dynamic polynomial models. *Chemical Engineering Research and Design* 149: 95–103. https://doi.org/10.1016/J.CHERD.2019.06.034.

45 Lahiri, S.K. and Khalfe, N. (2009). Process modeling and optimization of industrial ethylene oxide reactor by integrating support vector regression and genetic algorithm. *Canadian Journal of Chemical Engineering* 87: 118–128. https://doi.org/10.1002/CJCE.20123.

46 Panja, S., Chatterjee, A., and Yasmin, G. (2019). Kernel functions of SVM: a comparison and optimal solution. *Communications in Computer and*

Information Science 955: 88–97. https://doi.org/10.1007/978-981-13-3140-4_9/FIGURES/4.

47 Babangida, N.M., Ul Mustafa, M.R., Yusuf, K.W. et al. (2016). Evaluation of low degree polynomial kernel support vector machines for modelling pore-water pressure responses. *MATEC Web of Conferences* 59: 04003. https://doi.org/10.1051/MATECCONF/20165904003.

48 Chandra, S., Singh, H., Chauhan, D.S. et al. (2009). Detection of brain tumors from MRI using Gaussian RBF kernel based support vector machine. *International Journal of Advancements in Computing Technology*. https://doi.org/10.4156/ijact.vol1.issue1.7.

49 Debnath, R. and Takahashi, H. (2004). Kernel selection for the support vector machine. *IEICE Transactions on Information and Systems* E87-D: 2903–2904.

50 Haykin, S., York, N., San, B. et al. (2009). *Neural Networks and Learning Machines*, 3e. Simon and Schuster.

51 Chollet, F. (2021). *Deep Learning with Python*, 478. Simon and Schuster.

52 Zhou, J., Shi, T., Qian, Q. et al. (2023). Protocol for the design and accelerated optimization of a waste-to-energy system using AI tools. *STAR Protocols* 4: 102685. https://doi.org/10.1016/J.XPRO.2023.102685.

53 Zhou, J., Chu, Y.T., Ren, J. et al. (2023). Integrating machine learning and mathematical programming for efficient optimization of operating conditions in organic Rankine cycle (ORC) based combined systems. *Energy* 281: 128218. https://doi.org/10.1016/J.ENERGY.2023.128218.

54 Neocleous, C. and Schizas, C. (2002). Artificial neural network learning: a comparative review. *Lecture Notes in Computer Science (Including Subseries Lecture Notes in Artificial Intelligence and Lecture Notes in Bioinformatics)* 2308: 300–313. https://doi.org/10.1007/3-540-46014-4_27/COVER.

55 Wang, Y., Shi, J., Zhou, D., and Gao, F. (2006). Iterative learning fault-tolerant control for batch processes. *Industrial and Engineering Chemistry Research* 45: 9050–9060. https://doi.org/10.1021/IE060726P.

56 Nagy, Z.K. and Braatz, R.D. (2003). Robust nonlinear model predictive control of batch processes. *AIChE Journal* 49: 1776–1786. https://doi.org/10.1002/AIC.690490715.

57 Bonvin, D. and François, G. (2017). Control and optimization of batch chemical processes. In: *Coulson and Richardson's Chemical Engineering*, 4e (ed. S. Rohani and R. Chhabra), 441–503. Oxford, UK: Butterworth-Heinemann https://doi.org/10.1016/B978-0-08-101095-2.00011-4.

58 Kumar, D. and Budman, H. (2017). Applications of polynomial chaos expansions in optimization and control of bioreactors based on dynamic metabolic flux balance models. *Chemical Engineering Science* 167: 18–28. https://doi.org/10.1016/J.CES.2017.03.035.

59 Rani, K.Y. and Patwardhan, S.C. (2004). Data-driven modeling and optimization of semibatch reactors using artificial neural networks. *Industrial and Engineering Chemistry Research* 43: 7539–7551. https://doi.org/10.1021/IE0305521.

60 Zhan, Y., Zhu, J., Schrader, L.C., and Wang, D. (2023). Modeling and optimization of bioenergy production from co-digestion of poultry litter with wheat

straw in anaerobic sequencing batch reactor: response surface methodology and artificial neural network. *Applied Energy* 345: 121373. https://doi.org/10.1016/J.APENERGY.2023.121373.

61 Kumar, A., Bajaj, M., and Nath, V.P. (2013). Implementation of neural model predictive control in continuous stirred tank reactor system. *International Journal of Scientific and Engineering Research* 4: 1989–1996.

62 Justice, E.O. and Sunday, R.A. (2005). An optimum solution for a process control problem (continuous stirred tank reactor) using a hybrid neural network. *Journal of Theoretical and Applied Information Technology* 4: 906–915.

63 Sivapathasekaran, C. and Sen, R. (2013). Performance evaluation of an ANN–GA aided experimental modeling and optimization procedure for enhanced synthesis of marine biosurfactant in a stirred tank reactor. *Journal of Chemical Technology & Biotechnology* 88: 794–799. https://doi.org/10.1002/JCTB.3900.

64 Wang, H., Chaffart, D., and Ricardez-Sandoval, L.A. (2019). Modelling and optimization of a pilot-scale entrained-flow gasifier using artificial neural networks. *Energy* 188: 116076. https://doi.org/10.1016/J.ENERGY.2019.116076.

65 Jiang, J., Lu, Z., Ferreira Gomes, F.A. et al. (2024). Experiment and simulation of hydrogen oxidation in a high-pressure turbulent flow reactor. *Fuel* 357: 129714. https://doi.org/10.1016/J.FUEL.2023.129714.

66 Shokry, A. and Espuña, A. (2014). Sequential dynamic optimization of complex nonlinear processes based on Kriging surrogate models. *Procedia Technology* 15: 376–387. https://doi.org/10.1016/J.PROTCY.2014.09.092.

67 Ayub, Y., Zhou, J., Ren, J. et al. (2023). Plasma gasification based monetization of poultry litter: system optimization and comprehensive 5E (Energy, Exergy, Emergy, Economic, and Environmental) analysis. *Energy Conversion and Management* 282: 116878. https://doi.org/10.1016/J.ENCONMAN.2023.116878.

68 Zhou, J., Ren, J., and He, C. (2023). Wind energy-driven medical waste treatment with polygeneration and carbon neutrality: process design, advanced exergy analysis and process optimization. *Process Safety and Environmental Protection* 178: 342–359. https://doi.org/10.1016/J.PSEP.2023.08.040.

69 Kim, J.Y., Kim, D., Li, Z.J. et al. (2023). Predicting and optimizing syngas production from fluidized bed biomass gasifiers: a machine learning approach. *Energy* 263: 125900. https://doi.org/10.1016/J.ENERGY.2022.125900.

70 Waqas, S., Harun, N.Y., Arshad, U. et al. (2024). Optimization of operational parameters using RSM, ANN, and SVM in membrane integrated with rotating biological contactor. *Chemosphere* 349: 140830. https://doi.org/10.1016/J.CHEMOSPHERE.2023.140830.

71 Liang, R. and Yuan, Z. (2020). Computational shape optimization of microreactors based on CFD simulation and surrogate model driven optimization. *Computer Aided Chemical Engineering* 48: 925–930. https://doi.org/10.1016/B978-0-12-823377-1.50155-5.

72 del Rio-Chanona, E.A., Wagner, J.L., Ali, H. et al. (2019). Deep learning-based surrogate modeling and optimization for microalgal biofuel production and photobioreactor design. *AIChE Journal* 65: 915–923. https://doi.org/10.1002/AIC.16473.

73 Noble, R.D. and Agrawal, R. (2005). Separations research needs for the 21st century. *Industrial and Engineering Chemistry Research* 44: 2887–2892. https://doi.org/10.1021/IE0501475/ASSET/IE0501475.FP.PNG_V03.

74 Ibrahim, D., Jobson, M., Li, J., and Guillén-Gosálbez, G. (2017). Surrogate models combined with a support vector machine for the optimized design of a crude oil distillation unit using genetic algorithms. *Computer Aided Chemical Engineering* 40: 481–486. https://doi.org/10.1016/B978-0-444-63965-3.50082-9.

75 Lu, J., Wang, Q., Zhang, Z. et al. (2021). Surrogate modeling-based multi-objective optimization for the integrated distillation processes. *Chemical Engineering and Processing: Process Intensification* 159: 108224. https://doi.org/10.1016/J.CEP.2020.108224.

76 Quirante, N., Javaloyes, J., and Caballero, J.A. (2015). Rigorous design of distillation columns using surrogate models based on Kriging interpolation. *AIChE Journal* 61: 2169–2187. https://doi.org/10.1002/AIC.14798.

77 Kim, J., Son, M., Sup Han, S. et al. (2022). Computational-cost-efficient surrogate model of vacuum pressure swing adsorption for CO separation process optimization. *Separation and Purification Technology* 300: 121827. https://doi.org/10.1016/J.SEPPUR.2022.121827.

78 Nikolaidis, G.N., Kikkinides, E.S., and Georgiadis, M.C. (2017). Modelling and optimization of pressure swing adsorption (PSA) processes for post-combustion CO_2 capture from flue gas. In: *Process Systems and Materials for CO_2 Capture*, 343–369. https://doi.org/10.1002/9781119106418.CH13.

79 Beck, J., Friedrich, D., Brandani, S., and Fraga, E.S. (2015). Multi-objective optimisation using surrogate models for the design of VPSA systems. *Computers and Chemical Engineering* 82: 318–329. https://doi.org/10.1016/J.COMPCHEMENG.2015.07.009.

80 Tadepalli, A., Pujari, K.N.S., and Mitra, K. (2023). A crystallization case study toward optimization of expensive to evaluate mathematical models using Bayesian approach. *Materials and Manufacturing Processes* 38: 2127–2134. https://doi.org/10.1080/10426914.2023.2238051.

81 Miriyala, S.S., Pujari, K.N.S., Naik, S., and Mitra, K. (2022). Evolutionary neural architecture search for surrogate models to enable optimization of industrial continuous crystallization process. *Powder Technology* 405: 117527. https://doi.org/10.1016/J.POWTEC.2022.117527.

82 Dutta B. *Heat Transfer: Principles and Applications* 2023.

83 Liu, Y., Yang, M., Zhao, L. et al. (2021). Simultaneous optimization and heat integration of an aromatics complex with a surrogate model. *Industrial and Engineering Chemistry Research* 60: 3633–3647. https://doi.org/10.1021/ACS.IECR.0C05507/SUPPL_FILE/IE0C05507_SI_001.PDF.

84 Mastrippolito, F., Aubert, S., and Ducros, F. (2021). Kriging metamodels-based multi-objective shape optimization applied to a multi-scale heat exchanger. *Computers and Fluids* 221: 104899. https://doi.org/10.1016/J.COMPFLUID.2021.104899.

85 Ali, W., Khan, M.S., Qyyum, M.A., and Lee, M. (2018). Surrogate-assisted modeling and optimization of a natural-gas liquefaction plant. *Computers and Chemical Engineering* 118: 132–142. https://doi.org/10.1016/J.COMPCHEMENG.2018.08.003.

86 Koo, G.W., Lee, S.M., and Kim, K.Y. (2014). Shape optimization of inlet part of a printed circuit heat exchanger using surrogate modeling. *Applied Thermal Engineering* 72: 90–96. https://doi.org/10.1016/J.APPLTHERMALENG.2013.12.009.

87 Shi, H.-n., Ma, T., Chu, W.-x., and Wang, Q.-w. (2017). Optimization of inlet part of a microchannel ceramic heat exchanger using surrogate model coupled with genetic algorithm. *Energy Conversion and Management* 149: 988–996. https://doi.org/10.1016/J.ENCONMAN.2017.04.035.

88 Maakala, V., Järvinen, M., and Vuorinen, V. (2018). Optimizing the heat transfer performance of the recovery boiler superheaters using simulated annealing, surrogate modeling, and computational fluid dynamics. *Energy* 160: 361–377. https://doi.org/10.1016/J.ENERGY.2018.07.002.

89 Vilasboas, I.F., Dos Santos, V.G.S.F., Ribeiro, A.S., and da Silva, J.A.M. (2021). Surrogate models applied to optimized organic Rankine cycles. *Energies* 14: 8456. https://doi.org/10.3390/EN14248456.

90 Henao, C.A. and Maravelias, C.T. (2011). Surrogate-based superstructure optimization framework. *AIChE Journal* 57: 1216–1232. https://doi.org/10.1002/AIC.12341.

91 Palmer, K. and Realff, M. (2002). Optimization and validation of steady-state flowsheet simulation metamodels. *Chemical Engineering Research and Design* 80: 773–782. https://doi.org/10.1205/026387602320776849.

92 Palmer, K. and Realff, M. (2002). Metamodeling approach to optimization of steady-state flowsheet simulations: model generation. *Chemical Engineering Research and Design* 80: 760–772. https://doi.org/10.1205/026387602320776830.

93 Davis, E. and Ierapetritou, M. (2007). A Kriging method for the solution of nonlinear programs with black-box functions. *AIChE Journal* 53: 2001–2012. https://doi.org/10.1002/AIC.11228.

94 Davis, E. and Ierapetritou, M. (2009). A Kriging based method for the solution of mixed-integer nonlinear programs containing black-box functions. *Journal of Global Optimization* 43: 191–205. https://doi.org/10.1007/S10898-007-9217-2/METRICS.

95 Davis, E. and Ierapetritou, M. (2008). A Kriging-based approach to MINLP containing black-box models and noise. *Industrial and Engineering Chemistry Research* 47: 6101–6125. https://doi.org/10.1021/IE800028A.

96 Henao, C.A. and Maravelias, C.T. (2010). Surrogate-based process synthesis. *Computer Aided Chemical Engineering* 28: 1129–1134. https://doi.org/10.1016/S1570-7946(10)28189-0.

97 Caballero, J.A. and Grossmann, I.E. (2008). An algorithm for the use of surrogate models in modular flowsheet optimization. *AIChE Journal* 54: 2633–2650. https://doi.org/10.1002/AIC.11579.

98 Quirante, N., Javaloyes-Antón, J., and Caballero, J.A. (2018). Hybrid simulation-equation based synthesis of chemical processes. *Chemical Engineering Research and Design* 132: 766–784. https://doi.org/10.1016/J.CHERD.2018.02.032.

99 Fahmi, I. and Cremaschi, S. (2012). Process synthesis of biodiesel production plant using artificial neural networks as the surrogate models. *Computers and Chemical Engineering* 46: 105–123. https://doi.org/10.1016/J.COMPCHEMENG.2012.06.006.

100 Di Martino, M., Avraamidou, S., and Pistikopoulos, E.N. (2022). A neural network based superstructure optimization approach to reverse osmosis desalination plants. *Membranes* 12: 199. https://doi.org/10.3390/MEMBRANES12020199.

101 Cozad, A., Sahinidis, N.V., and Miller, D.C. (2014). Learning surrogate models for simulation-based optimization. *AIChE Journal* 60: 2211–2227. https://doi.org/10.1002/AIC.14418.

102 Cozad, A., Sahinidis, N.V., and Miller, D.C. (2015). A combined first-principles and data-driven approach to model building. *Computers and Chemical Engineering* 73: 116–127. https://doi.org/10.1016/J.COMPCHEMENG.2014.11.010.

103 Wilson, Z.T. and Sahinidis, N.V. (2017). The ALAMO approach to machine learning. *Computers and Chemical Engineering* 106: 785–795. https://doi.org/10.1016/J.COMPCHEMENG.2017.02.010.

104 Boukouvala, F. and Floudas, C.A. (2017). ARGONAUT: AlgoRithms for global optimization of coNstrAined grey-box compUTational problems. *Optimization Letters* 11: 895–913. https://doi.org/10.1007/S11590-016-1028-2/FIGURES/4.

Index

Applied AI Techniques in the Process Industry: From Molecular Design to Process Design and Optimization, First Edition. Edited by Chang He and Jingzheng Ren.

n